Exploring the Limits of Preclassical Mechanics

Peter Damerow   Gideon Freudenthal
Peter McLaughlin   Jürgen Renn

Exploring the Limits of Preclassical Mechanics

A Study of Conceptual Development in
Early Modern Science: Free Fall and Compounded
Motion in the Work of Descartes, Galileo, and Beeckman

With 151 Illustrations

Springer-Verlag
New York Berlin Heidelberg London Paris
Tokyo Hong Kong Barcelona Budapest

Peter Damerow
Max-Planck-Institut für
 Bildungsforschung
Lentzeallee 94
D1000 Berlin 33
Germany

Peter McLaughlin
Fachgruppe Philosophie
Universität Konstanz
Postfach 5560
D7750 Konstanz 1
Germany

Gideon Freudenthal
Cohn Institute for the History and Philosophy
 of Science and Ideas
Tel Aviv University
Ramat-Aviv
69978 Tel Aviv
Israel

Jürgen Renn
The Collected Papers of Albert Einstein
Boston University
745 Commonwealth Avenue
Boston, MA 02215
USA

Library of Congress Cataloging-in-Publication Data
Exploring the limits of preclassical mechanics / Peter Damerow . . . [et
al.].
      p. cm.
    Includes bibliographical references.
    ISBN 0-387-97602-7
    1. Motion—History.   2. Mechanics—History.   I. Damerow, Peter.
QC133.E97   1991
531—dc20   91-35196

Printed on acid-free paper.

Photocomposed copy prepared from the authors' Microsoft Word file.
Printed and bound by Edwards Brothers, Inc., Ann Arbor, Michigan.
Printed in the United States of America.

9 8 7 6 5 4 3 2 1

ISBN 0-387-97602-7 Springer-Verlag New York Berlin Heidelberg
ISBN 3-540-97602-7 Springer-Verlag Berlin Heidelberg New York

Preface

This book grew out of at a colloquium initiated by Wolfgang Lefèvre and Peter Damerow in Berlin many years ago. In this colloquium we attempted over a number of years to acquire an understanding of cognitive development in science that takes account of a number of factors usually considered in isolation: the cumulative or progressive aspects of science, its embedding in culture and society, the cognitive content of theories, and the relationship of science to contemporary philosophical systems. The results of our discussions are embodied in a number of books and other publications by various members of the colloquium – some programmatic in nature, some case studies of particular scientific developments. Among the way stations are: Wolfgang Lefèvre, *Naturtheorie und Produktionsweise* (Darmstadt: Luchterhand, 1978), *Die Entstehung der biologischen Evolutionstheorie* (Berlin: Ullstein 1984); Peter Damerow and Wolfgang Lefèvre, *Rechenstein, Experiment, Sprache* (Stuttgart: Klett-Cotta 1981); Gideon Freudenthal, *Atom and Individual in the Age of Newton* (Dordrecht: Reidel 1986); Peter McLaughlin, *Kant's Critique of Teleology in Biological Explanation* (Lampeter: Mellen 1990); a number of papers on Galileo by Jürgen Renn (see Bibliography) as well as his contributions as an editor of the *Collected Papers of Albert Einstein* (Princeton Univ. Press, 1987, 1989, 1992); and finally the numerous papers on cognitive development by Peter Damerow published mainly in German, a selection of which is to appear in English under the title, *The Cultural Conditions of Thinking*, as a volume of *Boston Studies in the Philosophy of Science* in the near future.

In the present volume we attempt to display some aspects of the cognitive development involved in the emergence of modern science. We study this development in a detailed analysis of three interrelated examples at the core of early theories of motion.

The book as a whole is both in its approach and in many of its individual findings a genuinely collective product and reflects a common understanding established in many years of close cooperation. Our study of early modern mechanics at first followed some preliminary theses introduced by Freudenthal but soon took on a life of its own and superseded our initial conceptions. The Introduction and Epilogue were written jointly, and all other parts, while written by individuals, have been scrutinized and criticized by the others in the group. Damerow and Freudenthal wrote Chapter 1; Freudenthal and McLaughlin wrote Chapter 2; Renn wrote Chapter 3. On McLaughlin fell the task of editing, sometimes heavily, each of the individual chapters in order to free shared ideas from idiosyncratic presentations.

The translations in the text and in the Documents Chapter have a checkered history and unclear parenthood: some were originally produced by Freudenthal as working translations for our internal discussions; translations from the Italian are

by Renn; others were first drafted by McLaughlin, with whom final respon-
sibility for all French and Latin translations lies. As the only native speaker of
English in the group, McLaughlin is also responsible for all grammatical errors,
Germanisms and stylistic solecisms that remain in the text as a whole.

Since our Berlin days the four authors have been dispersed to four different
cities on three different continents. Our continuing work would not have been
possible without the generous support of a number of institutions. The home
base and main stay of the project has always been Damerow's office in the
Center for Development and Socialization at the Max Planck Institute for
Human Development and Education in Berlin. We would especially like to thank
Director Wolfgang Edelstein for constant encouragement and a kind of support of
our long term interdisciplinary project going beyond academic hospitality. We
would also like to thank Yehuda Elkana of the Van Leer Jerusalem Institute, and
Jürgen Mittelstrass of the Center for Philosophy and Philosophy of Science at
the University of Constance whose support over the years have made our cooperative venture feasible. Furthermore, we are grateful to the Fritz-Thyssen
Foundation (Cologne), the Minerva Foundation (Bonn), and the Edelstein Center
for the History and Philosophy of Science (Jerusalem) for financial support at
various times, and to the Institute for Advanced Study in Berlin for bringing two
of us back to Berlin for a year.

We also thank Mr. E.J. Aiton, the British Society for the History of Science,
Cambridge University Press, Mrs. I.E. Drabkin, Mr. Stillman Drake, Kluwer
Publishers, Wall & Emerson, Inc., and the University of Wisconsin Press for
permission to use or modify figures or translations. In particular, we want to
thank the Biblioteca Nazionale, Florence, for making accessible its entire collection of Galileo manuscripts and for permission to reproduce some of them in
facsimile. We also want to thank Franca Principe-Saba and Maria Miniati of the
Institute and Museum of the History of Science (Florence), as well as its director
Paulo Galluzzi, for the help and hospitality accorded to Renn while he studied
the Galileo manuscripts. We should also like to thank the Springer Verlag's
anonymous referee for an unusually thorough and helpful commentary on the
original manuscript.

Finally we should like to acknowledge our debt to the scores of historians and
philosophers of science whose editions, interpretations, and translations of 17th
century scientists and philosophers we have used, abused, criticized and
cannibalized.

Berlin, Boston, Constance, and Jerusalem

Contents

Abbreviations

The following conventional abbreviations are used for works often cited in the text:

AT	Descartes, *Oeuvres* (ed. by Adam and Tannery)
Dialogo	Galileo, *Dialogue Concerning the Two Chief World Systems*
Discorsi	Galileo, *Two New Sciences* (Discourses and Mathematical Demonstrations concerning ...)
EN	Galileo, *Le Opere* (Edizione nazionale)
DM	Galileo, *De Motu*
GP	Leibniz, *Philosophischen Schriften* (ed. by.Gerhardt)
GM	Leibniz, *Mathematische Schriften* (ed. by.Gerhardt)
J	Beeckman, *Journal*
MS	Galileo, *Manuscripts*

Introduction

In this book we undertake to explain the 17th century transition from early modern conceptualizations of motion, to the theory of motion of classical mechanics. We present this transition not in an extensive narration of events but in the detailed examination of key texts dealing with two central topics: the free fall of bodies and the composition of motions and forces. The choice of these two topics is not arbitrary, for the theorems developed to account for them were the cornerstones of all studies of motion of the time, whether they already formed part of classical mechanics or still preceded its establishment.

The focus of our studies is hence the theory transformation resulting in classical mechanics. The emergence of classical mechanics has for many decades been a privileged topic of historical and philosophical studies since it represents not only a new theory but the constitution of a new kind of knowledge, modern science. It is therefore the prime example for discussions of the nature and determinants of science. In this study we shall be concerned exclusively with the conceptual aspects of the Scientific Revolution. Traditionally, interpretations of the Scientific Revolution have stressed either the continuity or the radical break with premodern intellectual traditions. According to the approach to conceptual development that we adopt and which we shall sketch in the following, continuity and break were neither alternatives nor even poles of a continuum. The studies presented here will show that it was by working within the conceptual system and so *exploring the limits of preclassical mechanics* that 17th century natural philosophers established classical mechanics.

Most of the documents to be analyzed have been published and are readily available to scholars, almost all have previously been interpreted. However, we have found most interpretations unsatisfying, and the explicit or implicit presuppositions on which they are based seem to us to be problematical and rather to obscure than to clarify the nature of the conceptual novelty which we see as constitutive of the transition to classical mechanics. While our different reading of the texts depends on a different approach to conceptual development, the explanatory power of the results of these studies does not depend on such presuppositions, but rather should demonstrate their fruitfulness.

For example, one standard theme in discussions of the emergence of classical mechanics is the proof of the law of free fall. Here one of the crucial questions is whether, when, and how Galileo (Descartes, Beeckman, or others) realized that the velocity of falling bodies is proportional to the time elapsed and not to the distance traversed in fall. Now, if the law is interpreted as the correct choice between these alternatives, then its discovery is naturally considered to be the result of examining the alternatives and making a well founded choice; the existence of the alternative prior to the discovery of the law is, of course, presupposed. However, from the point of view of Aristotelian natural philosophy – where no equivalent to a functional dependence of motion on a certain parameter exists, and where the velocity of a motion always refers to its overall extension in space *and* time – the alternative: velocity stands either in such a specified relation to space or else to time cannot sensibly be posed. It would be like demanding that a zebra have either white stripes or else black ones.

The existence of such an alternative itself presupposes the relevant concepts, "velocity" and "motion" as they figure in the law of free fall, i.e., with the meanings they have in classical mechanics. Thus this interpretation assumes that the concepts of classical mechanics are available before classical mechanics, and that only the correct empirical quantitative relations among the entities referred to need be discovered under the specific circumstances (free fall). This position either contains an implicit denial that conceptual development takes place at all, since the concepts remain the same as they were before, or else must claim that a conceptual development *preceded* the discovery of the central laws of a new science, and thus must have resulted from some other processes. The main discoveries of science would thus turn out to be uninteresting for the study of conceptual development in science.

We shall argue, on the contrary, that these classical concepts as well as the above alternative in the formulation of a law of free fall are rather the outcomes of the establishment of the law than its prerequisites. Thus we must attempt to explain how for instance the law of free fall, which implicitly defines some of the concepts of classical mechanics (such as velocity and acceleration), could be developed and formulated within the conceptual framework of preclassical mechanics, which did not possess these concepts. To accomplish this task is precisely to explain a conceptual development, i.e., the emergence of something genuinely new from within the old.

However, an explanation of genuine novelty that does not avoid the problem by simply giving it a name, so to speak introducing a *vis creativa* or some equivalent empty term, must confront a number of difficulties. The paradox involved in novelty is: how can concepts be genuinely new and nonetheless be derived on the basis of an old conceptual system? Alternatively, if the new concepts can be derived from the old system, were they not always implicitly contained in that system? And why were they not derived much sooner? And if the old system is capable of so much, why was a new one needed at all?

One important characteristic of our approach to explaining conceptual change is the strong emphasis we place on the role of inconsistencies of existing conceptual systems. It is due to such inconsistencies that conceptual development is not simply one continuous and cumulative process, but a development characterized by the far-going restructuring of conceptual systems. Shortcomings of a system can surface as puzzling consequences or even straightforward contradictions when concepts in the system are applied to new objects or to old objects under new conditions. However, while such arising inconsistencies can motivate or even compel changes in the system, they cannot themselves determine the direction of development or provide the means for their own resolution. It is also not necessary that inconsistencies in a conceptual system will always result in cognitive development. An inconsistency can be neutralized by excluding the problematic application of a concept, by introducing *ad hoc* hypotheses, or by similar ploys. This may in fact be the most common response, so that it should be clear that conceptual development cannot be necessitated simply by the appearance of an inconsistency. It may however be worthwhile to pursue some of the consequences of the simple truism that conceptual development will occur only where it is possible. This truism is far from empty, for it implies that since the recognition of an inconsistency and the desire to eliminate it do not guarantee success, success cannot be explained by reference merely to problems, goals, wishes, and the like. The saying, "Where there is a will, there is a way," is unfortunately wrong. Where the means to solve a problem are objectively not available, no amount of willing will do the job; and where the job was in fact done, we ought to look for the specific means which allowed it to be accomplished. The consideration of the means may help to provide a solution to the paradox of novelty by reformulating it as the question: what were the means (conceptual and otherwise) available in preclassical mechanics and what were the circumstances that allowed their application to supersede the conceptual system in which they were embedded? In the case of the Scientific Revolution of the 17th century this would mean that the cognitive means for the transformation from preclassical to classical mechanics or at least for its initiation must have been available in the conceptual system of preclassical mechanics and natural philosophy. A philosophical analysis of this historical transformation must locate these means and show how their application to particular, historically given tasks triggered a development which superseded some of the fundamental presuppositions on which it was based.

In preclassical mechanics we encounter many different kinds of conceptual means at work as well as some scientific instruments and experimental devices. Some of the conceptual means are universal or almost so (such as number and measure), some are common only to certain fields and may take on different forms according to the fields of their application (such as notions of causation), and some are specific to mechanics and kinematics (such as concepts of velocity, including the mathematical techniques for quantifying them). These means usually serve to solve problems formulated within the conceptual system, and a successful application does not lead to any inconsistencies but rather supports the validity of a theory. In fact it is very often through guided participation in such successful applications that students of a discipline appropriate the state of knowledge.

It may however happen that the attempt to solve a problem by applying the usual means produces new and unexpected results, inconsistencies, or both. Not only the extension to a new object may produce new results, but also the combination of previously disparate means when applied to the same problem may do much the same. If new results can be taken as a point of departure for a reconceptualization of the problem which removes the inconsistencies, then the application of particular means under the circumstances mentioned gives us a mechanism for transforming a conceptual system into a new and more powerful one. Thus we shall see in the work of Galileo how the application of the Aristotelian concept of motion as well as the traditional proportions among time, space, and velocity (which from the point of view of classical mechanics are valid only for uniform motions) to accelerated motion eventually led to a new concept of velocity. A new object may be introduced into a field of inquiry in very different ways: a peripheral phenomenon may move into the focus of interest (as did accelerated motion in the 17th century), or it may simply come from outside (as Galileo was confronted with Guidobaldo's experiment). We shall also see that the combination of a traditional doctrine of oppositions with the concept of "quantity of motion" resulted in the system of impact rules of Descartes' physics, while the interest in the previously marginal phenomenon of impact was clearly new and due to its central role in 17th century corpuscular philosophy.

The examples of conceptual *means* adduced so far, such as the Aristotelian concept of velocity, are clearly themselves the *products* of previous studies and can easily be made the *object* of future studies. This illustrates the fact that no entity is intrinsically a means or an object or a product; the same thing can in appropriate circumstances be taken in each role. Within the process of application, however, there are important differences which can be seen in any human activity: the object is changed into the product, but, at least ideally, the means remains the same.

Having stressed so much the role of the means as compared with goals in explaining conceptual progress, we should finally point out the determination of the goals themselves by the means available. While it may well be the case that a situation that does not satisfy some needs may give rise to a wish to alter the situation, this dissatisfaction cannot be transformed into a problem nor can a wish be transformed into a purpose, except by reference to the available means. We see this mechanism at work also in our studies, where the very questions asked by the natural philosophers of the seventeenth century result from their particular conceptualization of the natural phenomena investigated and would not be formulated in this way, if at all, by us. The same determination is conspicuous in the case of techniques for providing answers to these questions. Thus for example, whether the problem to be solved is to express the relation of the time elapsed to the space traversed in free fall by means of numerical examples or rather by proportions or even by a general function depends precisely upon the means for the solution of the problem. The wish to know how much space will be traversed after some time in free fall (in itself dependent on a specific conceptualization of "fall") is not yet a problem. Thus, we shall for instance see that Beeckmann's question, which initiated Descartes' study of free fall, presupposed an elaborate physical theory and was reformulated by Descartes with reference to the mathematical means that were to be applied for its solution.

Hence both the formulation of a problem and the way it is solved are determined by the means available in the conceptual system at hand.

We have stressed above that concepts can change their meanings in new circumstances of application. This argument may seem contrary to the widespread view, that objects are subsumed under a concept or they are not, but that concepts do not change their meanings in application. While this view may possess some initial plausibility and normative appeal for an ideally rigorous system with an almost universal empirical base, it is extremely implausible as a description of real cognitive activity in the real world. The conceptual systems dealt with in our studies are all formulated in a controlled technical language (*Fachsprache*), which displays characteristics of both a formal system and the natural language. While it is flexible enough to cope with a wide range of experience, it is also rigid enough occasionally to display inconsistencies when separate legitimate applications of a concept lead to extensions of meaning which turn out to be incompatible.

We shall try to show on concrete examples that the application of a preclassical conceptual system to new or newly important phenomena did indeed lead to changes in the meanings of the concepts and thus to changes in the structure of the system: but we shall also see that not every change in a conceptual system represents a development. We shall attribute conceptual development only to those changes that result in structural transformation leading to a new system better able to deal with the problems initiating the change. It will be one of the major theses of this book that even those changes in the conceptual system of preclassical mechanics introduced by Descartes and by Galileo, which constitute their lasting contribution to modern science, did not in themselves represent the conceptual development from preclassical to classical physics since – as Descartes and Galileo interpreted them – they did not lead to an abandoning of the conceptual system of preclassical mechanics. The conceptual development embodied in the transition to classical mechanics cannot be identified with any particular way station and is not to be found in any particular text. It is a process which begins with such figures as Descartes and Galileo and takes shape with the generation of their successors. These disciples or even adversaries read the old problems and arguments from the point of view of their new solutions, thus establishing classical mechanics, because their point of departure was now the concepts as they are implicitly defined within the derivations of the theorems, e.g., the law of free fall. Thus, while for the first discoverer, the law of free fall is achieved by applying and modifying an independently grounded, pre-existing conceptual system, for his disciples it is the law of fall that canonically defines key concepts in a new conceptual system. The very same reading of these theorems that establishes classical mechanics also obliterates the traces of its real historical genesis because the original problems and the concepts involved are now understood within a very different theoretical and semantic framework. But since the successors themselves derive the inherited theorems on the basis of the new concepts, they impute these concepts to the discoverers.

In the Epilogue we shall give examples of just this process and show how later readings of the results achieved by Descartes and Galileo and analyzed in the three main chapters misunderstood these results, attributing to them concepts that belong rather to classical mechanics. We shall illustrate this on the example of Descartes' final treatment of the law of free fall, which, although not announced as new, is incompatible with his previous presentations and is in fact the first interpretation of the traditional geometrical representation that uses only concepts that are "correct" in classical mechanics; we also illustrate this in the misinterpretations by Descartes' disciples of the problematical concept of "determination," which establish the parallelogram theorem for the compounding of motive forces as it is valid in classical mechanics. Finally, we shall see exactly the same mechanism in Cavalieri's and Torricelli's comments on Galileo's derivation of the parabolic trajectory of oblique projection, in which they use – and attribute to Galileo – inertia and superposition as general concepts.

The emphasis we place on concept development from preclassical to classical mechanics commits us of course to the position that, although classical mechanics is the contingent historical result of the development we investigate, it also has a privileged position for its interpretation. This does not, however, commit us to a theory of historical teleology. Classical mechanics is the privileged standpoint for analyzing this development not because it was the *goal* but because it was the *result*. It is sufficient that classical mechanics is the only consistent theory of mechanics that developed in the aftermath of the work we analyze, i.e., that fulfilled at least some of the objectives formulated by the proponents of preclassical mechanics, and that it adopted the contributions analyzed as part of its corpus. The situation is somewhat more difficult in the analysis of the philosophical foundations of Descartes' theory. There is no consensus on the philosophical presuppositions of classical mechanics comparable to that on textbook physics, and thus the point of view we take must be more controversial.

The first chapter, "Concept and Inference," analyzes the attempts of Isaac Beeckman and René Descartes to derive a law of fall within the conceptual framework of preclassical mechanics. Our presentation of these attempts stresses two points. On the one hand, we show the dependency of their inferences on the semantics of the concepts within their conceptual system as well as the reason for the compelling nature of such inferences; on the other hand, we show how the application of this conceptual system to a new object failed, since inferences were no longer unambiguous and necessary (in our case the relevant part of the system is the Aristotelian proportions of motion as applied to the accelerated motion of falling bodies). The resistance of this new case to integration into the given conceptual structure led to a proliferation of legitimate but incompatible inferences, some of which opened up new objective possibilities to restructure the entire conceptual system. We conclude the first analysis by showing that these objective possibilities were not realized by Descartes. It will be seen later (in the third chapter) that first steps towards realizing such possibilities were in fact taken by Galileo when he encountered precisely the same conceptual difficulties that surfaced in the application of the conceptual system of preclassical mechanics.

The second chapter, "Conservation and Contrariety," takes up the problem of the compounding of motions and forces and analyzes a very general conceptual system of the Aristotelian philosophy of the time, showing that it provided a general method of inference relatively independent of specific topics: the logic of contraries. This logic of contraries is shown at work both in specific topics such as Descartes' derivation of the inverse sine law of refraction in optics as well as in his derivation of the basic concepts of his entire physical system including its fundamental laws, i.e., inertia and the rules of impact. The chapter also examines Descartes' philosophical reflection on the conceptual foundations of such a deductive theory, which led him to introduce a kind of principles of conservation adequate not only to his system but also, with certain modifications, to later classical physics.

In the third chapter, "Proofs and Paradoxes," the same two central topics, free fall and compound motion, are pursued in the work of Galileo. It will be seen that Galileo, too, worked within the same conceptual system of preclassical mechanics, using the Aristotelian proportions, the logic of contraries, and the medieval mathematical techniques for representing qualities. The analysis of Galileo not only confirms the generality of the results of our analysis of Descartes but also illustrates (on the basis of Galileo's manuscripts) just what it means to explore the limits of a conceptual system. Galileo attempts again and again to derive the law of falling bodies and the parabolic trajectory for oblique projection with the conceptual means available in preclassical mechanics. In attempting ever new proofs of theorems he discovers ever new paradoxes within his conceptual system.

In the Epilogue the three preceding investigations are rounded out. We return here to the outcome of the case studies and show how afterwards the complex and difficult derivations of central theorems of classical mechanics by preclassical mechanics disappear once the results are taken not as results of operations within a conceptual system but as implicit definitions of concepts which form the basis of a new system. We shall see Descartes' precise and unproblematic use of the geometrical representation used to derive the relations of space and time in free fall – written after Galileo's final work. We shall sketch the productive misinterpretation of Descartes' concept of "determination" that enabled his disciples to read Descartes' complicated theorems as if they incorporated the parallelogram of forces for dynamics, thus formulating this theorem for classical mechanics. Finally we shall look at this kind of reconceptualization of a theory from the point of view of its results in the Galilean tradition, showing that it was Galileo's disciples who first derived the parabolic trajectory for oblique projection, although they present it merely as an explication of Galileo's *Discorsi*.

1
Concept and Inference: Descartes and Beeckman on the Fall of Bodies

So närrisch als es dem Krebse vorkommen muß,
wenn er den Menschen vorwärts gehen sieht.
(Georg Christoph Lichtenberg)

1.1 Introduction

The discovery of the law of free fall is usually considered to be a milestone in the development of modern physics and a major step in superseding medieval ways of thought. The subject of the law is the relation between the space traversed by a falling body and the time elapsed. The law states that under certain conditions the spaces traversed measured from rest are proportional to the squares of the times elapsed.

Discovering the law of free fall is partly an empirical problem. The simplest way to acquire information about the relation between the time elapsed and space traversed of a freely falling body is to estimate or measure these magnitudes. And indeed it is unlikely that the law of free fall would have been discovered in the 17th century without the new approach of seeking knowledge about nature by means of empirical investigation characteristic of this century.

Nevertheless, the empirical aspect of the problem played no major role in the discovery of the law of free fall or at least was not considered to be of major importance by scientists of the time. There are two reasons for this.

First of all, because freely falling bodies very quickly reach a relatively high velocity, the times to be measured are rather short and the spaces to be measured soon go beyond the limits of simple experimental setups. Therefore, it is not easy to acquire information about the relation between these short times and long distances that is precise enough to give an accurate picture of the mathematical nature of the relation between them. It is even difficult to achieve a correct qualitative understanding of the process of free fall. Intuitive perception is more likely to mislead than to provide a fundament for more elaborate investigation. The notion that the free fall is a continuously accelerated motion seems counterintuitive. The falling body actually seems to linger a moment at rest after being set free to fall and then seems to aquire a certain minimum speed instantaneously. Further-more, it seems intuitively more adequate to assume, that the falling body soon reaches a maximum speed than to assume that it accelerates indefinitely.

Second, the law of free fall as an empirical relation is rather meaningless as long as it is not interpreted in a theoretical framework, for instance the frame-work of classical mechanics. The mere statement of the law is only a contingent isolated fact without far-reaching theoretical implications. There is no reason to accept the restricting conditions for the truth of the new law (motion in a vac-uum), and the law cannot be generalized cosmologically so long as there is no theoretical reason for the second order importance of these conditions.

The *locus classicus* for the proof of the law of free fall in the 17th century is the "Third Day" of Galileo Galilei's *Discorsi* of 1638 (*Two New Sciences*); how-ever, several attempts to derive the law earlier in the century are known. Besides Galileo's own work on this problem dating back at least to 1604, the best known attempts were the cooperative studies of Isaac Beeckman and René Descartes. These studies, it is generally agreed, were independent of Galileo's, but they show astonishing similarities to Galileo's investigations. Beeckman and Descartes worked on this problem together towards the end of 1618, and each of them dealt with this topic in letters to Marin Mersenne in 1629; Descartes discussed the problem further in letters to Mersenne and others until at least 1643.

Isaac Beeckman (1588-1637), who was later to found a *Collegium mechanicum* for scholars and craftsmen devoted to the study of science and its technological application as well as the first metereological station in Europe (1628), first met and worked with Descartes in 1618 when he was conrector of the Latin school in Veere. They met again in 1628 and 1629 when Beeckman was rector of the Latin school in Dordrecht, and on this occasion Descartes also studied Beeckman's *Journal*; they met once more in 1634 when Beeckman visited Descartes in Amsterdam bringing with him a copy of Galileo's *Dialogo* (*Dialogue concerning the Two Chief World Systems, 1632*). We shall first discuss Descartes' and Beeckman's now famous dialogue of 1618 and somewhat later their encounter of 1629.

Ever since Koyré's influential interpretation the Beeckman-Descartes dialogue has been considered in the literature as a "comedy of errors."[1] The plot of the

---

[1] See Koyré 1978, p. 79 (originally published in French, 1939). Koyré's interpreta-tion of the documents is essentially shared by Clagett (1959), Hanson (1961), and Shea (1978). In the following we shall not criticize these interpretations on each oc-casion of diagreement, since in our view they in principle impute an anachronistic framework to the texts interpreted. The differences to our interpretation follow from their reading the geometrical figures used as rudimentary representations of velocity (in the sense of classical physics) depending on time respectively space. They thus reduce the difficulty in elaborating the concepts involved to deciding between the al-ternative of velocity being proportional to the time elapsed or to the space traversed in fall and then deriving the space-time function without modern integral calculus. Closest to our approach is Schuster's (1977) interpretation that stresses the applica-

comedy runs as follows: Beeckman, who had worked out a physical theory of motion, asked Descartes whether he could derive mathematically from this theory the quantitative relationship between space and time for a freely falling body. Descartes submitted to him a document which, as Koyré put it, "combines ... supreme mathematical elegance with the most hopeless physical confusion."[2] This document in hand, Beeckman wrote down in his *Journal* his own understanding of Descartes' answer: a proof of the law of free fall derived from the correct assumption of the proportionality between time and velocity. Descartes in turn wrote down his own understanding of his proof in his diary, but this version assumes a proportionality between space and velocity – and is consequently wrong. On top of this, it seems that neither of them realized that their understandings of the law of free fall were different. In later years Descartes gave more instances of his incorrect theorem without noticing his error. On the contrary: when Galileo, who had originally derived the correct law from the wrong assumptions, published his final proof in 1638, Descartes erroneously identified this proof with his own earlier attempts which he now rejected as fundamentally misguided.

This comedy raises two questions. First, how is it possible that the judgments on whether a specific argument is compelling or not are so divergent? We need to understand the conditions that make a proof convincing. Second, we have to acquire an understanding of the conceptual background for the question in the 17th century that led to accepting answers as adequate or rejecting them as inadequate. Therefore, we shall first summarize the medieval traditions that are related to concepts used by Descartes and Beeckman (section 1.2). We will then (section 1.3) interpret the documents of the dialogue between Descartes and Beeckman. This discussion will be restricted to those parts of the documents which directly show the deductive structure of the arguments. Afterwards (section 1.4), we shall interpret these documents again, this time in the broader context of the underlying theories of gravity, thus including parts of these documents which are not directly part of the proof but provide information about the relative importance of the assumptions and the consequences of the arguments. Finally (section 1.5), we shall discuss the deductive structure of the Descartes' proof and return to the dialogue between Descartes and Beeckman to draw consequences for the interpretation of the comedy.

---

tion of medieval concepts (Part 1, pp. 72-93). However, limited to the 1618/19 documents, his interpretation does not comprehensively account for the entire deductive structure of the conceptual system involved. Dijksterhuis (1961) also stresses the importance of the relation between the medieval concepts and the concepts of pre-classical mechanics. However, he deals only with Beeckman's proof of the law of falling bodies (pp. 329-333) because he assumes that Beeckman and not Descartes is the original author of this proof. This opinion seems to us to be incompatible with the sources.

[2] Koyré 1978, p. 83.

1.2 The Medieval Tradition

The documents we shall study were written not in symbolic but in natural language, French and Latin. Moreover, many of the terms used in the documents belong to the language used in everyday communication: distance, time, motion, moving force, etc.; these all refer to elementary human activities and belong to everyday language. As such these terms share the characteristic richness and context dependency of the semantics of currently spoken language.

Interpreting the documents according to the general knowledge of the language in which they were written would, however, necessarily lead to misunderstandings and grave mistakes. The reason is that these documents belong to a specific professional discourse in which the terms, even when not defined, have far more precise and restricted meanings than those of the same words in everyday and less controlled use of language. It is little wonder, therefore, that scholars at the time tended to formulate the core of a theory in Latin rather than in the vernacular: This had advantages not only for international communication and not only because Latin was the traditional language of scholarly discourse, but also because Latin was no longer a living language, and thus its semantics had become isolated from influences of nonscholarly discourse.

Hence, while on the one hand, the overlapping semantics of the terms common to both scholarly and everyday usage may have been of importance for the content of a theory and therefore also to its interpretation, the specific scholarly meanings of scholarly terms have to be elaborated by studying the semantic links between them as they appear in arguments conducted within a conceptual system and along such semantic connections. This method of studying the semantics of the language used does not commit us to any particular view on the indebtness of the theory studied to its medieval predecessors, in which the terms were first coined.

1.2.1 Proportions

At the time of preclassical mechanics natural language provided a rich terminology to describe the mutual dependencies of magnitudes. However, the conceptual means to give a precise scientific description of such dependencies were still limited. Whereas in classical mechanics such dependencies were conceived as functions, analytically represented by equations of variables, the medieval tradition[3] provided only the Euclidian technique of proportions, which had already been applied by Archimedes to various problems of mechanics. The linear function between spaces traversed and times elapsed in uniform motion, for instance,

---

[3] See Oresme, *De proportionibus proportionum,* in: Oresme 1968a.

was expressed by saying that the spaces are proportional to the times.[4] This technique, although limited, could also be used to represent certain nonlinear, in particular, quadratic functions.

According to the medieval theory of proportions, three magnitudes a, b, and c were said to be in "continued proportion," written

$$a : b : c$$

if (in modern notation)

$$a : b = b : c.$$

In this case the magnitude b was called the "mean proportional" between a and c, and a was said to be in the "double proportion" to c of that which a has to b, because it follows from the Euclidian theory of proportions in this case that

$$a : c = (a : b) \times (b : c) = (a : b) \times (a : b) = a^2 : b^2,$$

that is, a and c are in the same relation as the squares of a and b. Hence a quadratic relation could be expressed using the concepts continued proportion, mean proportional, and double proportion. To double a proportion $a : b$ (in classical mechanics: to square the variable that stands for them) means to continue the proportion $a : b$ by $b : c$ such that a proportion $a : b : c$ results, where $a : c$ is in double proportion of $a : b$, that is, $(a : c) = (a : b) \times (a : b)$. On the other hand, to divide a proportion $a : c$ (in classical mechanics: to take the square root of the corresponding variable) means to find a mean proportional b such that the proportion $a : b$ results, which is in *half* proportion of $a : c$, so that $a : b : c$ holds. This terminology was also used in a generalized way. The continued proportion

$$a : b : c : d,$$

for instance, denoted the equations

$$a : b = b : c = c : d.$$

In this case a was said to be in *triple* proportion to d and it follows that

$$a : d = a^3 : d^3.$$

More generally, if a, b, and c were not in continued proportion so that the ratio of a to b was not equal to the ratio of b to c, for instance,

$$a : b = m : n,$$

$$b : c = p : q,$$

then a and c were said to be in the compound ratio of $m : n$ and $p : q$ (in classical mechanics we would simply multiply two variables), for it follows that

$$a : c = (m : n) \times (p : q) = mp : nq.$$

---

[4] This was already tacitly used by Archimedes in the proof of Proposition 1 of "On Spirals"; see Archimedes 1953, p. 155.

1.2.2 The "Moments" of Motion and the Quantification of Motion by Proportions

Motion and its defining concepts (body, time, and distance) were conceived in the Aristotelean tradition as continuous magnitudes. In his *Physics* (Bk. VI) Aristotle discussed the relation between unextended points and extended continuous qualities, arguing that time is not composed of "nows" nor a line of points nor a motion out of indivisible elements of motion. Whereas the very short elements of time (whether conceived as divisible or not) had traditionally been called "momenta" in Latin, ever since William of Moerbeke's *nova translatio* of Aristotle's *Physics* (13th century) the term "momentum" was also used to designate the indivisible element of *motion*.[5] Both meanings of the term were compatible since according to Aristotle the same relation holds between the indivisible element and the continuous magnitude of time, line (space), and motion. The now or the moment of time has the same relation to time as the point to the line or the moment to motion. Hence it would have been (in principle) legitimate to designate the smallest element of space as momentum as well. And indeed, such a usage did in fact occur in the Middle Ages.[6] Finally, by the 16th century momentum had also acquired the meaning of the first tendency to motion of a suspended weight; this was later to become the technical term momentum in statics. Hence by the 16th century momentum had the connotation of the minimal magnitude of motion and of the entities by which motion was defined.[7] However, in the Aristotelean tradition, the quantification of motion did not refer to its momenta since these were conceived as limits or termini of the respective magnitudes rather than as their component parts. Motion was quantified according to the Aristotelian concept of velocity as the space traversed (in a given time)[8] and by proportions based on the Aristotelean definitions of the "quicker" of two moving bodies.

In his *Physics* Aristotle discusses the meaning of quicker, and in the course of a lengthy discussion there he makes two clarifications which later became the basis of a definition of "velocity" in terms of proportions. The first proposition states that the quicker of two bodies traverses more space in the same time, the second proposition states that the quicker of two bodies traverses the same space in less time.[9]

In Oresme in the 14th century we find a reference to Aristotle's discussion of the quicker in the *Physics*, embedded in a dissertation on complex proportions of

---

[5] Thus Moerbeke translates the (absurd) conclusion in *Physics* VI.1 deduced by Aristotle from the premises of his adversaries as: "... erit utique motus non ex motibus, sed ex momentis" (see Galluzzi 1979, p. 121).

[6] See Galluzzi 1979, p. 15, fn 41.

[7] For a detailed history of the concept momentum see Galluzzi 1979, pp. 3-149.

[8] See Maier 1949, pp. 118-120 and 1952, pp. 284-286.

[9] *Physics* VI: 232a23-232b15.

force, resistance, and velocity. The reference to Aristotle, which formulates the ratios of velocities in terms of proportions, is presented as a well-known proposition that needs no elaboration[10]:

> Now any ratio of magnitudes [or distances] would be just like the ratio of velocities with which those magnitudes [or distances] were traversed in the same time or in equal times. And a ratio of times is just like the converse ratio of velocities when it happens that equal distances are traversed in those times, which is clear from the sixth [book] of the Physics [of Aristotle].

In symbolic notation we can write

$$\text{If } t_1 = t_2, \text{ then } v_1 : v_2 = s_1 : s_2.$$

$$\text{If } s_1 = s_2, \text{ then } v_1 : v_2 = t_2 : t_1.$$

Note that the Aristotelian definition of quicker as well as the more precise formulations in the corresponding proportions do not allow a comparison of the velocities of motions that differ in *both* distance and time.

There is an interesting problem related to these proportions, which in fact (as we shall see in section 3.4.2) presented a number of puzzles when the Aristotelian concept of velocity was applied to accelerated motions. In uniform motion the spaces traversed are proportional to the times elapsed. Archimedes had already characterized uniform motions by this proportionality. In this particular case the ancient and the modern concepts of velocity coincide and the validity of the Aristotelian proportions is obvious. Moreover, a third proportion can be established, which was in fact explicitly introduced for uniform motions by Archimedes[11]:

$$\text{If } v_1 = v_2, \text{ then } s_1 : s_2 = t_1 : t_2.$$

Furthermore, in all three proportions the reverse implications are valid as well. If, for instance, the spaces traversed by two uniformly moving bodies are always in the same relation as the times, they necessarily have the same velocity. If the velocities are different, the spaces traversed cannot be in the same relation as the times.

Finally, uniform motions, as opposed to nonuniform motions, can always be compared with one another. Even if two uniformly moving bodies move for different times ($t_1 \neq t_2$) with different velocities ($v_1 \neq v_2$) and traverse different spaces ($s_1 \neq s_2$), a third body can be introduced that moves with the same velocity as the first ($v_3 = v_1$) for the same length of time as the second body ($t_3 = t_2$) thus traversing a certain space s_3. It follows that

$$s_1 : s_3 = t_1 : t_2 \text{ (proposition of Archimedes)},$$

$$s_3 : s_2 = v_1 : v_2 \text{ (Aristotelian proportion)}.$$

---

[10]   Oresme 1966, p. 303. We have simplified the translation. On the extensive use made of these proportions by Oresme, see Grant's introduction, pp. 52f and 84.
[11]   Archimedes, "On Spirals," Proposition 1 in Archimedes 1953, p. 155.

Thus the spaces traversed are in the compound ratio

$$s_1 : s_2 = (t_1 : t_2) \times (v_1 : v_2).$$

Therefore, in the case of uniform motions the relation between the velocities can always be inferred from the relations between times and spaces.

However, Aristotle's definition of velocity was not restricted to uniform motions. The crucial question arises whether the conclusions drawn in the *particular case* of uniform motions are also valid for nonuniform motions. The proof techniques of Greek mathematics did not allow this question to be answered. As we will see, one of the key problems of preclassical mechanics was to derive from Aristotle's definition of velocity consequences for nonuniform, accelerated motions. In particular, the attempts of Descartes and Galileo to derive the relation of times elapsed and spaces traversed for projectiles and falling bodies were intimately linked to this problem.

1.2.3 The Configuration of Qualities and Motions

In the 14th century yet another quantification of motion was invented which later proved to be of importance in the study of motion in the 17th century. This was the doctrine of the "configuration of qualities and motions."

The doctrine of the configuration of qualities and motions as developed in the 14th century has been viewed, since Duhem's influential interpretation, as the source of the geometrical representation by which Galileo and his contemporaries attempted to derive the law of free fall. There are, however, considerable differences of opinion as to how much this doctrine contributed to the derivation of the law.[12] These differences depend, on the one hand, on the interpretation of what is essential to the derivation of the law of free fall in the 17th century and, on the other, on the interpretation of just what is the core of the configuration theory, known primarily through the seminal work of Nicole Oresme. A further difficulty is presented by the rather well substantiated fact that some essentials of the configuration doctrine were not handed down to the 17th century. We shall, therefore, first discuss some essentials of the configurations of qualities and then those elements of it which formed part of the later tradition.

The doctrine of the configuration of qualities was conceived to depict and represent physical and moral intensive qualities. Intensive qualities are those qualities which have a quantity but not coexisting or successive parts (e.g., "heat," "whiteness," and "charity"). The intensity of a quality is expressed by degrees. The quantity of an intensive quality in a substance was thus conceived as dependent, on the one hand, on the intensity of the quality and, on the other hand, on the size of the subject "informed" by the quality. The subject was represented by

---

[12] See Maier 1952, pp. 354-384.

a horizontal line or plane (*longitudo* or *extensio*) and the intensity of the quality and its degree by lines drawn perpendicularly from various points of the subject line (*latitudo*, *intensio*, *gradus*). The proportions of the lines of intensity were conceived as corresponding to the proportions between the intensities of the quality at these points of the subject and the whole figure thus obtained represented and depicted the configuration and the quantity of the quality.[13]

If the subject is represented by a line and if all intensities of the quality are equal (i.e., if the quality is "uniform"), then the figure that represents the overall quantity of the quality is a rectangle; if the intensities are not all equal, the quality is said to be "difform." However, if the intensities change in proportion to the subject (i.e., if the differences between the intensities are proportional to the lengths of the corresponding parts of the line representing the subject) the quality is said to be "uniformly difform"; and its quantity is represented by a figure with a straight slope, either a right triangle or right trapezium (see Fig. 1.1) depending on whether the least intensity is of "no degree" or of some determinate degree.

While the line of "intensity" represents unequivocally the intensity of the quality, the "extension" of the subject could be interpreted either as the spatial extent of the substance which has the quality or else as the *time* during which the quality is present. It is also possible to represent both extensions by adding a further dimension to the figure. The whole figure represents the "quantity of the quality." Thus a rectangle could represent the overall quantity of a quality with an equally distributed intensity in the subject at a particular time and a box could represent the quantity of this quality over a certain period of time.

The doctrine of configuration of qualities was applied to various cases of "alteration" including local motion, and although the latter was of only minor importance in the 14th century, it is this application which is usually in the focus of scholarly attention, because the representational technique of the configuration of qualities was applied to derive the law of free fall in the 17th century. According to the two meanings of "extensio" in the configuration of qualities, in motion too there are two "subjects" to be considered: the moving body and the time of motion. Thus different parts of a moving body can have different degrees of velocity if the body is, for instance, rotating, and the same parts can have different degrees of velocity in successive times if the body is accelerating.[14]

In the application of the configuration of qualities to local motion the form of the figure depicts one of the motion's properties (uniform, uniformly accelerated,

---

[13] This twofold meaning of the configuration of qualities is the basis of the very different interpretations of Duhem and Maier. Whereas Duhem saw in this technique a rudimentary form of analytical geometry, Maier insisted on the "symbolic" character of the configuration (see Maier 1949, p. 125, note, and 1952, pp. 290-291, 306-307, 312, and 328). Clagett (1968a, pp. 15 and 25f) discusses both interpretations.

[14] See Maier 1949, pp. 116ff; 1952, pp. 277 and 314ff; see also Clagett's introduction to Oresme 1968a, pp. 31 and 35.

and difformly accelerated). The area of the figure represents, in general, the quantity of the quality and hence in the case of local motion the total quantity of velocity (*quantitas velocitatis totalis*). When the *extensio* represents time and not some spatial dimension of the body moved, Oresme interprets the area of the figure to represent the distance traversed. This is possible because of the concept of motion involved. In this case *velocitas* is conceived according to the Aristotelian definition of velocity as the space traversed in a given time.[15]

The most celebrated conclusion drawn by Oresme from his application of the configuration of qualities to motions is the validity of the "Merton Rule" for the configuration of a uniformly accelerated (uniformly difform) motion. The importance attributed to this consideration is, of course, due to its role in Galileo's derivation of the law of free fall, but for Oresme this is only one among many corollaries to the doctrine and is not of paramount importance. The Merton Rule (invented at Merton College, Oxford) refers first to uniformly difform qualities, in general, and notes that the right triangle (or trapezium) is equal in area to a rectangle whose height is the mean height of the triangle (or trapezium). Thus, the quantity of a uniformly difform quality can be compared to the quantity of a uniform quality of mean intensity (see Fig. 1.1). In Oresme's words, a uniformly difform quality "is of the same quantity as would be the quality of the same or equal subject that is uniform according to the degree of the middle point of the same subject." After some further elaborations of different applications, Oresme states that "one should speak of velocity in completely the same fashion as linear quality, so long as the middle instant of the time measuring a velocity of this kind is taken in place of the middle point [of the subject]."[16]

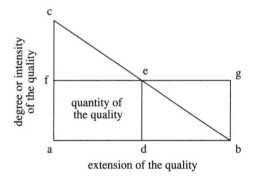

Fig. 1.1

---

[15] It should be noted that neither this concept of "total quantity of velocity" represented by the area nor the intensity of the velocity represented by the altitude corresponds to the modern concept of velocity, as will become clear below. For a detailed discussion of the concept of velocity in this period, see Maier 1952, pp. 285f, 293, 316f, 331, 337-342, and 351f; Maier 1949, p. 128f.

[16] *De Configurationibus* III.vii; Oresme 1968a, 409f. On the application of the "Merton Rule" in the *De Configurationibus*, see Maier, 1952, pp. 334f and 347f, and Clagett's introduction to Oresme 1968a, p. 46.

In his *Questiones super geometriam Euclidis* Oresme derives explicitly as a consequence of this application of the Merton Rule to uniform decelerated local motion that the total velocities, i.e., the spaces traversed can be calculated through their proportionality to the squares of the times[17]:

> ... in the case of every subject uniformly difform [in quality] to no degree, the ratio of the whole quality [*qualitatis totius*] to the quality of a part terminated at no degree is as the double proportion of the whole subject to that part [of the subject] ... the ratio of the qualities is as the double proportion of subject to subject [...]. And one ought to speak in this way of the quality of a surface with respect to its subject and of the quality of a body with respect to its subject, and of velocities in motions with respect to time.

The context of this argument leaves no doubt that in the case of velocities of motions the total quality is the space traversed, just as it is in the corresponding diagram of classical mechanics. However, there is also an essential difference: Whereas in classical physics this conclusion could be derived only through a summation of infinitesimals constituted by velocities (representing what Oresme called the degrees of velocity) conceived as uniform over infinitely small time intervals, in Oresme's configuration doctrine it followed *directly* from the interpretation as the area of the figure – and not from the summation of vertical lines – as velocity in the sense of Aristotle's definition. He made this perfectly clear when he applied the Merton Rule to a uniformly decelerating body[18]:

> If *a* is moved uniformly for an hour and *b* is uniformly decelerated in the same hour by beginning from a degree [of velocity] twice [that of *a*] and terminating at no degree, then they will traverse equal distances, as can easily be proved. Therefore, by the definition of velocity, it ought to be conceded that they were moved equally quickly for the whole hour. Therefore, the whole motion of *b* ought not to be said to be as fast as is its maximum degree [of velocity].

The tradition of Oresme's doctrine of the configuration of qualities presents a grave problem to those historians who interpret Galileo's derivation of the law of free fall as consisting in the assumption or insight that the motion in free fall is uniformally accelerated and in the application of the geometrical method of calculation (in addition to the Merton Rule) as developed by Oresme. This interpretation meets with the difficulty that the later tradition of the doctrine of configuration omitted precisely its use as a calculation device. In the most important books, the geometrical representation degenerated to an illustration of conceptual relations expressed in language. Thus, for instance, in the presumably most influential book *De latitudinibus formarum* (composed around 1390 and widely available in manuscript copies and at least three printings in the Renaissance), which was often erroneously attributed to Oresme himself and which summarizes the content of the first book of Oresme's *De configurationibus*, there is no trace either of Oresme's discussion of the *quantitas*

---

[17]   See Oresme 1968a, pp. 557 and 559.

[18]   See *Questiones super geometriam Euclidis*, in Oresme 1968a, pp. 559-561.

qualitatis or of the application of the Merton Rule. Anneliese Maier has succinctly summarized the character of the book, saying that in the configuration doctrine there "is not so much a method of calculation, but rather a method of graphical representation, of illustration, suited to provide an easier overview of some opaque and complicated concepts and relations."[19] The same is true of all other books printed in the 15th and 16th centuries in which this rudimentary form of the configurations illustrates in the margins the concepts discussed in the text itself. However, just as in Oresme, so too in the later literature, the graphical illustration and the conceptual relations involved were so intimately integrated that qualities were characterized by the geometrical figure representing them (e.g., "a right-triangular quality"), and the geometrical representations were named after the kinds of qualities they represented (e.g., "a uniformly difform surface").[20] The following text taken from Jacobus de Sancto Martino's *De perfectione specierum* shows clearly the function of the figures as representing conceptual links between terms rather than as calculation devices[21]:

For latitude is to be imagined by a plane surface and this in many ways. For a certain latitude is uniform:▭. Such a one is to be imagined by a rectilinear plane surface in which the lines bounding it that are opposite each other are equidistant and such lines are called parallel lines, or a surface in which the bounding lines meet so as to form four equal angles. But another one is called a difform latitude and there are two kinds of it. One is uniformly difform: ◺; the other is difformly difform: ◿. A uniformly difform latitude either begins from no degree and is terminated at a certain degree: ◺; or it is not terminated: the latter is to be imagined by a rectilinear surface whose lines form an acute angle [∠]. The other kind begins from a certain degree and is terminated at a certain degree: ◳; or it is not terminated and in which case it is to be imagined by a rectilinear surface whose lines are so ordered that they do not form a single angle but which if protracted would form a rectilinear, acute angle: ◁. Or a uniformly difform latitude is one which has equal excess between equally distant degrees [...].

A degree is also said to be a certain thing in the latitude which indeed exists imaginatively indivisible according to extension while being divisible according to intensity; and this is to be imagined by a straight line perpendicularly ascending and producing the whole latitude in respect to intensity. From this it is evident that in any latitude a degree to which a longer line corresponds is said to be more intense and one to which a shorter line corresponds is said to be more remiss. It also follows from this and the given description of uniformly difform that in such a uniformly difform latitude, two degrees that are equally intense cannot be found. From this it is further inferred that any degree more intense than another degree exceeds it in some determined ratio; double, or triple or three-halves, or at least in some irrational ratio.

---

[19] See Maier 1952, p. 371; see pp. 359, 368ff, and 383f.
[20] *De configurationibus* I.viii, Oresme 1968a, p. 185. See Clagett's introduction to Oresme 1968a, pp. 19 and 82. See also Lewis 1980.
[21] Jacobus is presumed to be the author of *De latitudinibus formarum*. Text and translation by Clagett in Oresme 1968a, pp. 89f.

1.2.4 "Impetus" as the Cause of Motion and of Acceleration

The theory of impetus can be considered primarily as a theory of causality. It undertakes to explain how a projectile (*proiectum separatum*) can be moved when it is separated from the projecting agent (*proiciens*) without appealing to action at a distance. The answer provided by the impetus theory is that the projecting agent impresses a force (*vis impressa*) onto the mobile and that this force continues to act when the projectile has been separated from the primary agent. This impressed force is the impetus, which mediates between the primary cause and the effect, since it is produced by the primary cause and produces the effect even when the body is no longer attached to the primary agent.[22] It is important to remember, however, that the original agent (say the hand) produces the impetus only through the mediation of motion (*mediante motu*): If the hand does not move the projectile as long as it touches it, then it also does not produce in it an impetus which will act when the mobile is separated from the hand. Hence the motion of the projectile is explained as follows: the exterior agent (*movens*) produces both an initial motion of the mobile and an impetus which will inhere in the body and continue to act when the mobile is separated from the agent.

These essentials of the impetus theory are subject to very different interpretations about the cases to which the theory may be applied (besides the motion of projectiles) and to the concrete progression of the motion thus produced. We shall not consider these questions here since they lie beyond the scope of our present interest and have been discussed in great detail by Anneliese Maier and others. We shall simply note the fact that in the 14th century the impetus theory was applied to the natural motion of heavy bodies falling on the surface of the earth; but here, too, different interpretations were possible. We shall concentrate on the views of Buridan and Oresme, who apparently conceived the two alternatives later considered by Descartes.

These alternative interpretations are inherent in the essentials of the theory, though they need not have been clear as alternatives to Buridan and Oresme. The impetus was thought to explain the motion of a body after separation from the exterior agent which had set the resting body in motion; and the impetus was conceived to have been produced together with the initial motion. But this initial action of the exterior cause could be conceived either as imparting a *velocity* to the body (and thus also imparting an impetus proportional to velocity) or as *accelerating* it from rest to a certain velocity (and thus imparting an impetus proportional to acceleration). Hence the impetus could be conceived as dependent

---

[22] The significance of the impetus theory as a theory of causality was stressed in various places by Maier. See, for example, her "Die Impetustheorie," in Maier 1951, pp. 113-314, and "Das Problem der Gravitation," in Maier 1952, pp. 143-254. Wolff (1978) has also emphasized this idea showing that proponents of the impetus theory also employed the same notion of causality as mediated by the transference of an entity from the agent to the object acted upon in other contexts as well, e.g., in the theological theory of the sacraments or in the economic theory of value production.

either on velocity (Buridan) or acceleration (Oresme) and in turn as producing either velocity (Buridan) or acceleration (Oresme). However, the unclear characterizations offered by Buridan and Oresme show that reference to the impetus produced with the initial motion did not necessarily entail recognition that these possibilities are clear alternatives.

Buridan explains the accelerated motion of heavy falling bodies [23]:

> ... it follows that one must imagine that a heavy body not only acquires motion unto itself from its principal mover, i.e., its gravity, but that it also acquires unto itself a certain impetus with that motion. This impetus has the power of moving the heavy body in conjunction with the permanent natural gravity. And because that impetus is acquired in common with the motion, hence the swifter the motion is, the greater and stronger the impetus is. So, therefore, from the beginning the heavy body is moved by its natural gravity only; hence it is moved slowly. Afterwards it is moved by that same gravity and by the impetus acquired at the same time; consequently, it is moved more swiftly. And because the movement becomes swifter, therefore the impetus also becomes greater and stronger, and thus the heavy body is moved by its natural gravity and by that greater impetus simultaneously and so it will again be moved faster; and thus it will always and continually be accelerated to the end.

As far as these thoughts of Buridan's allow for a quantitative interpretation of acceleration, it seems that acceleration is produced by the increasing impetus (in addition to gravity). Impetus is produced by gravity but its magnitude is dependent on the speed of the falling body (perhaps proportional to the speed) and the impetus as such produces a speed (perhaps proportional to the impetus), not an acceleration.

Oresme, on the other hand, conceives the impetus as produced by the mover through the mediation of the initial motion (*a motore mediante motu*) and as producing on its part acceleration. The impetus or the "accidental gravity" (in distinction to the essential, constant gravity) grows "due to the acceleration of motion by which a certain ability or impetus and a certain accidental fortification to move faster is acquired."[24] Now, the reason that a falling body moves faster at the end than at the beginning of the fall is that "since it is accelerated in the be-

---

[23] John Buridan, *Quaestiones super libris quattuor de caelo et mundo*. The Latin text was edited by E.A. Moody (Buridan 1942); this passage is also quoted in Maier's "Das Problem der Gravitation" (Maier 1952, pp. 201-202). The translation is taken from Clagett 1959, pp. 560-561. Clagett observes that from Buridan's wording it is not clear whether the velocity is proportional to time or to distance of fall, and he even doubts whether the stated dependency of speed on fall should mean proportionality at all. Clagett rather suggests that Buridan "made no clear distinction between the mathematical difference involved in saying that the velocity increases directly as the distance of fall and saying that it increases directly as the time of fall" (Clagett 1959, p. 563, cf. p. 552).

[24] "Ex velocitatione motus per quam acquiritur quaedam habilitas vel impetus et quaedam fortificatio accidentalis ad velocius movendum." Quoted from Oresme's Latin commentary to *De caelo* in Maier 1951, p. 244.

ginning, it acquires such an impetus and this impetus coassists in producing motion. Thus, other things being equal, the movement is faster."[25] The impetus, which in the French commentary to *De caelo* is called "impetuosité," is caused by the "enforcement of the increase in speed [*par l'enforcement de l'acressement de l'isneleté*] whether in natural or in violent motion and is the cause of motion of objects thrown from the hand or an instrument."[26] It is by this impetuosity that Oresme explains why a falling body would not immediately come to rest in the center of the universe but oscillate back and forth beyond the center of the earth. Discussing again this problem, Oresme gives what is perhaps the most clear formulation of the relation between impetus and acceleration, when he says that the impetuosity is acquired by the growth of the speed of motion of the body.[27]

But if the impetus is dependent on the acceleration of the body (and not on its speed) and further accelerates it, then the acceleration in fall is not uniform but growing.

1.3 Descartes' Proof of the Law of Fall

There are essentially three documents related to Beeckman's original question about the quantitative relationship between space and time for a falling body; all three date from 1618, although two of them were preserved only in later copies. These three documents are not equally elaborated, but each one contains an explicit version of Descartes' answer to Beeckman's question. The documents are Descartes' original answer to Beeckman's question (see section 1.3.1), the note which Beeckman wrote down in his *Journal* in reaction to this answer (see section 1.3.2), and a note which Descartes wrote down immediately after the event in his diary (see section 1.3.3).[28] More than ten years later Descartes referred again to his original answer in letters to Mersenne. These letters elaborate several details of the problem even more specific (see section 1.3.4). Finally Descartes reconsidered his answer on certain occasions, in particular when Galileo published his *Dialogo* and his *Discorsi* (see section 1.3.5).

---

[25] Latin commentary to *De caelo*; see Maier 1951, p. 246.

[26] Bk. I, chap. 18; Oresme 1968b, p. 144.

[27] Explaining the oscillation to and fro at the center of the Earth Oresme says: "Et la cause est pour l'impetuosité ou embruissement que elle [la pierre] aquiert par la cressance de l'isneleté de son mouvement jouste ce que fu dit plus a plain ou. xiiie chapitre"(Oresme 1968b, p. 572). See Maier 1951, pp. 252f and 1952, pp. 203-206.

[28] The documents are printed in Beeckman's *Journal*, J IV, 49-52, J I, 260-265, 360-361; and in Descartes' *Oeuvres*, AT X, 75-78, 58-61, and 219-220. All three are given in translation in Chapter 5 below; see documents 5.1.1, 5.1.2, and 5.1.3.

1.3.1   Descartes' Initial Document (1618)

The first version of Descartes' argument on the law of free fall is contained in a paper he drafted in answer to Beeckman's inquiry, probably during their cooperative work at Breda in November and December of 1618. It has been handed down to posterity only because Beeckman had it copied into his *Journal* in 1628, i.e., ten years after its original composition.[29] The paper as copied does not contain Beeckman's original question, but some information about the question can be inferred from the context.

We know from Beeckman's own note in his *Journal*, which we shall discuss later, that he had a qualitative explanation of acceleration in free fall. This explanation is based on the assumption that "every object once put in motion will never come to rest unless because of an external impediment," which Beeckman had already endorsed years before.[30] A falling body would hence retain its motion if not acted upon from without, so that the influence of gravitation increases its motion. The question presented to Descartes concerned the quantitative relation between the time elapsed and space traversed. According to Beeckman's note[31] it was as follows:

> Whether one could know how much space an object traverses in a single hour if it is known how much it traverses in two hours according to my principles, viz. that *in a vacuum, what is once in motion will always be in motion*, and supposing that there is a vacuum between the Earth and the falling stone.

Descartes' initial document begins with the statement that the "force" of the falling object increases as "the transverse lines *de, fg*, *hi*, and the infinite other transverse lines that can be imagined between them" (see Fig. 1.2). Thus he assimilates the dynamical concept force to the geometrical representation of the kinematics of motion by identifying force with intensity[32] :

> In the proposed question, in which it is imagined that at each single time a new force is added to that with which the heavy body tends downward, I say that this force increases in the same manner as do the transverse lines *de, fg*, *hi*, and the infinite other transverse lines that can be imagined between them.

Descartes then gives a detailed justification for this identification of force with intensity. This justification is carried out in two steps. In the first step he derives the quantity of the increasing force "added" to the falling body by the "attractive force of the earth" under the assumption that this force is added in discrete minima.

---

[29] See the note by Cornelis de Waard in J IV, 49; compare also the"Advertissement" by Charles Adam in AT X, 26f.

[30] J I, 24; see J I, 10 and *passim*. Beeckman's principle is not restricted to rectilinear motion: "*Id, quod semel movetur, in vacuo semper movetur*, sive secundum lineam rectam seu circularem ..." (J I, 353).

[31] J I, 263; AT X, 60.

[32] AT X, 75-78; J IV, 49-52.

To demonstrate this I take as the first minimum or point of motion, caused by the first attractive force of the Earth that can be imagined, the square *alde* [Fig. 1.2]. For the second minimum of motion we have the double, namely *dmgf*; the force, which was in the first minimum persists and a new, equal force is added to it. Thus in the third minimum of motion there will be three forces, namely those of the first, second, and third time minima, and so on. This number is triangular, as I will perhaps explain more fully elsewhere, and it appears that the figure of the triangle *abc* represents it.

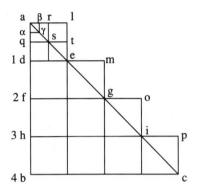

Fig. 1.2 (AT X, 76)

In the second step Descartes discusses the consequences of taking ever smaller minima until these are reduced to "true minima," i.e., points.

But, you will say, there are parts *ale*, *emg*, *goi*, etc., which protrude outside of the figure of the triangle. Therefore, the figure of the triangle cannot explain this progression. But I reply that these protuberant parts originate because we have ascribed latitude to the minima which must be imagined as indivisible and as consisting of no parts. This is demonstrated as follows. I divide the minimum *ad* into two equal ones at *q*; then *arsq* will be the [first] minimum of motion, and *qted* the second minimum of motion, in which there will be two minima of force. In the same manner we divide *df*, *fh*, etc. Then we will have the protuberant parts *ars*, *ste*, etc. Clearly they are smaller than the protuberant part *ale*.

Furthermore, if I take a smaller minimum such as *aα*, then the protuberant parts will be yet smaller, such as *aβγ*, etc. If, finally, I take as this minimum the true minimum, i.e., the point, then there will be no protuberant parts, for they clearly could not be the whole point, but only a half of the minimum *alde*, and there is no half of a point.

The argument is based on establishing that the area of the geometrical figure represents the quantity of motion, in short: motion. This motion is conceived of as consisting of "minima of motion," which through the entire document are consistently conceptualized in the following way (compare our Fig. 1.3): The "minimum of motion" is the area of a rectangle, one side of which represents the total of successively added "minima of force," the other the "minimum of time."

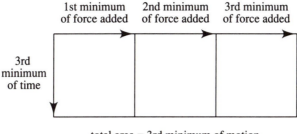

total area = 3rd minimum of motion

Fig. 1.3 Reconstruction of Descartes' concept of minimum of motion

Thus, the minimum of motion (the area) is produced by the accumulated minima of force acting in a minimum of time. Since attraction is supposed to add equal minima of force in successive discrete minima of time, it follows that the minima of motion increase in arithmetic progression. The resulting area, representing the total motion, has a steplike slope and converges to a triangle by reducing the minima of time to points. With this, the justification of the identification of force and intensity of motion is complete.

The term minimum as used in this argument proves to be synonymous with the term moment as used in Beeckman's note in his *Journal* and in Descartes' diary note. Applied to the extension (line *ab* in Fig. 1.2) minimum is as indifferent to a temporal or spatial interpretation as the concept of moment. Although minimum of extension was explicitly interpreted by Descartes in a temporal sense in the argument thus far presented,[33] it should be noted that the argument is not dependent on this interpretation. The ambiguity is thus not yet resolved, and later Descartes switches over to a spatial interpretation. This is manifested in the following sections of the argument and especially in his diary note.

In the subsequent short passage Descartes draws the conclusion in answer to Beeckman's question.

> From which it clearly follows that if we imagine, for example, a stone which is attracted in a vacuum by the earth, from *a* to *b* [Fig. 1.5], by a force which flows equally from the Earth while the first persists, then the first motion at *a* will be to the last at *b* as the point *a* is to the line *bc*. The part *gb*, which is half, will be covered by the stone three times more quickly than the other half *ag*, because it will be drawn by the Earth with three times the force. The space *fgbc* is three times the space *afg*, as is easily proved. And one can say this of the other parts proportionately.

---

[33] Descartes says, "those [forces] of the first, second, and third time minima"; this is overlooked by Koyré (1978, p. 84) and Hanson (1961, p. 48); Shea (1978, pp. 141f) even sums up the argument incorrectly by omitting the part of the argument cited above. Only Schuster (1977, pp. 72-84) discusses Descartes' denotation of the minima as "minima temporis." Accordingly, he severely criticizes the standard interpretation which can be traced back to Koyré's mistake (pp. 84-88).

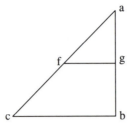

Fig. 1.4 (AT X, 77)

Descartes argues that, since the relation between the area associated with the upper half of the extension and the area associated with the lower half (and hence the relation of the respective quantities of motion) is one to three, it follows that "the part gb, which is half, will be covered by the stone three times more quickly than the other half ag." If we keep Beeckman's question in mind, then according to Descartes' temporal interpretation of the extension ab it follows that the halves ag and gb represent the first and the second hours, respectively. According to the relevant Aristotelian proportion (in equal times velocities are proportional to spaces traversed) it follows that in the two consecutive times the relation between the spaces traversed is one to three.

This is in fact the correct result. If nothing else were known, it might have seemed beyond doubt that Descartes had formulated the law of free fall. And indeed, this interpretation of Descartes' argument was endorsed by Beeckman, who furthermore expressed the result obtained by Descartes in a general form as a "double proportion" between time and space. But because of the ambiguity of the concepts *moment* and *minimum*, Descartes' concluding remark can be interpreted as stating that the stone is falling three times quicker in the second half of the space traversed. As we shall see later, this interpretation was in fact endorsed by Descartes himself in his diary note.

Note that in our interpretation the implicit inconsistency of the spatial and temporal interpretations of minimum results necessarily from the application of the given conceptual framework to a new area of investigation and thus paves the way for conceptual development to the concepts of classical mechanics. In light of Descartes' diary note, Koyré (and other scholars following him) interpreted this section of the initial document as also committed to the spatial interpretation in spite of the explicit time minima of the text and thus overlooked the ambiguity discussed above. Believing that Descartes interpreted extension consistently as space and that Beeckman's temporal interpretation is hence independent of Descartes, Koyré read the difference in interpretation as a comedy of errors.

In the note in his *Journal* on Descartes' initial document, Beeckman refers only to the parts quoted above. We shall discuss the rest of Descartes' document later.

1.3.2 Beeckman's Note in his *Journal* (1618)

The part of Beeckman's note in his *Journal* which we will discuss in this subsection ends with the remark: "This was demonstrated thus by Mr. Peron [Descartes]."[34] Indeed the text closely follows Descartes' initial document.

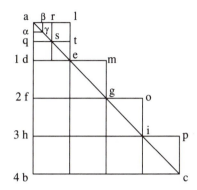

Fig. 1.5 Descartes' figure (AT X, 76)

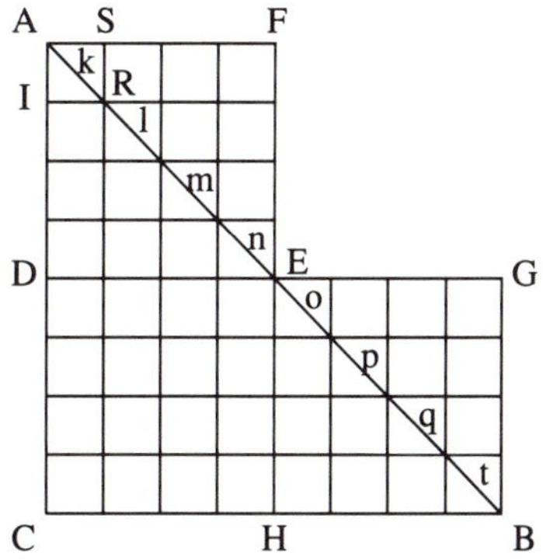

Fig. 1.6 Beeckman's figure (J I, 262)

Descartes:

In the proposed question, in which it is imagined that at each single time a new force is added to that with which the heavy body tends downward, I say that this force increases in the same manner as do the transverse lines *de, fg, hi,* [Fig. 1.5] and the infinite other transverse lines that can be imagined between them.

To demonstrate this I take as the first minimum or point of motion, caused by the first attractive force of the Earth that can be imagined, the square *alde*.

For the second minimum of motion we have the double, namely *dmgf*; the force, which was in the first minimum persists and a new, equal force is added to it.

Thus in the third minimum of motion there will be three forces, namely those of the first, second, and third time minima, and so on.

Beeckman:

Objects are moved downward toward the center of the Earth, the intermediate space being a vacuum, in the following manner:

In the first moment [Fig. 1.6] so much space is traversed as can be by the attraction of the Earth [*per Terrae tractionem*].

[The object] continues in this motion in the second moment and a new motion of attraction is added, so that in the second moment double the space is traversed.

In the third moment, the doubled space is maintained, to which is added a third space resulting from the attraction of the Earth so that in the one [i.e., the third] moment a space triple the first space is traversed.

---

[34] J I, 260-265; AT X, 58-61 (in part).

The first important difference between the two is that Beeckman completely omits Descartes' dynamical reasoning. Instead of beginning with the addition of forces, he considers the acceleration as resulting directly from the addition of "motions of attraction." Referring to the geometrical representation, Beeckman explicitly identifies the area with space traversed in fall. He does not justify this identification, but it follows from his question and from the Aristotelian definition of velocity as space traversed in a certain time. Beeckman looked for the spaces traversed in consecutive equal times, and the appropriate proportion (for equal times velocities are proportional to spaces traversed) permits an interpretation of the area as representing the quantity of motion or the space traversed. Since Beeckman already switches over from motion to space traversed at the very beginning of his note, the temporal interpretation of extention is immediately given and consistently applied throughout the entire argument.

From this, the second difference to Descartes' initial document follows: Descartes inferred that the stone falls "three times more quickly" in the second half of the fall than in the first from the relation of the corresponding areas; Beeckman on the other hand states explicitly that the spaces traversed are in "double proportion" of the times elapsed. This, in fact, is the way to express a quadratic relation by means of proportions.

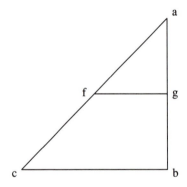

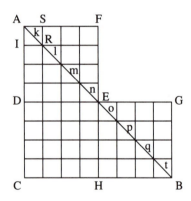

Fig. 1.7 Descartes' figure (AT X, 77)   Fig. 1.8 Beeckman's figure (J I, 262)

Descartes:

This number is triangular, as I will perhaps explain more fully elsewhere, and it appears that the figure of the triangle *abc* represents it ...

Beeckman:

Since these moments are indivisibles [*individua*], you will have a space such as *ade* [Fig. 1.8] through which a body falls in one hour. **The space through which it falls in two hours doubles the proportion of time, i.e., *ade* to *acb*, which is the double proportion of *ad* to *ac*** [...].

From which it clearly follows that if we imagine, for example, a stone which is attracted in a vacuum by the earth, from a to b [Fig. 1.7], by a force which flows equally from the Earth while the first persists, then the first motion at a will be to the last at b as the point a is to the line bc. **The part gb, which is half, will be covered by the stone three times more quickly than the other half ag,** because it will be drawn by the Earth with three times the force. The space $fgbc$ is three times the space afg, as is easily proved. And one can say this of the other parts proportionately [emphasis added].

It remains, therefore, that the space through which something falls in one hour is related to the space through which it falls in two hours as the triangle ade is to the triangle acb [emphasis added].

The intermediate part of Beeckman's note, skipped over above, closely follows Descartes' discussion of the consequences of taking ever smaller minima until these are reduced to points:

Descartes:

But, you will say, there are parts ale, emg, goi, etc., which protrude outside of the figure of the triangle. Therefore, the figure of the triangle cannot explain this progression. But I reply that these protuberant parts originate because we have ascribed latitude to the minima which must be imagined as indivisible and as consisting of no parts. This is demonstrated as follows. I divide the minimum ad into two equal ones at q; then $arsq$ will be the [first] minimum of motion, and $qted$ the second minimum of motion, in which there will be two minima of force. In the same manner we divide df, fh, etc. Then we will have the protuberant parts ars, ste, etc. Clearly they are smaller than the protuberant part ale.

Furthermore, if I take a smaller minimum such as $a\alpha$, then the protuberant parts will be yet smaller, such as $a\beta\gamma$, etc. If, finally, I take as this minimum the true minimum, i.e., the point, then there will be no protuberant parts, for they clearly could not be the whole point, but only a half of the minimum $alde$, and there is no half of a point.

Beeckman:

For let the moment [*momentum*] of space through which something falls in one hour be of some magnitude, say $adef$. In two hours it will go through three such moments, namely $afegbhcd$. But $afed$ consists of ade with afe, and $afegbhcd$ consists of acb with afe and egb, i.e., the double of afe. Thus, if the moment be $airs$, the proportion of space to space is as ade with $klmn$ to acb with $klmnopqt$, i.e., double $klmn$. But $klmn$ is much smaller than afe.

Since, therefore, the proportion of space traversed to space traversed will consist of the proportion of triangle to triangle with some equal additions to each term, and since these equal additions continually become smaller as the moments of space become smaller, it follows that these additions would be of no quantity [*nullius quantitatis*] when a moment of no quantity is taken. But such is the moment of space through which an object falls.

On the basis of the remaining part of Beeckman's note containing his commentary on Descartes' proof we shall later discuss the question of whether Beeckman's result can be considered as the discovery of the law of free fall.

1.3.3 Descartes' Diary Note (1618)

Not only Beeckman but also Descartes made a note in his diary on his solution of the problem posed by Beeckman.[35] In the following we shall quote and discuss that part of this note which addresses the topics also considered by Beeckman, and we shall come back later to its concluding remark, which we omit here. Descartes wrote in his diary:

It happened a few days ago that I came to know an extremely clever man who posed me the following problem:

A stone, he said, descends from A to B in one hour [Fig. 1,9]; it is perpetually attracted by the Earth with the same force without losing any of the speed impressed upon it by the previous attraction. According to him, that which moves in a vacuum will move always. He asks in what time such a space will be traversed?

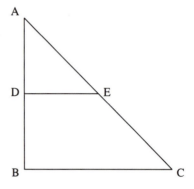

Fig. 1.9 (AT X, 219)

I have solved the problem. In the right-angled isosceles triangle, space ABC represents the motion; the inequality of the space from point A to the base BC [represents] the inequality of the motion. Therefore, AD will be traversed in the time represented by ADE; and DB in the time represented by DEBC; it being noted that the smaller space represents the slower motion. AED is one-third of DEBC; therefore it [the stone] will traverse AD three times more slowly than DB.

Descartes' diary note, written shortly after his initial document, is only a very short summary of his argument. The differences between the initial document and the diary note indicate that Descartes did not have the original when he wrote the note (presumably because it was already in Beeckman's possession). However, the diary note is so short that we can understand it only with the help of information contained in the initial document. Consider, for example, the question itself. "A stone, he said, descends from A to B in one hour ... He asks in what time such a space will be traversed?" This somewhat peculiar question

---

35   AT X, 219-220; J I, 360-361; this document, too, was preserved only in a later copy – in this case made by Leibniz 70 years later.

indicates that Descartes has already formulated his report of the question in light of his argument in the initial document. There, the area of the geometrical representation was identified with the quantity of motion, so that by applying the second possible proportion (for equal spaces velocities are inversely proportional to times elapsed), the spatial interpretation of the extension follows immediately, and hence the magnitude looked for is shifted from space to time. Since Descartes at the same time quotes the question as really put by Beeckman, the problem as stated becomes nonsensical.

The short answer following the question essentially contains the statement that AD, the first half of the extension AB, is traversed three times slower than the second half DB because the area of the figure represents the motion and the partial area ADE constructed on AD is only one-third of the partial area DEBC constructed on DB. Descartes infers that the motion ADE needs three times as much time as the following motion DEBC. This is exactly what we claimed to be the second possible interpretation of the concluding remark in the initial document. The extension AB is now unambiguously interpreted as space traversed. Consequently, the second medieval proportion is applied (for equal spaces velocities are inversely proportional to times elapsed) so that the areas represent the times (the relation between the areas represents the inverse relation between the times). What looks like hopeless confusion if the concept of velocity of classical mechanics is presupposed is hence perfectly reasonable in the logical framework of the concepts involved.

1.3.4 Descartes' Letters to Mersenne (1629-1631)

In the spring and summer of 1629 Beeckman and Mersenne corresponded on the phenomenon of falling bodies. In this context, Mersenne posed some questions concerning this topic to Descartes as well. Descartes discussed the problem of falling bodies in two letters (13 Nov. and 18 Dec.), the latter of which asserted that the solution is the same as the one contained in his diary note of 1618 which he consulted when writing the letters[36]:

> ... the speed of motion in a vacuum always increases in the proportion which I mentioned above and which I sought eleven years ago when the issue was presented to me and noted in my notebooks at that time.

And indeed the first letter contains an elaboration of the very short diary note. Contrary to Koyré,[37] we consider Descartes' present argument to follow closely his previous one of 1618, if the latter is interpreted on the background of its conceptual framework. The argument of 1629 follows the diary note in interpret-

---

[36] AT I, 69-75 and 82-104; here p. 91. A later letter (Oct./Nov. 1631; AT I, 226-232) reiterates the position. See documents 5.1.5, 5.1.6, and 5.1.7.

[37] See Koyré 1978, p. 86.

ing extension as space and in its almost verbatim repetition of the derivation of the implied (erroneous) result. But this time, the dynamical justification given in the initial document (which was omitted in the diary note) is reproduced.[38]

> As to your question concerning the principle on the basis of which I calculate the time in which a weight, which is attached to a chord of 2, 4, 8, and 16 feet, descends, although I will insert it [the principle] into my Physics, I do not want to make you wait for it until then, and I will try to explain it. First I assume that the motion once impressed into a body remains there always, unless it is removed from it by some other cause, i.e., that which has once started to move in a vacuum moves always and with equal speed. Suppose, then, a weight at A, pushed by its gravity toward C. I say that if its gravity leaves it as soon as it has begun to move, it would nonetheless continue with the same motion until it arrived at C. But it would descend neither more quickly nor more slowly from A to B than from B to C. But since this is not the case, and [the weight] retains its gravity which pushes it downward and which, at each of the

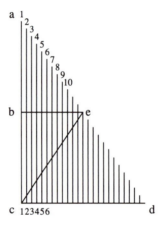

Fig. 1.10 (Bibl. nat., f. fr., nouv. acq. 5160, fol. 48r)

> moments adds new forces to descend, it happens consequently that it traverses the space BC much faster than AB, because in traversing it retains all the impetus which moved it through the space AB and, in addition, it adds to this a new [impetus], because of the gravity which propels it anew at each of the moments. As for the proportion in which this speed increases, this is demonstrated by the triangle ABCDE; for the first line denotes the force of speed impressed in the first moment, the second line the force impressed in the second moment, the third the force conferred in the third moment, and so on. Thus is formed the triangle ACD which represents the increase of the speed of motion of the body in its descent from A to C, and ABE which represents the increase of speed in the first half of the space traversed by this weight; and the trapezium BCDE which represents the increase in speed in the second half of the space traversed by the weight, namely BC. And since, as is clear, the trapezium BCDE is three time as large as the triangle ABE, it follows from this

---

[38] AT I, 71-73. See document 5.1.5 and the note there on the source of the figure.

that the weight will descend three times more quickly from B to C than from A to B ...

Noteworthy among the additions to the diary note is the following. The areas of the geometrical representation are explicitly said to represent velocity (*celeritas*) thus supporting our interpretation that the medieval proportions were applied to the motions represented by areas. The concept of force designating in the initial document the "attractive force of the earth" as well as the "added force" accumulated in the falling body is now differentiated into "gravity" and "impressed force" or "impetus," respectively. This version hence is even more clearly related to the medieval conceptual framework than the documents of 1618. So is the geometrical representation itself into which the successive added impetuses were drawn.

Additional information given in the letter under discussion sheds light on the difficulties involved in expressing the functional dependence of one magnitude on another with the mathematical means of the time.[39] In the diary note, Descartes had computed numerical values for only one case: A body falling from rest through two equal spaces traverses the second in a third of the time needed for the first. In the exposition for Mersenne he tried to go one step further towards characterizing mathematically the implied relation between space and time.[40]

> ... i.e., that if it descends from A to B in three moments it will descend from B to C in a single moment; that is to say, in four moments it will traverse twice the distance as in three and, consequently, in 12 moments twice as much as in 9, and in 16 moments four times as much as in 9, and so on.

Observing that for spaces as 1 and 2 the corresponding times are as 3 and 4, he assumed erroneously that this would give a general rule for the relation between spaces and times. Based on this assumption he inferred for distances as 1, 2, and 4 times as 9, 12, and 16. If Descartes had applied his geometrical representation for calculating the time for 4 spaces traversed he would have been able to realize that his generalization was wrong. For spaces as 1 and 4 he would have received times as 105 and 176, which are not in the relation 9 to 16.[41]

In fact there is no simple way to derive the functional relation implied by Descartes' argument. He could at best have derived values of time for the discrete segments of space obtained by successively adding up equal space units. The corresponding values for time would increase in the sequence:

---

[39] This part of the letter has been a particular source of confusion for historians of science who interpret the documents anachronistically from the viewpoint of classical physics and thus overlook the fact that the values Descartes gives are not consistent with the figure but rather are calculated using the factor $4/3$; see section 1.5.4.2.

[40] AT I, 73.

[41] The areas in Descartes' figure for the first four spaces traversed are as 1 to 3 to 5 to 7. According to Descartes' argument the corresponding times are as 1 to $1/3$ to $1/5$ to $1/7$. The total of these times is $176/105$. The internally correct result is therefore that the times for distances as 1 and 4 are as 1 to $176/105$ or as 105 to 176. This result, too, is of course incompatible with the correct law of fall.

$$1, \; 1 + \tfrac{1}{3}, \; 1 + \tfrac{1}{3} + \tfrac{1}{5}, \; \ldots$$

This is so because the assumption of equal spaces is necessary for the application of the Aristotelian proportion. Descartes probably realized this problem since he remarked in a later letter to Mersenne, dated October, 1631[42]:

> there would be no way to explain the speed of that motion by means of other numbers than those which I have sent you, at least [not] by rational [numbers], and I do not even see that it would be easy to find irrational.

1.3.5   Later References to the Original Proof

It seems that even later on, Descartes did not realize that spatial and temporal interpretations of moments would yield different conclusions as to the relations of time and space in free fall.

In August 1634 Beeckman visited Descartes in Amsterdam bringing him a copy of Galileo's *Dialogo*[43]:

> Mr. Beeckman came here on Saturday evening and brought me the book of Galileo. However, since he took it [with him] to Dort this morning, I had it in my hands for only about 30 hours. I did not fail to leaf through the whole book. I found out that he philosophizes quite well about motion, although there are only very few things he says which I find completely true; what I did notice is that he is more deficient where he follows accepted views than where he departs from them ... I would like, however, to say that I found in his book some of my own thoughts, among others two of which I believe to have written to you before. The first is that the spaces traversed by heavy bodies when they fall, are to one another as the squares [Fig. 1.11] of the times which they employ to descend, i.e., that if a ball employs three moments to descend from A to B, it will employ but one to continue from B to C, etc. And I say that with many reservations, since in fact it is never completely true as he thinks he has demonstrated.

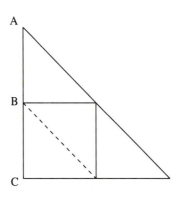

Fig. 1.11 (AT I, 304)

---

[42] AT I, 222. See section 1.5.4.1 for a discussion of the consequences of a functional interpretation of Descartes' figure.

[43] Letter to Mersenne, August 18, 1634; AT I, 304-305.

Descartes repeated the numerical values of his diary note as well as the corresponding geometrical representation. At the same time he identified erroneously Galileo's correct law of fall with his own findings. He did not notice the incompatibility between the quadratic relation and his own figures which he refered to in the same sentence:

> the spaces traversed ... are to one another as the squares of the times ..., i.e., that if a ball employs three moments to descend from A to B, it will employ but one to continue from B to C, etc.

This is obviously a mistake. His example implies a relation of times as 3 to 4 for the spaces AB and AC, respectively, which are in the relation as 1 to 2. According to Galileo's law of fall, however, for times as 3 to 4 the spaces traversed should be as 9 to 16 and not as 1 to 2. Thus the identification of his result developed from a spatial interpretation of the moments of fall with Galileo's result which uses a temporal interpretation is obviously based on an error, and it seems clear that Descartes (presumably just as Beeckman) was not yet aware of the alternative between a spatial and a temporal interpretation of extension, which represents primarily nothing more than the continuum of infinitely small momenta which seemed to be indifferent to spatial or temporal interpretation.

The same holds true even later after Descartes had conceptualized the new theory of gravity to be discussed below. In a draft contained in the so-called "Excerpta Anatomica" he interchanged spatial and temporal interpretations of moments as in the initial document, only this time in the reverse direction: In the initial document Descartes had shifted from a temporal to a spatial interpretation, but in the Excerpta Anatomica he shifted from a spatial to a temporal interpretation. One section begins with a spatial interpretation[44]:

> If any body is activated or impelled into motion always with an equal force ... and moves *in vacuo*, it always takes three times as long to travel from the beginning of its motion to the middle as it does from the middle to the end, and so on.

But by the end of the same document, Descartes had changed to a temporal interpretation, utilizing a geometrical figure (see Fig. 1.12) to infer the speed of accelerated bodies and concluding[45]:

> ... that the speed during the first [interval] of time is to the speed during the second as the area *abc* is to the area *aced*.

---

[44] AT XI, 629. The Excerpta Anatomica comprise a number of manuscripts by Descartes, later in the possession of Clerselier, many of them anatomical in content; they were transcribed by Leibniz in Paris in the 1670's. This particular manuscript is dated around 1635 by Gabbey (1985), whose translation we have adopted. See document 5.1.9.

[45] AT XI, 631.

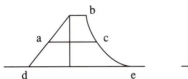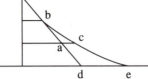

Fig. 1.12 (AT XI, 631)

The speeds represented by the areas are here explicitly correlated to the times elapsed, not to the spaces traversed.

1.4 The Law of Fall Derived within the Context of the Theory of Gravity

1.4.1 Beeckman's Commentary on Descartes' Initial Document

Let us return to Beeckman's first reaction to Descartes' initial document. After stating his understanding of Descartes' proof, Beeckman continued his note, commenting extensively on two points. The first pertains to the assumption of continuous acceleration which underlies Descartes' proof, the second pertains to the problem of resistance in a medium.

Beeckman first discussed the consequences that ensued if he replaced Descartes' assumption of continuous acceleration with the assumption of discontinuous acceleration implied by his own theory of gravity.[46]

> But if the minimum moment [*momentum minimum*] of space be of some quantity, the progression will be arithmetic [...] we see that the space of fall in one hour is to the space of fall in two hours as *ade* to *acb*, namely if we consider that in arithmetic progression, all the numbers contained under half of the terms are to the numbers of all terms not at all as 1 to 4, even though the proportion grows perpetually. This way the progression of two terms, which is 1, 2 is as 1 to 3. Thus 1, 2, 3, 4, [5], 6, 7, 8, is as 10 and 36. Thus these 8 terms are to 16 [terms] are as 36 to 136 [the sum of the first 8 terms is to the sum of the first 16 terms as 36 to 136], which is not at all as 1 to 4. If, then, the stone descends in distinct intervals because the Earth is attracting it by means of material spirits, then these intervals or moments will be nevertheless so small because of the multitude of the particles that their arithmetic proportion will not be sensibly smaller than 1 to 4. The demonstration [by means] of the triangle should therefore be retained [...].

> The proportion of the triangle appeals to us not because there is really no certain mathematically divisible physical minimum of space in which a minimal

---

[46] J I, 263-264. See document 5.1.2.

physical attractive force moves an object (this force is namely not truly continuous but discrete, and, as one says in Flemish: *sy trect met cleyne hurtkens*, and, therefore, the aforesaid increases consist in a true arithmetical progression); but, I say, it is appealing because this minimum is so small and insensible, that because of the multitude of the terms of the progression, the proportion of the numbers does not differ sensibly from the continuous proportion of the triangle.

According to Beeckman's theory the attractive force, "is not truly continuous but discrete" and "pulls with small jerks" (*sy trect mit cleyne hurtkens*), and therefore, the increase mentioned (i.e., of spaces traversed) is in a "true arithmetic progression." Because the increase of motion is caused by a large number of particles and because the successive moments in which the body traverses successively larger spaces are very small, the quadratic relation between times and spaces is an approximation of the true arithmetic progression of spaces traversed; and therefore, "the demonstration by means of the triangle should be retained." Beeckman observes that in the case of discrete acceleration the relation of the spaces traversed depends not only on the relation of the times elapsed but also on the number of the moments considered. For the relation of 1 to 2 segments representing the moments elapsed the relation between the corresponding areas representing the spaces traversed will be 1 to 3; for the relation of 4 to 8 moments the corresponding relation of spaces will be 10 to 36; for the relation of 8 to 16 it will be 36 to 136. Thus, the more moments are considered the closer the relation of the spaces traversed approaches to the relation 1 to 4 of the triangles *ade* and *acb* without, however, ever reaching it.

The quadratic relation celebrated by historians is hence not Beeckman's law of free falling bodies but rather contradicts it bluntly if taken as a physical theorem. The results obtained according to the quadratic relation are merely suited to compute by an approximation the relations determined by Beeckman's law.

Beeckman goes on to discuss the consequences that ensue if the assumption of free fall in a vacuum is replaced by the assumption of a fall in a resisting medium which is essential to his theory of gravity.[47]

The point of equilibrium is sought, that is where the speed of a stone falling in the air is no longer increased [marginal note].

In the same way as the space is multiplied [*multiplicatur*] the impediment is multiplied too if you consider it [the fall] in air or water, i.e., in a plenum. The falling object describes an oblong figure, all lines of which are parallel. Since the object falls more quickly in the second hour and traverses more space, the proportion of the figure it describes in the first hour to that which it describes in the second hour is the same as the space traversed in the first hour to that traversed in the second hour. If, therefore, the falling object were not impeded by an impediment, it would encounter as much more air in the second hour as the parallelopiped of the second hour is larger than that of the first hour. But since it is certain that a falling object is impeded by the air − experience showing that the speed of every falling object is not always augmented, but

---

[47] J I, 263-264.

that there is a certain place where, when reached, [the object] will move through the remaining space uniformally – let us see in what way it happens.

[…] if an object which falls for one physical minimum moment of time (in which it traverses one physical spatial minimum) encounters as much air as it itself consists of body, then it will not move more quickly, but will persist in that motion, i.e., if the parallelopiped, which is described in such a moment contains as much corporeality as the object itself contains, then the attractive force of the Earth will not be able to add to the motion of the object, since the gravity of the body in which it moves, i.e., the air, equals the gravity of the object …

According to Beeckman, the resistance of medium increases in the same proportion as the spaces traversed because the quantity of matter encountered by the body is proportional to these spaces. The motion is augmented until the quantity of matter displaced by the falling body in one "physical minimum moment" equals the quantity of matter of the body itself. From this moment on "the attractive force of the earth cannot add to the motion of the body … since the gravity of the body equals the gravity of the body in which it is placed."[48] From this point of equilibrium [*punctum aequalitatis*] onwards the body will continue in uniform motion. Beeckman's theory of gravitation is hence conceived in analogy to hydrostatics and Archimedes' principle is applied to determine the point of equilibrium in free fall.

Beeckman's concept of gravitation thus has the following consequence. The fact of increased motion presupposes a medium (material spirits), the particles of which impart new motion to the falling body. The medium, on the other hand, causes a retardation of the falling body until the latter reaches the point of equilibrium, from which point on it proceeds in uniform motion. The concept of uniform increase of the motion of a falling body thus demands that there should be a medium behind and none in front of the body, and this is taken into account in the first condition as put down in Beeckman's note in his *Journal*: "Bodies are moved downward toward the center of the Earth, the *intermediate* space being a vacuum."[49]

Thus, the very concept of gravitation is not compatible with the assumption stated nor is the quadratic relation valid for ideal conditions but is merely an approximation.

1.4.2 Descartes' Alternative "Law of Free Fall" in his Initial Document

As for Descartes, the status of inferred relations between times and spaces of fall is more complicated than for Beeckman. In his initial document Descartes

---

[48] J I, 264.

[49] J I, 260; AT X, 58; emphasis added.

accepted at first Beeckman's assumptions and assimilated them to a dynamical theory; thus he was on the one hand able to reconstruct Beeckman's assumption (the so-called principle of inertia) and his qualitative explanation of acceleration in free fall in a dynamical conceptual framework. On the other hand, this framework enabled him to justify the assumption that gravity acts uniformly and hence that moments of motion grow in arithmetic progression for discrete moments of time.

The relevant conceptual framework is probably the medieval theory of impetus, although Descartes refers explicitly to this theory in this connection only in 1629. The medieval impetus theory, however, offered more than one possible explanation for acceleration in free fall – each with different quantitative consequences. One of these possibilities was to conceive acceleration as resulting from impressed impetus added to the impetus already impressed earlier. The previously impressed impetus conserves the motion and the newly added impetus increases it. This is essentially the argument of Buridan, who was the first to apply impetus theory to free fall, and it is Descartes' argument in the part of his initial document already discussed. But, as indicated above, another possibility was developed in the Middle Ages as well. Oresme conceived the impetus as not only conserving the motion but as increasing it, too. In consequence, acceleration itself increases with increasing motion. Hence, projected on the conceptual system of classical mechanics, it would seem that Buridan conceives of impetus as proportional to velocity while Oresme conceives of it as proportional to acceleration. This latter possibility is essentially Descartes' argument in a part of the initial document and of the diary note which we have yet to discuss.

Having shown that on the basis of Beeckman's assumptions the falling body will traverse the second half of its fall "three times more quickly" than the first half Descartes' initial document continues with the following consideration[50]:

> This question can be posed in another, more difficult way in the following manner. Imagine that the stone is at point a [Fig. 1.13] and that the space between a and b is a vacuum; and that all at once, say today at nine o'clock, God should create in b a force attracting the stone; and that afterwards he should create in every single moment a new force equal to the one he created in the first moment; and that together with the force created before it would attract the stone ever more strongly, because in a vacuum that which is once in motion moves always; and finally the stone which was at a will reach b today at ten o'clock. And if it is asked, in how much time it will traverse the first half of the space, i.e., ag and in how much [time] the remaining [space]; I answer that the stone fell through the line ag in $7/8$ of an hour; but through the space gb in $1/8$ of an hour. Now, a pyramid should be constructed on a triangular basis, so that its altitude is ab, which is in whatever manner agreed upon divided along with the whole pyramid by parallel horizontal transverse lines [planes]. The stone will traverse the lower parts of the line ab as much more quickly as the sections of the whole pyramid are larger.

---

[50]   AT X, 77-78; J IV, 51. (In the original "$7/8$" and "$1/8$" are mistakenly exchanged.)

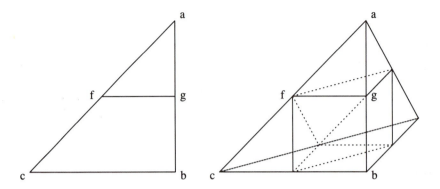

Fig. 1.13 Descartes' figure (AT X, 77)   Fig. 1.14 Corresponding pyramid

Descartes' own diary note ends with the following corresponding remark[51]:

> But the problem could be posed in a different way, such that the attractive force of the Earth be equal to that in the first moment: and, while the first remains, a new one is produced. In this case the problem would be resolved by a pyramid.

Descartes introduced the argument saying that, "this question can be asked in another, more difficult way." As the text shows, a more difficult way is a different and more complicated principle than the one assumed by Beeckman. According to this principle, a new equal force is created in each moment "by God" in addition to the gravity of the body so that it will be attracted "by an ever increasing force." For the derivation of the quantitative relations Descartes now uses a three-dimensional pyramid instead of the two-dimensional triangle used before, the increase of volume representing the increase of motion. The quantitative relation derived is that the body traverses the second half of a space in $1/7$ of the time employed to traverse the first (instead of $1/3$ according to the first assumption, see our diagram, Fig. 1.14). This result is obtained by an inference which exactly parallels the two-dimensional case. The quadratic relation between length and area results in the relation of $1/4$ to $3/4$ of the areas representing the motion in subsequent intervals; the cubic relation between length and volume results in the corresponding relation of $1/8$ to $7/8$. (Note, however, that here Descartes has already shifted in the initial document from a neutral usage to a spatial interpretation of extension.)

This additional argument shows that Descartes did not subscribe at this point to a specific quantitative interpretation of his dynamical theory. Descartes' still undetermined interpretation of free fall as well as his reservations concerning the concept of gravity as elaborated in a document from the same period submitted to

---

[51] AT X, 220; J I, 361.

Beeckman[52] raise doubts whether his argument at this time really presents a "law" of free fall.

1.4.3 Descartes' Revised "Law of Fall"

It seems that for a long time to come, Descartes did not integrate his argument concerning free fall into a consistent theoretical framework. However, in the context of his first elaboration of a comprehensive physical theory, which was presented in *Le Monde* (completed ca. 1633), Descartes reconsidered the problem of fall and discussed it in letters exchanged with Mersenne. As we have already shown, Descartes still adhered in these letters of 1629 to his previously inferred relations of fall. However, he now discusses for the first time Beeckman's theory of resistance of a medium. He accepts, "that the quicker a certain body falls the more does the air resist its motion," but rejects the idea that the falling body will reach a point of equilibrium after which it will proceed in uniform motion. He points to a mistake of Beeckman's and attempts to prove that from Beeckman's assumptions it follows that the falling body approaches this point of equilibrium asymptotically without ever reaching it.[53]

By 1631, however, a change in Descartes' conception of fall becomes manifest. He is still convinced by his inference of the relations in free fall, but he now doubts whether the physical assumptions underlying the proof can be retained. In an answer to a letter from Mersenne concerning his computation of the speed of the falling weight, Descartes writes[54]:

> Concerning the manner of computing that speed which I have sent you, you should not make much of it since it supposes two things which are certainly wrong: namely that a space can be found which is completely void, and that the motion which takes place there is in the first instant, when it begins, the most slow that can be imagined, and that it is later always augmented equally. But if all that were true, there would be no way to explain the speed of that motion by means of other numbers than those which I have sent you.

Thus the following basic presuppositions are rejected here: that the assumption of fall in a vacuum is reasonable and that acceleration is continuous and uniform.

The reasons for this change in judgment concerning gravity become clear if seen in connection with the principles of the comprehensive system of physics elaborated by Descartes in *Le Monde*. A manuscript (cited above, section 1.3.5) dated about 1635 clarifies in particular the issue of gravitating bodies.[55]

---

[52] AT X, 67-74. See document 5.1.4.
[53] See AT I, 92-93.
[54] Oct. 1631[?], AT I, 221-222.
[55] Excerpta Anatomica, AT XI, 629-630. See document 5.1.9.

If any body is activated or impelled into motion always with an equal force, i.e., a force imparted to it by a mind [*mens*] (no other such force can exist), and moves *in vacuo*, it always takes three times as long to travel from the beginning of its motion to the middle as it does from the middle to the end, and so on [Fig. 1.15]. But because no such vacuum can be granted, there being space [*spatium*] of some sort, it always resists in some way, and does so in such a way that the resistance always increases in geometric proportion to the speed of the motion, so that eventually it reaches a point where the speed no longer increases appreciably; and one can determine a particular finite speed which it will never equal.

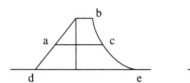 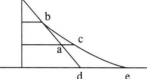

Fig. 1.15 (AT XI, 631)

As for those bodies impelled by the force of gravity, since gravity does not always act equally, like a mind [*anima*], but is effected by some other body which is already in motion, this can never impel a heavy body as fast as it itself moves. But even *in vacuo* the impulse will always diminish in geometric proportion.

Descartes stated that "gravity does not always act equally" since it is caused by impact of bodies (i.e., particles) and a body "never can impel a heavy body as fast as it itself moves" so that "even *in vacuo* the impulse will always diminish" with increasing velocity of the impelled body. As Descartes himself later pointed out, the statement that gravity does not always act equally is tightly connected to the core of the physical theory of *Le Monde*. In a letter to De Beaune,[56] Descartes first explains that gravity is

... nothing other than that all the subtle matter which is between here and the moon and which turns very quickly around the Earth chases toward it all the bodies which cannot move as quickly.

Since the acceleration of the falling body depends on the quantity of motion imparted to it by the particles of the vortex, and since this depends also on the difference in their speeds, it follows that the quicker the body is falling the less it will be accelerated.

Finally, when it happens that they [the bodies] descend as quickly as it [the subtle matter] moves, it will not push them at all anymore, and if they descend more quickly, it will resist them.

Descartes then remarks that it was this dependency on so many contingent empirical facts that led him to give up his attempts to derive a law of fall. He also

---

56 April 30, 1639; AT II, 544. See document 5.1.0.

discussed this a year later in two letters to Mersenne explaining why he no longer considered his former proof of the law of fall to be valid.[57]

> I cannot determine the speed with which each heavy body descends in the beginning, because this is purely a question of fact, and it depends on the speed of the subtle matter ... The subtle matter pushes the descending body in the first moment and gives it one degree of speed; then, in the second moment, it pushes it somewhat less and gives it only almost one degree of speed, and so on in the other [moments]; this occurs in almost double proportion [fere rationem duplicatam] at the beginning of the descent of bodies. But this proportion is completely lost when they have fallen several fathoms, and then the speed does not increase anymore, or hardly anymore.

Abstracting from the medium as the cause of resistance could hence not have been considered reasonable in this conceptual framework since the concept of gravity itself involves the existence of a medium. It is explicitly stated in the manuscript of about 1635 cited above that continuous acceleration could be affected only by a nonphysical cause such as the activity of a mind. Moreover, Descartes believed that from this necessary presupposition of a medium it follows necessarily that every body will encounter resistance to its motion.

The revolution in Descartes' thought on the problem is well illustrated by his different reactions to Galileo's two publications of the law of free fall. In 1634, referring to Galileo's *Dialogo*, Descartes claimed (erroneously) that Galileo had discovered the same law as he himself had in 1618 and only expressed reservations as to whether the law is completely true; but the situation had changed dramatically by 1638, when Galileo published his proof of the law of free fall in the *Discorsi* and Descartes wrote a long commentary on the book for Mersenne.[58] In his comments on the proof presented there, Descartes now bluntly rejected the assumptions on which the proof is based, although he "formerly believed" them.[59] In particular, he asserts that everything Galileo said about the velocity of bodies falling in vacuum "is built without foundation."[60] Galileo should first have determined what gravity is, and "if he had known the truth, he would have known that it is nothing in the void." Moreover, "he supposes that the speeds of falling weights always increase equally, which I formerly believed as he does, but I now believe I know by demonstration that it is not true."

A derivation, whether correct or erroneous, of the relation between times and spaces for a uniformly accelerated body in a vacuum could thus not be stated as a law for gravitating bodies in Descartes' new conceptual framework. This seems to be the reason why Descartes did not further elaborate the implicit inconsistency resulting from the ambiguity of the concept of moment. The development

---

57 March 11, 1640; AT III, 36-38. Further explanations were added in a letter of June 11, 1640; AT III, 79. See documents 5.1.11 and 5.1.12.

58 October 11, 1638; AT II, 380-402. See 3.1, 3.8, and document 5.3.1.

59 AT II, 386.

60 AT II, 385-386.

towards his relatively closed system of physics would not have been disturbed even by the adoption of the Galilean assumption of the proportionality between velocity and time of fall: as we shall see in the Epilogue, this was indeed Descartes' last contribution to the "comedy of errors."[61]

1.5 The Conceptual and Deductive Structure of Descartes' Proof of the Law of Fall

Our analysis of documents related to the Beeckman-Descartes dialogue has yielded detailed information about difficulties of discovering and proving the law of free fall. These results may be of some interest for the historiography of the discovery of this law. There is, however, also a much more general aspect to our analysis. In our opinion the dialogue displays some fundamental problems in the advance of knowledge which are not restricted to the discovery of the law of free fall – in particular, the dependence of inferences on conceptualizations and the role of contradictions for reconceptualizations.

The different versions of Descartes' proof show remarkable failings and inconsistencies. But they also show a coherent pattern of argument which is closely linked to the semantic structure of some of the key concepts involved. This pattern differs from the pattern of the modern proof of the law of fall. A major problem posed by the dialogue is therefore the question, whether the differences between Descartes' proof and the modern one lie simply in the fact that Descartes' proof still exhibits serious deficiencies and erroneous arguments, or whether the difference is also one of content.

If the first alternative were true, we would have only to identify the key mistakes which account for incorrect or physically meaningless arguments. Our claim, however, is that the second alternative is true. We claim that the difference between Descartes' proof and the modern proof is due to incompatible conceptualizations and therefore cannot be traced back simply to mistakes in Descartes' argument.

We shall first show this incompatibility by analyzing the structure of Descartes' proof and systematically comparing it to the structure of the modern proof. This analysis will be carried out in three steps. First we shall compare the conceptualization of the problem in Descartes' proof and in the modern proof. We shall then compare the different functions of the geometrical representation of the problem. Finally, we shall compare the deductive structure of the modern proof with the coherent common pattern of inference found in the different documents in which Descartes laid down his argument.

---

61   See Descartes' letter to Huygens, February 18/19, 1643 (AT III, 617-630; document 5.4.1) and the Epilogue (Chapter 4) below.

We shall then argue that the advance of knowledge is intimately linked to the development of conceptual structures.

1.5.1 Conceptualization as a Condition for the Proof

There are some fundamental differences between the conceptualizations of the problem in Descartes' proof of the law of free fall and in the modern proof. The differences pertain primarily to the concepts motion, velocity, and force. But other concepts like acceleration, which depend on these fundamental concepts, are also involved.

1.5.1.1 Conceptualization of Free Fall in Classical Mechanics

In classical mechanics only acceleration (or deceleration) but not motion itself needs a cause. The acceleration of a falling body is caused by an attracting force between the body and the earth, called gravity. Newton's law of gravitation implies that this force depends only on the masses of the falling body and of the Earth and on the distance between their centers of gravity. The force is nearly constant as long as the space traversed does not alter this distance considerably. Newton's second law furthermore implies that the velocity of the falling body is increased by the constant force of gravity with constant acceleration, so that the velocity increases in proportion to time. The concepts of velocity and acceleration involved in this inference presuppose that every instant of the motion and every point of the trajectory can be ascribed numerical values for velocity and acceleration and that the velocity function and the acceleration function are mathematically connected in a specific way.

The conceptualization by means of these concepts and propositions embeds the concept of motion in a semantic structure (partly constructed by them) which represents fundamental theoretical assumptions of classical mechanics. If the law of free fall is described in these terms it is no longer indifferent to theoretical premises but becomes dependent on (and partly constitutes) a particular physical theory.

1.5.1.2 Conceptualization of Free Fall in Descartes' Proof

In Descartes' proof motion is conceptualized as being itself caused by force. Force is not only an external cause. It can be imparted to a body (Excerpta Anatomica), produced in a body (diary note), can persist in a body (initial document), can be added up there (initial document, letter of Nov. 13, 1629), and can thus increase in the body (initial document). The force imparted to a body is the cause of its motion; the force impells the body (letter of Dec. 18, 1629, Excerpta Anatomica).

This concept of force is not only analogous to the medieval concept of impetus. The letter of Nov. 13, 1629 shows that the medieval concept is its origin. In this letter the concept of force is used synonymously to the concept of impetus. It is stated there that the impetus is impressed into the body, that the body retains all the impetuses impressed into it, and that these impetuses move the body through space.

There is, however, an important difference between Descartes' use of the concepts force, impetus, and motion and their original use in the medieval tradition. In this tradition impetus and motion were clearly distinct concepts. Descartes seems sometimes to use motion and impetus or force as synonymous terms. On the one hand, he stated that the impressed force persists in the body (initial document) and that the body retains all the impetus impressed (letter of Nov. 13, 1629). Hence, it seems that in contrast to classical mechanics, Descartes adhered to the opinion that a body moves only as long as it is impelled by an impressed force. On the other hand, Descartes conceived of motion, too, as being impressed into a body and remaining there, impelling the body to continue in motion (letter of Nov. 13, 1629).

This latter statement, which Descartes probably considered to mean the same as the first, is to some extent compatible with the conceptualization of motion in classical mechanics. The concepts of force or impetus on the one hand and of motion on the other hand, as they were used by Descartes and probably also by others at his time, evidently share a common conceptual structure. Based on this common structure the law of inertia of classical mechanics could also be expressed in medieval terminology as a law of conservation of impetus, i.e., by assuming that an impetus impressed into a body persists as long as there is no resistance. Newton himself said that motion is maintained because of a "vis inertiae." In a similar way in Descartes' argument the sharp distiction between the medieval concept of impetus as the cause of motion and the classical concept of force as the cause of acceleration is already obliterated.

In Descartes' proof motion is further assumed to consist of minima (initial document) or moments (letters of Nov. 13 and Dec. 18, 1629) of motion. These minima are indivisible, containing no parts. They are directly related to the points of the motion (initial document). In every moment the minima of motion are produced by the imparted forces. Especially in free fall the force of gravity which exists in a body impells it anew in each moment (letter of Nov. 13, 1629).

There is a major difference between these minima of motion and the instants of motion or points of its trajectory in classical mechanics. Whereas in classical mechanics a velocity is assigned to each instant or point, in Descartes' conceptualization the concept of velocity is applied not to minima of motion but to finite intervals (letter of Nov. 13, 1629, Excerpta Anatomica). This conceptual structure again can be traced back to a medieval one, namely, the structure of the concept of motion in the doctrine of the configuration of qualities and motions discussed above (section 1.2.3). It corresponds to the idea that motion is characterized not by one quantitative concept, namely velocity, but by two such concepts. These are, on the one hand, the intensity or degree of the motion, which is assigned to each minimum or moment of motion and represents something like a qualitative difference of motions with different velocities, and, on the other hand, the quantity of motion, which is a global characteristic of motion representing the effect of the motion in terms of space traversed. The latter was identified with the concept of velocity as defined in the Aristotelian tradition. Descartes' concept of velocity corresponds to this concept of velocity as representing the effect of motion measured in space traversed. Obviously, Descartes identified the concept of intensity or degree of a motion with the imparted force representing the cause of motion.

Thus the conceptual differences between Descartes' proof and the proof in classical mechanics concern primarily the concepts of force and velocity. No direct comparison is possible between Descartes' concept of force or impetus and, say, the Newtonian concept of force as proportional to acceleration.[62] Descartes' concept of the force imparted to a falling body, which persists and is accumulated there, is much more similar to the concept of velocity or to the concept of kinetic energy than to the concept of force in classical mechanics. The same holds true for Descartes' concept of velocity. Velocity is the effect of motion measured in space traversed relative to a given time interval. Velocity is thus a global qualification of motions and not a qualification assigned to the individual minima or moments of motion. From the modern standpoint this concept of velocity is at best a secondary concept characterizing the average velocity for an interval. But for Descartes it was still primary and was the only concept to which formal rules for infering statements about relations between times and spaces applied.

1.5.2 Geometrical Representation of Free Fall

The difference between Descartes' conceptualization of motion and the conceptualization in classical mechanics is reflected in the different use of geometrical representations of the motion of a falling body. Both representations are geometrically identical, but if we compare the use made of the figure [see Fig. 1.16], they turn out to serve in different ways to derive consequences from the respective assumptions.

---

[62] The concept of force in classical physics is not something that can survive its effect. Newton's axioms already determine a conceptual structure which is incompatible with anything like an idea of "conservation of forces" as it is inherent in the medieval concept of impetus.

1.5.2.1 Meaning of the Figure in Classical Mechanics

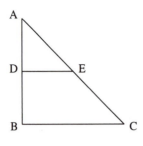

In classical mechanics line AB refers to the time elapsed and horizontal lines as DE and BC refer to the velocities at times D and B, respectively. Thus line AC represents the proportional increase of velocity in relation to time. Finally the main conclusion of the argument is that the area represents the space traversed which therefore increases proportionally to the squares of the times elapsed.[63]

Fig. 1.16

1.5.2.2 Meaning of the Figure in Descartes' Proof

Descartes, on the other hand, takes the increase of the area from the top (A) to the bottom (BC) as representing the increasing motion or – more precisely – the velocity of motion in the Aristotelian sense. The horizontals as DE and BC represent the minima or moments of motion which are added up to the total area. They are identified with the forces imparted to the body and which produce in each moment the motion.

To clarify further this conceptual structure, Descartes also used an auxiliary figure where the moments of motion were represented by thin stripes with a finite width instead of by horizontal lines, so that the figure was modified from a triangle to a stepped figure. Furthermore in his letter of Nov. 13, 1629 he added to his basic triangular figure vertical lines representing the impetuses imparted to the body in each moment (uppermost points of the vertical lines) and remaining in the body during the further motion (vertical lines themselves).

Thus Descartes' figure refers to different concepts than does the same figure in the conceptual framework of classical mechanics, and thus the elements of the figure have a different meaning. But besides these differences in meaning the figure is also used technically in a different way.

In the proof of classical mechanics the horizontal lines DE, BC, etc., and the vertical AB represent the given magnitudes. The area of the figure represents the unknown, and the inference that this area represents the space traversed is part of the argument. In Descartes' proof, however, the horizontal lines DE, BC, etc., and, in particular, the area represent the given magnitudes. For Descartes it is only when the Aristotelian proportions are applied that the vertical AB acquires its meaning and represents the auxiliary assumption of equal time or space intervals for the comparison of the corresponding areas. The unknown is not rep-

---

[63] In contemporary analytical geometry it is customary to represent the independent variable by the horizontal dimension. Therefore, figures like Descartes' geometrical representations of free fall are normally drawn rotated 90 degrees counterclockwise.

resented directly at all but only indirectly as proportional or inversely proportional to the areas.

1.5.3 Deductive Structure of the Proof

Let us now compare the structure of the proof itself in classical mechanics and in Descartes' argument.

1.5.3.1 The Structure of the Proof in Classical Mechanics

There are two different aspects of the proof of the law of free fall in classical mechanics. As it is usually understood, the proof itself comprises only the derivation of the law from the assumption that the velocity of the falling stone increases proportionally to the time elapsed. The justification of this assumption is not considered as part of the proof of the law of fall because it is an implication of the theoretical context of the proof. However, it has to be included here in the analysis because as we have seen above this is an essential part of Descartes' proof.

Justification of the linear increase of velocity

The assumption of an increase of the velocity proportional to the time elapsed holds true only under certain conditions. The main condition is that the body is falling freely, i.e., that there is no resistance of a medium. In a vacuum this condition is perfectly realized but for a body falling in air the law holds only approximately. The approximation is fairly good as long as the velocity of the falling body is relatively small. The resistance of the air increases with increasing velocity so that the velocity of the falling body asymptotically approaches a final constant velocity. Although under normal conditions every falling body shows the effect of the resistance of the air, there are good reasons according to classical physics not to conceive this effect as an inherent feature of gravity itself.

Another condition for empirical adequacy of the law of free fall is the homogenity of the field of gravity. Near the surface of the earth this condition is nearly realized. The law, however, cannot be generalized to a cosmological level because the force of gravitation decreases with increasing distance from the center of the earth. It is therefore a tacit assumption that the law of free fall is implied by the law of gravity only under conditions which are nearly realized for every falling body on the surface of the earth but which never holds true in an absolute sense.

Thus the justification of the assumption that the velocity of a falling body increases proportionally to the time elapsed – which is the main assumption of the proof – is allocated by the conceptual framework to other parts of the deductive system of classical mechanics. The justification is partly directly connected to fundamental propositions like the law of gravity and the existence of a vacuum so that a specific justification in the framework of a particular proof like the proof of the law of free fall appears to be neither necessary nor even possible. It is clear that once Newton's law of gravity was established it was out of the question that velocity might increase proportionally to space instead of time.

Integration of velocity over time

If we were to answer Beeckman's question about the quantitative relations of space and time on the basis of classical mechanics we would have to integrate the assumed linear function of velocity over time to get the desired relation between time elapsed and space traversed. This is an immediate consequence of the conceptualization in classical mechanics. The space traversed is by definition the integral of the velocity function over the interval of time. Of course, this answer is completely anachronistic if applied to the dialogue between Descartes and Beeckman, i.e., to a scientific situation before the invention of the infinitesimal calculus with its technique of integration.

However, in a special case – as in the case of free fall – where the function to be integrated is linear, there is also a simple geometrical solution – a geometrical method of integration, which cannot be generalized to other cases. The figures used by Descartes in his proof need only an adequate interpretation to yield such a proof. If we interpret the vertical in Descartes' triangular figure as a representation of the time elapsed in free fall and the horizontal lines as representations of the velocity of the falling body, then the area represents the space traversed, and it is obvious that the space traversed like the area of the triangle increases with the square of the time elapsed, represented by the vertical side.

This way of proving the law of free fall is evidently much closer to Descartes' proof. The main problem of this proof is to find out and prove that the area of the triangle represents the space traversed. The solution to this problem in classical mechanics is always connected to the fundamental notion of integrals. It is assumed that motion can be decomposed into infinitely small "differentials" whose velocities can be considered as constant. The areas of these differentials represent infinitely small spaces traversed. The continuously accelerated motion is thus conceived as the limit of a sequence of motions which are discontinuously composed of uniform motions. Finally it is inferred that the area of the triangular figure represents the space traversed because all the areas of the step-like figures representing the approximating discontinuous motions also represent the spaces traversed.

This proof is substantially based on the presupposition that the velocity increases in proportion to time and not space. Without going into the details, it should be mentioned that in the conceptual framework of classical physics it follows from the assumption that the velocity increases proportionally to the space that the space traversed increases exponentially with respect to time. If that implication had been known, it would have been obvious that this assumption could not apply to falling bodies because every falling body would already have been falling for an infinite time. There is, however, no simple way to derive this implication without the mathematical means of the infinitesimal calculus. Consequently it played no role in the attempts to derive the law of fall at the time of Descartes. The situation, however, changed a short time later once the calculus was developed. For instance, in his *Mechanica sive motus scientia analytice exposita* of 1736, Leonhard Euler was already able to publish a general solution to the problem.[64]

1.5.3.2 The Structure of Descartes' Proof

We know from Descartes' initial document that he was very close to the modern argument, according to which the law of fall is derived by determining the limit of an infinite sequence of discontinuous motions consisting of increasing numbers of decreasing intervals of uniform motions. Thus, it is not only tempting, it might even seem compelling, to read Descartes' proof primarily as an obscure version of this proof, obscured by a still inadequate conceptualization. As already mentioned, most scholars, following Koyré, have interpreted Descartes' proof as a more or less deficient version of the modern proof. Nevertheless, the documents as they were presented above show that it is completely misleading to interpret Descartes' proof this way.

Descartes' proof solves two problems. One of these problems is to justify the assumptions from which the law of free fall is derived. As we have seen, the assumptions of the proof in classical mechanics are intimately linked to core theorems of this physical theory so that they could not have been evident at all before the existence of this theory. The second problem which is solved by Descartes' proof is to find a way to derive the law of free fall by means of conceptual relations implicit in its conceptualization and its relation to contemporary theories at his disposal.

---

[64] Euler 1912, p. 21 (Chapter 1, Proposition 4).

Justification of the linear increase of force

The core of Descartes' proof consists of two steps. The first step is to derive from Beeckman's assumptions an explanation for the acceleration in free fall and to justify the triangular shape of the figure representing the motion.

Beeckman's assumption that "that which moves in a vacuum will move always" (diary note) is interpreted in Descartes' conceptual framework as "the force ... persists" (initial document), or the body "retains all the impetus which moved it." Descartes adds the assumption that in each moment of the motion "a new, equal force is added to it" "by the attractive force of the earth" (initial document), or "it adds to this a new [impetus], because of the gravity which propels it anew at each of the moments" (Letter, Nov. 13, 1629). Hence in the case of finite moments of motion the increase of forces can be described by a step function as sketched by Descartes in his initial document. If the moments of motion tend to become infinitely small this step function converges to a linear function. A substantial part of Descartes' initial document is devoted to prove this convergence. This argument completes the justification of the triangular shape of the geometrical representation for the motion in free fall.

Application of Aristotelian proportions

The second step of Descartes' proof is more sophisticated, although it is completely formal and shows no gap to be bridged by an intuitive understanding of motion. It is this very formalism that makes it somewhat difficult to understand the argument.

The Aristotelian proportions for a comparison of different velocities state that, if the times elapsed are equal, then the velocities are proportional to the spaces traversed, and that, if the spaces traversed are equal, then the velocities are inversely proportional to the times elapsed. The relation between the velocities of two time or space intervals is given by the relation between the corresponding areas of the figure. It is only necessary to identify equal intervals of time or space in the figure to derive the corresponding spaces or times.

The crucial question for this kind of derivation of a conclusion about space or time from the relations between velocities represented by Descartes' figure is: What does the vertical AB represent? Obviously every point of this line determines one of the minima of motion which is represented by the horizontal line beginning in this point. The line itself represents the force in this minimum of motion. Thus the corresponding point on the line AB is an abstraction from this motion, representing a minimum of motion but without any reference to the velocity. The points of the line AB represent the moments of this motion as far as they are determined temporally or spatially.

Hence the fact that the vertical line represents either the times elapsed or the spaces traversed is an implication and not a presupposition of Descartes' geometrical representation of the motion of free fall. This opens up a twofold possibility to apply the Aristotelian proportions. If we interpret two equal intervals on the vertical as equal times elapsed then the spaces traversed in these time intervals are proportional to the corresponding areas. If, on the other hand, we interpret the two equal intervals as representing equal spaces traversed then the times elapsed are inversely proportional to the corresponding areas.

Within the framework of Descartes' conceptualization there seems to be no substantial difference between these two possible applications. It seems to be . equally justified to interpret the vertical line AB as representing either the moments of time or the moments of space – and in both cases a seemingly acceptable quantitative result is achieved. It can be concluded either that for two equal time intervals starting with the beginning of the fall, the corresponding spaces traversed are as 1 to 3, or that for two equal space intervals starting with the beginning of the fall, the corresponding times are as 3 to 1. Both inferences seemingly lead to the same consequence, which Descartes repeated over and over again as his final statement of the proof: The falling body passes the second (time or space) interval "three times more quickly" (initial document) than the first one.

We shall argue at length below (section 1.5.4) that not only from the point of view of classical mechanics but also within the conceptual framework of Descartes' proof it is an error to assume that these two inferences are equivalent. For the present we shall only point out that it is not at all immediately clear that these alternatives for applying the Aristotelian proportions lead not only to different but even to contradictory conclusions.

Aside from this implicit contradiction, Descartes' proof is formally correct. There is nothing puzzling in Descartes' argument if we assume that he shares the medieval conception of velocity. We shall neither go into the details why this formal reasoning in the Aristotelian tradition also coincides with an intuitive understanding of motion nor discuss in detail the failings of this inference from the viewpoint of a modern conceptual framework. The incompatibility of Descartes' argument with classical physics is obvious. As seen from the modern concept of velocity, it makes no sense at all to identify the area of the geometrical representation of free fall with the velocity. The consequence, that even in uniform motion – represented by a rectangle – the velocity increases with time or space seems absurd. Furthermore, even if such an identification did make sense, it would remain puzzling that in Descartes' application of the Aristotelian definition of velocity the dependence of the area on the vertical does not represent the dependence of the velocity on the time elapsed or the space traversed.

It appears strange that Descartes rather assumes that only a proportion – namely the proportion of times or spaces to velocities in equal intervals of space or time represented by areas corresponding to equal intervals of the vertical – is represented which cannot be generalized to a geometrically represented function. But from the viewpoint of the medieval conceptual tradition there is no necessity for such a generalization. Motions are conceived as caused by forces and as having displacements in space as their effects. The effect depends on time and on some determination of the motion called velocity. This velocity is conceptualized as a space traversed in a given time. In this conceptual framework it does not make sense to compare these spaces if the times are not equal. From this viewpoint a comparison of the areas of the figure, i.e., the velocities, for arbitrary intervals makes no sense at all. Such a comparison, however, underlies our intuitive resistance against the medieval definition of a global and general concept of velocity as space traversed in a certain time, as it is adopted in Descartes' proof.

Conclusions drawn from the analysis of Descartes' proof

Summing up: In his answer to Beeckman's question, Descartes integrated two medieval theoretical traditions. Descartes and his contemporaries were familiar with these traditions not necessarily through knowledge of the sources in which these ideas were first expressed but rather through the knowledge of the standard works of their time influenced by this tradition. Descartes interpreted motion according to one of these traditions as resulting from force or impetus impressed upon a body. He quantified motion according to another tradition, namely the doctrine of configurations of qualities and motions, by means of a two-dimensional geometrical representation, one dimension of which represents the intensity, the other the extension of motion; the area represents the quantity of motion. This quantity of motion was traditionally understood as the space traversed in a certain time.

When the two traditions were linked together by Descartes in the course of his application of them to free fall, the intensity of motion was considered to represent the impetus impressed, the extension – which represented time in the medieval tradition – remained to be interpreted.[65] The quantity of motion was conceived as an accumulation of intensity in the smallest units or moments of motion over extension. The quantity of motion is thus dependent on its elements, the minima or moments in which the difference between time and space seems to vanish. Extension is thus understood as consisting of indivisible moments which could just as well be interpreted as moments of time or space.

In a comparison of two different motions, two Aristotelian definitions could be applied which medieval authors had already codified in the following proportions: *When times are equal then the velocities are proportional to the spaces traversed. When spaces traversed are equal then the velocities are inversely proportional to the times elapsed.* Descartes applied these proportions to the accelerated motion of free fall by identifying the quantity of motion with velocity in the Aristotelian sense. According to the twofold interpretation of extension he thus

---

[65] This seems to be a strong indication that, despite the widespread use of geometrical figures to represent motions, Descartes and his contemporaries probably had no direct access to genuine sources on Oresme and the Merton tradition.

attained either the result that for two equal time intervals starting with the beginning of free fall, the corresponding spaces traversed are as 1 to 3, or the result that for two equal space intervals, the corresponding times are as 3 to 1. This is essentially the argument of Descartes' proof, an argument which makes no sense in classical mechanics but is up to this point formally correct in Descartes' own conceptual framework.

Therefore, the main conclusion we draw from this analysis is that Descartes' derivation of a statement concerning the relation between times elapsed and spaces traversed from an assumption about the acceleration of a body falling due to the attraction of the earth is not deficient or obscure at all: It appears to be deficient only if it is projected into the conceptual framework of classical mechanics.

1.5.4 The Internal Contradiction in Descartes' Argument

As already mentioned, however, Descartes' proof is not only deficient from the point of view of classical mechanics but contains even in Descartes' own conceptual framework an implicit inconsistency which may yield apparent contradictions. Descartes' argument allows for a temporal as well as for a spatial interpretation of the moments of motion and thus of the vertical dimension of his geometrical representation. In the temporal interpretation, the impetus increases in proportion to the time elapsed; in the spatial interpretation, it increases in proportion to the space traversed. As we have shown, Descartes constantly shifted from one interpretation to the other as if they were logically equivalent. However, time elapsed and space traversed in free fall are not proportional because the motion of free fall is not uniform but accelerated. Thus the impetus of the falling body cannot be proportional to time and space alike. The temporal and spatial interpretations of moment are mutually exclusive.

This consequence was realized neither by Descartes nor by Beeckman.[66] One reason is probably the difficulty of expressing the derived law of fall in a general way. We are accustomed to conceive of motions as functions of time, space, and velocity. But the mathematical means of the 17th century did not yet contain a general concept of function. Relations between quantities like space and time were expressed by a variety of heterogeneous means, in particular by proportions, numerical examples, and numerical sequences. How could these means contribute help to reveal the difference between a temporal and a spatial interpretation of the moments of motion and hence to the interpretation of the vertical dimension in Descartes' geometrical representation? How could he elaborate the consequences of his argument by means of the mathematics of his time?

---

[66] It was exactly this contradiction that triggered one of Galileo's most important insights; see section 3.3.3.

We shall answer these questions in three steps. *First*, we shall briefly discuss some implications of the generalization inherent in the common misinterpretation of Descartes' proof, according to which his geometrical figure represents a function. *Second*, we shall discuss the generalization of Descartes himself which implicitly underlies the numerical values given in his letter of Nov. 13, 1629 and similar figures in a letter of Oct./Nov. 1631.[67] We shall argue that this generalization, which is usually interpreted as the legitimate outcome of Descartes' argument, is in fact the erroneous abstraction of a rule which contradicts certain implications of Descartes' geometrical procedure. *Finally*, we shall discuss these implications and show that they involve the contradiction mentioned above, which was perceived neither by Descartes nor Beeckman, although it could have been detected by deriving more numerical values from the geometrical representation used by Descartes. We shall argue that this contradiction originates from an extention of the use of the medieval concept of velocity, i.e., its application in the study of accelerated motion, and could have been eliminated only by a reconstruction of the conceptual framework.

1.5.4.1 Functional Interpretation of the Geometrical Representation

From a modern point of view it is tempting to interpret Descartes' figure as the representation of a function. We are used to immediately identifying the alleged representation of time or space by the vertical and of velocity by the area as representation of the dependency of velocity on time, respectively, space. But as we already pointed out when we analyzed the application of the Aristotelian proportions, such a functional interpretation is incompatible with the sources.

The areas of geometrical figures, as they were used by Descartes and his contemporaries to represent motions, are in fact functionally dependent on the length of the sides of these figures. But this does not necessarily mean that such figures were actually always used to represent functions. Every mathematical model contains properties which are meaningless and have no semantic reference.

In the case of Descartes' argument the functional relation between the length and the area of his geometrical figure is such a property without reference. Although geometrical figures in the traditon of the medieval doctrine of the configuration of qualities and motions implicitly represent in a fairly general way what in modern mathematics is conceived of as functions, a functional interpretation of such geometrical figures was neither necessary according to the conceptualization of relations between quantities in Descartes' time nor even supported by it. Accordingly, Descartes used the figure not to represent functional relations among time, space, and velocity but as geometrical representation of his conceptualization of free fall and thus as a means of deriving consequences.

---

[67] See section 1.3.4.

It is a consequence of the rise of classical physics, that only a short time later geometrical representations similar to those used by Descartes and his contemporaries to represent motions were not applied any longer to represent anything else but functions. Therefore, we will briefly elaborate the implications of the preclassical argument from the point of view of classical mechanics. Conceiving of Descartes' geometrical representation of uniformly accelerated motion as representing a function we obtain completely different results from a spatial and a temporal interpretation of its vertical dimension.

In a spatial interpretation, Descartes' geometrical argument, which resulted in the numerical example according to which a body moves three times as fast during the second interval as during the first yields a contradictory result. In this case the vertical line represents the space traversed, but the area has to represent the increase of the space traversed as well, although it increases in a different manner, namely proportionally to the square of the length of the vertical. This absurd consequence is probably the reason that Descartes' proof is usually considered to be crudely defective. It seems that up to now nobody has taken Descartes' proof literally as a formal proof.[68] As we have shown, however, the absurd conclusion is only possible if it is assumed that the figure represents a function. As soon as we stick to the Aristotelian condition that the velocities can be compared only for equal time or space intervals, an interpretation of the figure as a representation of a continuous function is excluded and no contradiction occurs.

In a temporal interpretation, however, Descartes' argument can easily be generalized to determine such a function. The vertical line represents the time elapsed, and the area which increases for geometrical reasons proportionally to the length of the vertical represents the continuous increase of the total velocity, i.e., the increase of the space traversed in this time. It thus turns out that these spaces increase proportionally to the squares of the times. The result is essentially the modern proof. Thus the generalization in the case of a temporal interpretation leads directly to a fundmental theorem of classical mechanics.

---

[68] The influence of the paradigmatical use of geometrical representations in classical physics seems to be a major obstacle to an adequate understanding of the attempts to derive the law of fall by Descartes and his contemporaries. Although the philological indications for the identification of the area of the representing figure with the total velocity of the represented motion are well known, not only in the work of Descartes but particularly – as we shall see in chapter 3 – also in the work of Galileo, this identification was neglected up to now by most modern historians of science. The similarity between the medieval tradition of representing motions as configurations of qualities and the representation by modern analytical geometry is so striking that the strong implicit assumption in analytical geometry that any graph is the representation of a *function* can indeed easily be overlooked.

1.5.4.2 Descartes' Erroneous Abstraction of a General Rule

As we have shown in section 1.3.4, Descartes himself, in a letter to Mersenne of November 13, 1629, implicitly generalized his result with regard to the functional relation between time and space when he tried to derive more numerical values (see Fig. 1.10)[69]:

> ... if it [the body] descends from A to B in three moments it will descend from B to C in a single moment; that is to say, in four moments it will traverse twice the distance as in three and, consequently, in 12 moments twice as much as in 9, and in 16 moments four times as much as in 9, and so on.

Although Descartes' procedure is not specified and must be reconstructed from the values, there can be no doubt about how these values were derived.

Descartes first changed the reference of his values to obtain a common point of reference. Instead of speaking about two different equal intervals with corresponding quantities of motion as 1 to 3, he related these intervals to the beginning of the free fall. Thus he obtained the result that for the intervals 1 and 2 measured from the beginning of fall the corresponding quantities of motion are 1 and 4. Depending on whether a temporal or a spatial meaning is ascribed to the moments of motion represented by the vertical dimension of the figure, the derived values are 1 and 4 for spaces traversed and 1 and $4/3$ for times elapsed.[70]

Second, Descartes assumed that the quantitative relations between these figures are valid for all numerical values that could be derived and used this assumption instead of his original geometrical procedure to derive further numerical values. But again there is a difference between the results depending on whether a temporal or a spatial meaning is ascribed to the vertical dimension of the figure.

In the first case the generalization yields the rule that double times always correspond to quadruple spaces, which again is correct from the point of view of Descartes' argument as well as from the point of view of classical physics. This is exactly the consequence Beeckman drew when he wrote down his understanding of Descartes' proof in his *Journal* ascribing a temporal meaning to the vertical. Descartes himself, although adhering to a spatial interpretation when deriving his numerical values, might have drawn this consequence intuitively as well. This can be seen from the fact that he never protested against Beeckman's formulation of the law of fall that "the space ... squares [*duplicat*] the proportion of time." Moreover he considered Galileo's law, published in the *Dialogo*, "that the spaces passed over are to each other as the squares of the times" to be compatible with his own argument.[71]

---

[69] AT I, 73.

[70] The quantity of motion in the second interval is 3. As the space traversed is proportional, the time inversely proportional to the quantity of motion, the corresponding values are 3 and $1/3$, respectively. The total for both intervals is therefore $1 + 3 = 4$ or $1 + 1/3 = 4/3$.

[71] See Galileo 1967, p. 222. See section 1.3.5.

In the second case the generalization yields the rule that double spaces always correspond to four-thirds times elapsed. Descartes generalized his numerical example in this way when he stated in his letter to Mersenne that "in four moments it [the body] will cover twice the distance covered in three and, *consequently*, in twelve moments twice as much as in nine, and in sixteen moments four times as much as in nine, and so on" (emphasize added). This interpretation of Descartes' numerical values is supported by similar calculations in another letter to Mersenne written about two years later[72]:

> You ask me, thirdly, how a stone moves *in vacuo*; but because you have forgotten to insert the figure, which you suppose to be in the margin of your letter, I could not well understand what you propose, and it does not seem at all that the proportions which you put forward agree with those which I once sent you, where instead of etc. as you have written me, I put $1/3$, $4/9$, $16/27$, $64/81$, etc., and this yields very different consequences.

There is no explicit information in the letter concerning the nature of these numbers and the values from which they are calculated. However, these numerical values show precisely the same relations among one another as the numerical values in his earlier letter: each value is $4/3$ of the preceding. It is therefore likely that they also represent a sequence of times corresponding to a sequence of spaces traversed where each space doubles the preceding one[73]:

spaces:	a	$2a$	$4a$	$8a$	$\ldots$
times:	$1/3$	$4/9$	$16/27$	$64/81$	$\ldots$.

Thus there is evidence that Descartes indeed generalized his originally computed example, assuming that it is a general property of the relation between time and space in free fall, that the space traversed is always doubled in $4/3$ of the time. By modern mathematical means we are able to determine the set of functions with this property. They all have the following form[74]:

$$s = \lozenge t^{\dfrac{\log 2}{\log 4 - \log 3}}.$$

Descartes' generalization therefore imputes a function between time and space where again space increases with a power of time but the exponent of the

---

[72] Oct./Nov. 1631; AT I, 230-231. See document 5.1.7.

[73] Descartes' sequence of figures begins with $1/3$ instead of 1 for the first time interval. Although the choice of the time unit is arbitrary, this choice is somewhat puzzling, in particular because this unit of time for the *first* interval is the same as that for the *second* time interval in Descartes' standard example, which does not make any sense. Although the result is correct, Descartes' choice of the time unit possibly depends on an error.

[74] That these functions have the property that the space s is doubled if the time t is multiplied by $4/3$ can easily be checked. The proof that all continuous functions with this property necessarily have this form requires more advanced mathematical methods and will not be given here.

function is not 2 as in the first case but $\log 2/(\log 4 - \log 3)$, i.e., approximately 1.41. We observe that the result is not the same as in the first case, but the mathematical means available to Descartes did not allow for an adequate representation of this function. Thus it is not at all astonishing that Descartes overlooked the difference.

Moreover, the generalization itself is wrong in the sense that the procedure by which Descartes derived the numerical values for the first two intervals defines a function between time and space which does not correspond to the rule assumed by Descartes.[75] This will become clear below when we discuss the implicit inconsistency of this procedure. For the present it should only be noted that it was a serious problem in Descartes' time to express functions in a way suitable for deriving general conclusions.

1.5.4.3 Generalization of Descartes' Deductive Argument

Thus, the most promising way to study the consequences of an argument like Descartes' proof was indeed to calculate actual numerical values and to compare them for different assumptions.[76] In particular, such a procedure would have also been the easiest and most direct way for Descartes to recognize that the temporal and spatial interpretations of the moments of motion, and hence of the vertical dimension of his figure, are mutually exclusive. Using methods he had often

---

[75] This has been overlooked up to now by historians of science. One of the editors of Descartes' works, Paul Tannery, in a note (AT I, 75) on the series of figures given by Descartes, explained them correctly as representing spaces traversed which are a function of powers of times elapsed with an exponent $\log 2/\log 4 - \log 3$ instead of 2. However, he erroneously considered this incorrect law of fall to be implied by Descartes' interpretation of the vertical side of his triangular geometrical figure as representing the spaces traversed. This mistake was reproduced by de Waard in his edition of the correspondence of Mersenne (vol. 2, p. 320). He even explicitly formulated the contradictory statement that in Descartes' figure the areas of 1, 3, 5, etc. represent the average velocities and that this is the same as to say that the spaces traversed are proportional to the powers of the times elapsed with an exponent of $\log 2/(\log 4 - \log 3)$ instead of 2. If they had really calculated the resulting values they would have recognized the contradiction between both statements immediately. It seems that historians of science who have adopted the interpretation never questioned this conclusion or calculated the corresponding values. Thus the error can still be found in recent publications. See, for instance, Shea (1978, p. 145), who, closely following a footnote of Koyré (1978, p. 87 fn. 99), stressed the incompatibility of Descartes values with the correct law of fall but overlooked the incompatibility of these values with Descartes' own proof. Marshall (1979) correctly reconstructed Descartes' incorrect generalization on p. 127, and correctly reconstructed the implications of Descartes' proof on p. 128, but did not notice the contradiction between the two.

[76] A method of testing theoretical assumptions by calculating numerical values is not an anachronistic suggestion: it was in fact actually applied by Galileo. See Renn 1990.

applied, he could have elaborated consequences of either interpretation. As we mentioned already when we discussed Descartes' numerical example in his letter of Nov. 13, 1629 (section 1.3.4), he could easily have inferred more values from his geometrical representation. If his figure is extended from two to several equal time or space intervals, the relations between the corresponding areas representing the quantity of motion are as 1, 3, 5, 7, etc. Deriving further numerical values from this sequence, Descartes would have noticed that the application of the Aristotelian proportions yields quantitatively different though qualitatively similar results according to which proportion is applied. He would also have immediately noticed that his derived rule, according to which for spaces as 1 to 2 the times always had to be as 3 to 4, is wrong.

In the first case, the temporal interpretation of the vertical, i.e., the assumption of equal time intervals, the spaces inferred increase in proportion to the areas, i.e., as the series

$$1 + 3 + 5 + 7 + 9 + \dots \ .$$

The sequence of partial sums

$$1, \ 4, \ 9, \ 16, \ 25, \ \dots$$

representing the spaces traversed increases in proportion to the squares of the times elapsed and allows one to infer directly the correct law of fall.

In the second case, the spatial interpretation of the vertical, i.e., the assumption of equal space intervals, the times inferred increase as the series

$$1 + \frac{1}{3} + \frac{1}{5} + \frac{1}{7} + \frac{1}{9} + \dots \ .$$

In this case there is no simple algebraical expression for the sequence of partial sums representing the times elapsed:

$$1, 1 + \frac{1}{3}, 1 + \frac{1}{3} + \frac{1}{5}, 1 + \frac{1}{3} + \frac{1}{5} + \frac{1}{7}, \ 1 + \frac{1}{3} + \frac{1}{5} + \frac{1}{7} + \frac{1}{9}, \dots \ .$$

Hence it is not immediately obvious that the sequence does not increase as the inverted function inferred above, i.e., as the square root of the spaces traversed. However, if Descartes had had any doubt that both inferences lead to an identical result, he could have checked this assumption as well as his own generalized rule of a relation as 3 to 4 for the relation of the times elapsed in the case of a double space traversed simply by comparing the calculated numerical values.

Checking the rule of time intervals as 3 to 4 for double space traversed in the first case, he could have compared the spaces for 3 and 4 time intervals and would have obtained the relation of 9 to 16 spaces, which clearly is not the double space. Checking the rule in the second case, he could have compared the times for 2 and 4 space intervals and would have obtained the relation of

$$1 + \frac{1}{3} \quad \text{to} \quad 1 + \frac{1}{3} + \frac{1}{5} + \frac{1}{7}$$

spaces which again violates his rule according to which the relation should be as 3 to 4.

In the same way Descartes could have checked whether the temporal and the spatial interpretation lead to identical results. According to the temporal interpretation, spaces in the relation 1 and 4 should correspond to times as 1 and 2. According to the spatial interpretation, they correspond, however, to times as

$$1 \text{ to } 1 + 1/3 + 1/5 + 1/7$$

which is clearly different from the relation of 1 to 2. The following table presents an overview of corresponding figures which are implied by a temporal interpretation, by a spatial interpretation, and by Descartes' generalization which was discussed in the preceding section:

Temporal interpretation		Spatial interpretation		Descartes' generalization	
time	*inferred space*	*space*	*inferred time*	*space*	*inferred time*
1	1	1	1	1	1
—	—	2	$1+1/3$	2	$4/3$
—	—	3	$1+1/3+1/5$	—	—
2	4	4	$1+1/3+1/5+1/7$	4	$4/3$ $4/3$

Summing up: It turns out that, on the one hand, Descartes, by elaborating the consequences of his argument and calculating and comparing numerical examples, could have realized that his generalized rule contradicts his argument. This contradiction is the result of a mistake by Descartes. The consequence would probably have been merely that this mistake would simply have been corrected. On the other hand, Descartes could have realized that there was in his conceptualization of the problem an internal contradiction which cannot be attributed simply to a mistake. This is the much more interesting aspect of the dialogue between Beeckman and Descartes, because in our view it refers to fundamental characteristics of the advance of knowledge. If not a mistake in the common sense of the word, what else can be the source of the contradiction? How could the contradiction have been eliminated given the epistemological situation of Descartes and his contemporaries?

1.5.4.4 The Source of the Contradiction

The analysis of Descartes' proof has shown that the internal inconsistency of his argument insofar as it cannot be attributed to a simple mistake, has its origin in an ambiguity of application of the Aristotelian proportions when deriving the relation between time and space by means of the geometrical representation of motion. The application of the proportions allows both temporal and spatial interpretations of the vertical dimension of this figure. The results however are not identical.

From a modern point of view, it may be plausible simply to decide which interpretation is right and which is wrong. However, in the framework of

Descartes' conceptualization, the two conflicting arguments are strictly sym-metrical. The reason is that instead of the modern concepts of motion and velo-city which make proportional increase of velocity in relation to time or space, a meaningful alternative Descartes had at his disposal only the concepts of motion and velocity in the Aristotelian tradition.

We know not only from Descartes' proof but also from other contemporary sources – especially the work of Galileo – that the Aristotelian conceptualization of velocity was still presupposed as evident: velocity was conceptualized as space traversed in a certain time and thus related not to individual instants or points but to finite intervals of a motion. We, today, accept the implications of this con-ceptualization, for instance that velocities are equal if equal spaces are traversed in equal times, as valid only for uniform motions. In this case – because the ve-locity is constant – there is indeed no difference between the consequences of the modern concept of instantaneous velocity and the Aristotelian concept. In all other cases we would accept the Aristotelian concept only when referring to the average velocity in a certain interval, i.e., an average over an infinite number of points which can generally be calculated only by means of the infinitesimal cal-culus.

Historically, it was impossible to view the Aristotelian concept this way, not only for technical reasons, i.e., that the calculus was not yet invented, but pri-marily because it is impossible to conceptualize something as an average of something else which itself cannot yet be conceptualized. Hence there was no reason to restrict the Aristotelian concept and the consequences based on it to uniform motions.

This consideration provides an answer to the question under what conditions the internal contradictions could occur. As long as there was no quantitative study of accelerated and decelerated motions, the Aristotelian concept could not lead to contradictions of the kind which become evident in Descartes' proof. As long as such motions are not resolved into different parts, the global characteriza-tion is completely adequate.

Therefore, it is historically no accident that contradictions originating in the Aristotelian concept of motion surfaced in the 17th century, when for the first time accelerated motions were studied and calculated in detail. It will be seen in Chapter 3 that contradictions similar to those discussed here can also be found in Galileo's work on mechanics. The general scheme of development is that new problems and experiences are conceptualized by means of concepts which were appropriate for their former area of application but do not sufficiently hold for the new situation. Thus, the fundamental concepts of classical mechanics were not the preconditions of the "new sciences" in the 17th century but rather their result. At first, motion was conceptualized completely in the framework of Aristotelian and related medieval conceptual traditions. This conceptualization was the basis for deriving consequences and constructing proofs. The immediate consequence was an extension of the meanings of the concepts involved as well as the establishment of new relations to other concepts. Finally, however, the concepts showed serious deficiencies if applied to these new experiences. These deficiencies became obvious when contradictions like the one discussed here surfaced and could not be attributed to simple mistakes that would be easily located and corrected.

According to a common understanding of deductivity, a deductive system which implies contradictions is simply incorrectly constructed. Concept development as a consequence of the appearance of contradictions would thus not be a genuine developmental process, but would consist merely in the elimination of mistakes. Accordingly, the common interpretation of Descartes' analysis of free fall is based on the assumption that Descartes was simply mistaken in assuming the proportion of the velocity in free fall to the space traversed instead of to the time elapsed. However, the result of our analysis of the structure of Descartes' proof is not compatible with such an interpretation. This interpretation overlooks the fact that in the beginning of the 17th century the modern concept of velocity had not yet been constructed and that only this concept makes the alleged alternative meaningful in an obvious way. It was only the restructuring of the conceptual framework in which motion was conceptualized, i.e., the refinement and partial revision of the internal relations among several concepts which finally led to the concepts and proof structures of classical physics.

If, therefore, Descartes had recognized the internal contradiction in his proof there would have been no simple solution to the emerging problem. The problem would have forced him to reconstruct his conceptual system and such a process is not bound to a predetermined end. As we have shown by presenting the documents related to Descartes' reconsideration of his proof in the course of the development of his theory of gravity, he did not arrive at the same result as Galileo, who studied free fall at about the same time, i.e., a restructuring of the concept of motion which might have brought the proof of the law of free fall closer to the proof in classical mechanics. Therefore, we shall in conclusion revisit the dialogue between Descartes and Beeckman in order to discuss why they did not arrive at a conceptualization of free fall which is compatible with the conceptualization in classical mechanics.

1.5.5   The Beeckman-Descartes Dialogue Revisited

Neither Beeckman nor Descartes was primarily interested in solving the particular problem of free fall. Beeckman posed his question in order to clarify implications of his atomistic theory of gravity. He wanted to know what could be inferred from his principle, "that in vacuum, what is once in motion will always be in motion." He interpreted Descartes' answer in the framework of his atomistic theory of gravity in which free fall is conceptualized as caused by means of "material spirits" or – more explicitly – a "multitude of ... particles" (*Journal* note, 1618). Descartes' answer to Beeckman reconceptualized the question in the framework of a dynamical theory using the medieval concept of impetus. It was a major point of his answer – which Beeckman did not pick up at all – that the quantitative relation between time and space in free fall are not determined as long as there is no special assumption introduced about how the attractive force of the earth influences the motion of a falling body. Apparently, as long as this proof could not be integrated into a broader conceptual framework, he was undecided about the status of the proof.

Hence, Descartes' initial document and Beeckmans' *Journal* note about Descartes' answer to his question are written from different perspectives; but this difference in perspective influenced the context of the argument more than the argument itself. According to the common interpretation, the main difference between Descartes' argument and the version reproduced by Beeckman is that Descartes erroneously assumed that the velocity of free fall increases proportionally to the space traversed and that, in his note, Beeckman tacitly corrected this mistake. Disagreeing with this interpretation, we have shown that Beeckman's note in his *Journal* closely follows Descartes' initial document with two qualifications.

The first qualification is that Beeckman omitted the dynamical explanation given by Descartes. We claim that Beeckman did not quote Descartes' theory because he adhered to a different dynamical explanation of free fall and did not substitute his explanation for Descartes' because it was inconsistent with Descartes' argument. Considering those parts of the manuscript which have usually been ignored by scholars, we were able to show that from Beeckman's dynamical theory it follows that acceleration in free fall is not continuous, and from this it follows that velocities are in arithmetical progression in subsequent times. Therefore, Beeckman regarded Descartes' solution as merely an approximation of the exact relations.

The second difference between Beeckman's note and Descartes' initial document is that Beeckman consistently follows – at least in this note[77] – one of the two possible interpretations of the concept of moment and explicitly states the quadratic relation between times and spaces of fall. Because of its importance within the framework of classical mechanics, this result became the focus of most established interpretations. However, it should not be overlooked that neither in Beeckman's note nor in Descartes' document is a generalized proportionality between velocity and time or space mentioned at all.

Descartes' diary note is merely a very short version of his initial document. The only difference is that in this version Descartes, too, consistently follows one of the possible interpretations of the concept of moment. But he applied the spatial interpretation, so that – as we have shown above – the derived numerical values are in fact not compatible with the quadratic relation stated by Beeckman.

When Descartes later returned to the subject again, his perspective had changed. But he never mentioned any modification of the argument itself. He referred, if at

---

[77]   See Schuster (1977, p. 67) on Beeckman, J III, 133f.

all, to his original answer always claiming that he was only repeating his argument. From our analysis of the related documents we have found no indication of a substantial modification of his argument which would cast any doubt on Descartes' assertion. In particular, we have shown that the letter of Nov. 13, 1629, which has been interpreted as representing a fundamentally new argument,[78] on the contrary supplies the key to an understanding of the medieval conceptual background of his initial document.

It remains to be explained why Beeckman and Descartes never noticed that their different perspectives on the argument resulted in differences between their conclusions even on a technical level. Although they referred to each other's interpretations several times (at least indirectly in their correspondance with Mersenne), and although Descartes studied Beeckman's *Journal* twice, in 1618 and 1628, neither of them realized that their differing interpretations of the concept moment would have led to a contradiction if explicit alternative functional relationships between time and space in uniformly accelerated motion had been derived.

The answer to this question is partly the same as the answer to the question why Descartes himself did not notice the contradictory consequences of a temporal and a spatial interpretation of the moments of motion. Moreover, the conditions for the exchange of ideas in this dialogue and the limited extent to which the conceptual and theoretical base of the arguments were explicitly raised were impediments that prevented Beeckman and Descartes from exploring the consequences of their differing understandings. In fact, the differing interpretations did not lead to contradictions in the inferences as actually drawn by Beeckman and Descartes.

This is the case, first of all because the problem as stated by Beeckman was not to derive a general relation between time and space in free fall nor is such a relation contained in Descartes' explicit answer. Rather, at issue was a procedure for calculating specific values. As such, the calculated values were not incompatible. It is thus misleading to read Beeckman's question and Descartes' answer as pertaining to a law of free fall. Moreover, one should keep in mind that the mathematical means for expressing a functional dependence of one magnitude on another were still very limited.

The more fundamental reason, however, that Beeckman and Descartes did not realize that their arguments implied different laws of free fall is that neither of the alternatives could have been a theoretical statement in their respective physical theories. Since this was the case, they did not elaborate these implicit consequences. For Beeckman, the impact of small particles on the falling body was the cause of gravitation, implying a discontinuous acceleration; Descartes' argument could at best be considered by Beeckman as a quantitative approximation, the assumptions of which were incompatible with his theory. For Descartes, on the other hand, the assumptions contained in Beeckman's question were arbitrary

---

[78]   Koyré 1978, p. 86.

and could be replaced by others. According to his principles of mechanics, which he stated at the same time,[79] gravitation is dependent on conditions not accounted for by Beeckman's presuppositions. In later years his previous lack of commitment to Beeckman's presuppositions gave way to an explicit rejection of similar ones stated by Galileo. When he had finally worked out a comprehensive theory of nature, with an explanation of gravity as caused by the impact of particles, he consequently rejected the whole idea that there could be a determined quantitave relation between time and space in fall, since fall is dependent on the existence of a medium. This conception had two immediate consequences: that acceleration is not uniform and that the resistance is not accidental but rather an essential aspect of the existence of acceleration.

What has been represented as a comedy of errors thus turns out to be a coherent long-term discussion of an internally consistent argument. Beeckman understood Descartes' argument very well although he did not accept it. He noted it correctly in his *Journal* leaving out what was not directly related to his question and commenting on it from his point of view. Descartes consistently stuck to his argument because for understandable reasons he noticed neither the implied internal contradiction nor the differences to Beeckman's note insofar as these differences were not attributable to the differing physical theories of mechanics. His rejection of the law after Galileo's publication was well founded in a comprehensive theory of gravitation which could not be convincingly refuted for a long time to come.[80] The dialogue between Descartes and Beeckman is therefore striking, not as a process of mutual misunderstandings, but as an example of the influence of global conceptual schemes on the perception, development, and evaluation of the truth of an argument.

---

[79] Aquae comprimentis in vase ratio reddita, AT X, 67-74. See document 5.1.4.

[80] For an impressive account of the difficulties involved in the notion of action at a distance, see the 68th letter of Euler's *Letters to a German Princess*. For a general account see Aiton 1972.

2
Conservation and Contrariety: The Logical Foundations of Cartesian Physics

> Profiteorque mihi nullas rationes satisfacere in ipsa
> Physica, nisi quae necessitatem illam, quam vocas
> Logicam sive contradictoriam, involvant.
> (Descartes to Henry More, Feb. 5, 1649; AT V, 275)

2.1 Introduction

The general theory of matter presented by Descartes in the second book of the *Principia Philosophiae* is the first well founded systematic physical theory of modern science; for it explicitly introduces the logical presuppositions necessary for a system of causal explanations of physical phenomena using equations. While it is true that Descartes himself takes very little advantage of the possibilities created by the introduction of these prerequisites (there is, for instance, very little mathematics, no formal equations, and few proportions in the *Principia* itself), he nonetheless determines basic requirements of such a system of explanations and provides conceptual instruments adequate for the formation of such a physical theory.

These requirements consist: (1) *first and foremost* in the formulation of conservation principles which define the nature of the systems studied, guarantee their identity over time, and put strong constraints on the kinds of assertions that can even be recognized as legitimate attempts at causal explanation.

(2) *Second*, the development of a physical explanation of qualitative and quantitative changes within a system or of the interactions between systems requires a conceptualization of the nature of change and interaction. Descartes conceives physical systems to consist ultimately of particles whose interactions are exclusively collisions. Therefore, he must determine the nature of this fundamental interaction and explain the behavior of particles in collision and in the absence of collision. Descartes explains the latter by a logical principle of inertia and in one-dimensional two-body cases he explains the former by a logical calculus of contrary predicates. These principles guarantee that the asserted causal connections in the material world correspond to the semantic connections of the terms in his conceptual system.

(3) *Third*, in order to apply the theory not merely to interactions in the same line, which are accounted for by the logic of contrary predicates, but also to two- and three-dimensional Euclidean space, a method is needed to reduce all possible interactions (e.g., oblique collisions of real bodies or those of point masses moving in intersecting lines) to the pure forms of direct opposition and lack of opposition. Descartes accomplished this reduction by the dynamical interpretation and application of a geometrical operation, the parallelogram rule for the compounding of motions. In the 17th century this technique was used to give a geometrical solution to a number of problems now solved with the help of the vector calculus.

As will be shown below, the conceptual means developed by Descartes to meet these requirements (1. *conservation*, 2. *interaction*, 3. *reduction*) are consistent with one another, but their application leaves the fundamental interactions of bodies which were to be explained by his physics still underdetermined. As we shall argue, the reason for this underdetermination is that Descartes asserted the conservation of "vector" quantities only for individual bodies (not for the system as a whole) and only in the *absence* of interactions. To achieve complete determinism, Descartes had on occasion to specify further the logic of contrary predicates through a minimum principle as well as implicitly to appeal to considerations of symmetry not explicitly introduced into the system of explanations.

In the following sections we shall discuss the role of conservation laws (2.2) in the foundations of science and their function in Descartes' system, the role of logic (2.3), especially the logic of contrary predicates in providing Descartes with the conceptual means for dealing with physical interactions, Descartes' application of the logical principles to the interaction of two particles in the same line (2.4), i.e., the derivation of the laws of impact, and Descartes' use of the parallelogram rule to reduce interactions in two-dimensional space to one-dimensional interactions (2.5).

2.2 Principles of Conservation and Conservation Laws

2.2.1 The Foundational Function of Conservation Laws

By a systematic physical theory, we understand a network of laws, usually formulated as equations, the transformations of which predict and explain the outcome of interactions among the phenomena to which the theory applies. We will argue below that every physical interpretation of such equations as stating causal equivalences presupposes some principles of conservation, and hence that the logical foundations of causal explanations in a deterministic physical theory are principles of conservation for certain physical entities. We shall then show that Descartes explicitly introduced conservation laws[1] appropriate to the theory he developed: these laws specify the conservation of matter, and of the quantity of motion, which fulfill much the same function as the laws of conservation of mass and energy in classical physics. The role of momentum in classical physics is played by contrariety and "determination" in Cartesian physics.[2]

Physics, and every science, presupposes the invariance or conservation of some entities, but not only specifically scientific knowledge relies on the conservation of some of its objects: if the invariance in manipulation of some objects could not be presupposed or if their variations were not at least predictable, then regular human activities as we know them, as well as the formation of concepts to describe or explain them, would be impossible. Alice in Wonderland cannot plan her next stroke if the croquet mallet unexpectedly turns into a flamingo and the ball becomes a hedgehog. Playing croquet, as well as forming concepts of the game's equipment or formulating the rules of the game, is impossible in such a situation. But the conservation laws of physics need not refer to entities identical with the objects of human practice, nor to the time spans relevant to our activities. Moreover, the formation of the concepts of the conserved entities in daily life and in science rests on different processes, which we will briefly discuss below.

In physics, the change of a material system under investigation can be defined only in relation to a determined state of the system considered unchanged over time. Searching for an external cause to explain change presupposes that the system is invariant and would not have changed without interference from outside. Thus in order to have a physical theory of change in a material system, one must already at least implicitly have defined what "staying the same" means in physical terms.[3] Hence the quest for a causal explanation of change presupposes not only that a determinate description of a system in different states at different times is possible – and thus that the standards of measurement are invariant over

---

[1] We shall speak of "*principles* of conservation" as the logical or philosophical prerequisites of a physical theory, and of "conservation *laws*" as the specific fulfillment of the requirements in a particular physical theory.

[2] An extensive discussion of the relation between conservation principles and the principle of causality in mechanics (with special reference to Descartes' and Leibniz's contributions) can be found in Wundt 1910, pp. 84-114. The principle of equivalence of cause and effect is the last of the six axioms or hypotheses on which mechanics is founded (according to Wundt). For a discussion of conservation principles with many historical examples, but from a point of view very different from ours, see Meyerson 1962.

[3] The conceptual link between change and invariance is discussed by Kant under the heading "The Permanence [of Substance]" in the *Critique of Pure Reason* (A182-189). He later explicitly refers to this discussion in his proof of the law of conservation of mass in the *Metaphysical Foundations of Natural Science* (1786). The law states: "First law of mechanics: With regard to all changes of corporeal nature, the quantity of matter taken as a whole remains the same, unincreased and undiminished" (*Gesammelte Schriften,* vol. 8, p. 541; Kant 1985, p. 102).

time – but also that the entities in the system which define its identity over time are conserved in the absence of interactions with other systems.

For example, in order to conceive force as the cause of the acceleration of a body and acceleration as a change in its velocity, one needs a conservation principle for the state of motion, i.e., a principle of inertia and a definition of state of motion, e.g., in terms of velocity or speed and rectilinear direction. Force as cause of change in general is considered as external to the system defined by conservation laws. If the cause, too, is considered as a physical entity, then a change in the system under observation implies an interaction with another system, and if the interaction is to be expressed in a causal law, then there must be a complementary change in the other system. A causal law in physics which expresses (in an equation) the possibility of transforming a system through interaction with another, implies a qualitative and quantitative invariance in such a transformation, i.e., that a change in one system will have a constant qualitative and quantitative relation to the change in the other. If both interacting systems are taken together as one, then their combination will be a closed system to which the conservation laws for the magnitudes defining the system apply.

Insofar as a physical law is expressed in an equation it fulfills the philosophical requirement that the cause should be equal to the effect (*causa aequat effectum*). However, the fact that two concrete systems are causally equivalent, in respect to a particular effect based on certain physical magnitudes, does not mean that they are causally equivalent in respect to every possible effect they may produce on the basis of these magnitudes.

The insight into this possibility on the example of the motions of bodies, which can be quantitatively equivalent in different manners according to different effects (measured by mv and mv^2), led Leibniz to distinguish between two kinds of equivalence and between two kinds of measurement: measurement by *congruence* and measurement by *equipollence*.[4] "Congruence" denotes the simplest kind of measurement, in which a standard is iterated and compared with the object measured, as a yardstick is used to measure the length of an object. In measurement by congruence, two systems are compared with respect to some property possessed by both, and the quantitative relations of the two in this regard are determined. As the name suggests, this type of measurement is originally derived from the *geometrical* properties of systems, where the additivity of the physical magnitudes consists in spatial juxtaposition of parts (i.e., to what were traditionally called "extensive" quantities). Leibniz also considered *counting* other standardized units to be measurement by congruence, so that a body (i.e., a standard unit of mass) or a spring (i.e., a standard unit of force) can be iterated, and the number of such units compared.

---

[4] "Equipollence" is a synonym for equivalence used in logic at least up to the time of Carnap. Leibniz gives it the particular technical meaning that we have adopted in a letter to l'Hospital (Jan. 15, 1696). See Leibniz, GM II, 305-307; see document 5.2.1. See also GP IV, 370-72; and in general "Initia rerum mathematicarum metaphysica," GM VII, 17-29.

Equipollence, on the other hand, was also used to measure intensive qualities such as velocity and force.[5] Here a phenomenon is measured by transforming it (or a representative sample of it) causally into an extensive effect, which can then be measured by congruence and counting. Leibniz measured the force (*vis viva*) of a body in motion by letting it ascend in the gravitational field of the Earth and measuring (by congruence) the height it could reach or (by counting) the number of standard (congruent) springs it could compress.

Equivalence or equality in the sense of congruence is hence given when the left- and right-hand sides of the equation are expressed in terms of the same physical magnitudes. If, however, the two phenomena to be equated (or compared by measurement) are not or cannot be expressed by means of the same physical magnitudes, they can only be equated as *cause and effect*. In this case they can be said to be equipollent or causally equivalent. The equipollence of two phenomena means that both can be produced by the same causes or that both can produce the same effects. The equation of *work* and *heat,* for instance, expresses an equipollence.

It is of course possible that an equation expressing a causal equivalence seems to be a simple congruence, i.e., if the cause and effect happen to be expressed in the same magnitudes. This form of representation disguises the fact that considerations of equipollence are ultimately the basis of the definition of the magnitudes. Moreover, even when an equation actually displays an equipollence, the *invariance* in transformation that it expresses can be interpreted as the conservation of a new entity and a new concept can be introduced for it. The concept does not refer to a new empirical phenomenon but to what is invariant in the transformation of phenomena, and the original phenomena can now be conceived as instances or embodiments of the same entity, e.g., energy, which is not an observable phenomenon or even a definable entity, but is identified in and distinguished from many specific phenomena, e.g., heat, work, electricity, etc.[6]

It should be stressed that, in spite of the operational simplicity of congruence, the conceptually really fundamental form of measurement for physical explana-

---

[5] Traditionally, velocity was the most important intensive magnitude; and as late as Kant *mass* was also sometimes taken as an intensive magnitude.

[6] "The law of energy directs us to coordinate every member of a manifold with one and only one member of any other manifold, in so far as to any *quantum* of motion there corresponds one *quantum* of heat, to any *quantum* of electricity, one *quantum* of chemical attraction, etc. In the concept of *work*, all these determinations of magnitude are related to a common denominator. If such a connection is once established, then every numerical difference that we find within one series can be completely expressed and reproduced in the appropriate values of any other series. The unit of comparison, which we take as a basis, can arbitrarily vary without the result being affected. If two elements of any field are equal when the same amount of work corresponds to them in *any* series of physical qualities, then this equality must be maintained, even when we go over to any other series for the purpose of their numerical comparison." (Cassirer 1923, p. 191) Cassirer calls the one-to-one correspondence of values in different series "equivalence" (p. 197).

tion is equipollence. For, physical equivalence refers essentially to the substitutibility of the entities equated. Substitution is of course relative to various purposes, and various forms of equivalence can be imagined; but at least for scientific explanation causal equivalence is decisive. Congruence can have the semblance of being a priori, since it is obvious that $5x = 2x + 3x$. But the concept of the magnitude whose units are called x's, and the physical meaning of equivalence and addition, is ultimately determined by equipollence.

Our discussion of the concepts equivalence and equipollence suggests that the scientific manipulation of physical entities and the transformation of propositions about concepts referring to them, are the basis of the formation of concepts of further physical entities which do not directly refer to objects of daily experience but to invariances stated in controlled manipulations of physical objects in experiment and in transformations of propositions about previous scientific concepts. If this approach is applied to concept formation in general, then on each level invariance in manipulations of objects and in transformations of propositions precedes and is prior to the notion of identity and equivalence on the next higher level.[7]

These considerations on causal equivalence and conservation imply two relevant consequences as to the relation of the principle of causality to conservation laws in science. Scientific conservation laws cannot be derived from a philosophical principle like *causa aequat effectum* or *ex nihilo nihil fit. Nil fit ad nihilum*, since the actual scientific problem is to determine *what* remains invariant in transformations supposing that something is conserved.[8] On the other hand, it is the conservation laws in a given physical theory that determine the specific meaning of the principle "cause equals effect" by defining what counts as a cause and an effect and how their measures are to be expressed. These laws guarantee the observance of the causal principle in that any candidate for the status of natural law must conform to them, if the whole framework of the discipline is not to be reworked.[9]

For the philosophical foundations of science it is not so important to ask what particular magnitudes are determined as invariant but rather *whether* they fulfill the requirements mentioned; the conservation laws may change and do change with the development of science. But as long as science states causal laws in equations, the abandoning of one set of conservation laws will necessarily lead to

---

[7] We have intimated that the formation of concepts for objects that are invariant in our manipulation of them is similar in structure to the formation of concepts of scientific entities and conservation laws. We cannot, however, further discuss the extent and significance of the similarity between these two processes of concept formation.

[8] Robert Mayer, for instance, argued that the conservation of energy follows from this philosophical principle. See Freudenthal 1983.

[9] As E.P. Wigner maintained in his Nobel Prize Lecture of 1963, the most important function of invariance principles is "to be touchstones for the validity of possible laws of nature. A law of nature can be accepted as valid only if the correlations which it postulates are consistent with the accepted invariance principles" (Wigner 1967, p. 46).

an attempt to formulate others because of their foundational role in scientific theory.[10]

The fact that conservation laws define an isolated system means that they define the system in reference to which causal laws apply. Applying conservation laws to the Universe as a whole expresses the claim that the applicability of causal laws is universal, i.e., that a change in any particular material system is to be explained by its interaction with other material systems and that no other causes are admissible. This claim was of paramount importance in the 17th century for the establishment of a scientific world view in opposition to the feudal-religious view dominant at the time, which allowed a transcendent immaterial deity to act physically in the material world.[11]

2.2.2 Descartes' Conservation Laws

Before we turn to Descartes' physics and examine whether the conservation laws he specifies can, in principle, serve as the foundation of a deterministic quantitative physical theory, we must first determine what *kinds* of conservation laws are necessary and which *particular* laws proved historically to be sufficient for this purpose.

In classical mechanics, in order to provide a deterministic description of a system of elastic particles in motion which act on each other by *impact only*, one needs conservation laws for mass, momentum, and kinetic energy as well as information on the positions and velocities of all particles in one instant. On the basis of this knowledge, Laplace's demon could calculate the evolution of the system and pinpoint the state of each and every particle at any given future time. There are two significant differences between the concepts of kinetic energy and momentum. On the one hand, their measures are different: $1/2mv^2$ for kinetic

---

[10] "We can put the question: what would be changed in physics if a *perpetuum mobile* were to be discovered today? Our conviction of the universal subjection of nature to law would not be shaken ... the validity of the law of the conservation of energy would be restricted to certain limits, and perhaps we could hope to recognize it ultimately as a special case of a still more general law" (von Weizsäcker 1952, p. 64-5).

[11] This does not mean, however, that any particular scientist had to realize that science demands a scientific world view nor that he suscribe to such a view. But even scientists who did not conceive of the universe as a closed system determined by conservation laws had nonetheless to presuppose (in practice) a constant quantity of the relevant magnitudes. In such a case, only inconsistency between the philosophy espoused and the science actually practiced allowed science to be pursued on the basis of unscientific presuppositions. Newton provides a good example of this state of affairs; see Freudenthal 1986, pp. 44-76. The politically and ideologically motivated attempt to construe the universe as a closed system, the states of which can be related only to physical entities, in conjunction with the attempt similarly to construe society was, according to Lefèvre (1978, pp. 45-79), an important factor in the construction of modern science.

energy and *mv* for momentum. On the other hand, kinetic energy is a *scalar* magnitude, while momentum is a *vector*, i.e., for our purposes, a magnitude with a determined direction. Mass, too, is scalar in classical physics.

Assuming that all interactions within a system can be reduced to two-body impacts, the information needed to determine the state of a system consists in the positions, masses, and velocities of the particles. Assuming, furthermore, the conservation of mass in each particle in interaction, the state of any given particle depends on its velocity. If we want to determine the velocities of *two* particles after collision, we need (for purely formal reasons) at least two independent equations. Furthermore, at least one of these (or a new equation) must guarantee the invariance of the system as a whole, i.e., that whatever the changes within the system, the magnitudes defining it remain the same. And at least one of the equations must be able to deal quantitatively with directions. Thus the *minimal* requirements for a deterministic idealized impact theory of elastic particles (assuming the conservation of the mass of each particle) are two equations, of which at least one defines the system as closed and at least one operates with a vector magnitude or some other entity capable of expressing direction. In classical mechanics these are conservation laws for kinetic energy (scalar) and for momentum (vector).

The difference in status between scalar and vector conservation laws in classical mechanics and the philosophical relevance of this difference has, to our knowledge, never been seriously studied, and we cannot go into detail on the differences.[12] However, we should point out that the attribution of a certain (vector) momentum to a system *of given mass* entails the attribution of a particular amount of (scalar) energy but not vice versa since (directionally) different momenta are compatible with the same amount of kinetic energy. In the 17th century, the system was *defined* by the "real and positive" (i.e., scalar) magnitudes not by those that are "merely modal" (e.g., directional), even though the latter are necessary for calculating the trajectories of colliding particles.

According to Descartes, material reality consists in "extended substance" (matter) and its "modes" (shape and motion or rest) (II, §25).[13] Descartes con-

---

[12] Max Planck began his book, *Das Prinzip der Erhaltung der Energie* (1913, p. 1) by isolating two conservation laws as somehow more fundamental than other laws: "There are two propositions which serve as the foundation of the current edifice of exact natural sciences: the principle of the conservation of matter and the principle of the conservation of energy. They maintain undeniable precedence over all other laws of physics however comprehensive; for even the great Newtonian axioms, the law of inertia, the proportionality of force and acceleration, and the equality of action and reaction apply only to a special part of physics: mechanics – for which, moreover, under certain presuppositions to be discussed later they can all be derived from the principle of conservation of energy." Planck seems to mean that the fundamental character of the conservation of energy and matter has to do with their being scalar magnitudes and thus not limited to mechanics.

[13] Numbers in parentheses refer to part and paragraph of the *Principia Philosophiae* (AT VIII); a translation of bk. II, §§36-53 is given in document 5.2.2.

ceived of substance in traditional terms as something that can exist of itself (*quae per se apta est existere*);[14] this definition applies strictly only to God, and all created substances depend for their existence on God's concourse (I, §51). The conservation of matter seems to be presupposed by Descartes as evident, i.e., as immediately following from its being a substance. He stresses the conservation of the *modes* of matter and mentions the conservation of matter only in passing (II, §36). However, the proposition that no matter is created or annihilated by natural causes was hardly in dispute. Since Descartes has defined matter as the extended in three dimensions, the quantity of matter is determined by volume (II, §4). Although this measure is different from the measure of mass in classical physics, it is nevertheless obvious that both concepts refer to the same experiences with material objects and serve comparable functions in the respective theories, namely to account for the resistance of a body to a change in its state of motion.

The conservation of motion on the other hand, a scalar magnitude measured by $|mv|$, is explicitly stated and argued by Descartes both for the world as a whole and, in the absence of interactions, for every single body. The conservation of the total quantity of motion and rest (as modes of matter) in the world, is introduced after the fundamental concepts of the theory of matter have been developed but before the laws of nature are discussed. Since Descartes attempts to ground his physics on principles whose validity is independent of the particular theory they are introduced to ground, he cannot justify these principles by appealing to the success of their instantiations. The origin and conservation of both matter and motion are logically prior to the physical theory, and therefore the arguments that Descartes adduces in support of conservation principles are not physical. As presuppositions of physical theory they need a qualitatively different kind of justification. The justification that Descartes actually gives is that the *constancy* and *immutability* of God demands that he conserve unchanged what he originally created, and that it is "most consonant with reason" that the creator be constant and immutable (II, §36). The fact that the "different kind" of argument takes on a theological form is not important here; for the subsequent analysis it is however important that the argument rests neither on any physical proposition, nor on logical or mathematical necessity.

The further specification, that the quantity of motion for *every single body* is conserved in the absence of interaction with others, is introduced as the *First Law of Nature* and is dependent on a logical principle (to be discussed in section 2.3.4 below). The law states that every thing, insofar as it is *simple*, will persist in its state, i.e., conserve its modes, if not acted on from without. Thus it will retain its shape (and presumably its volume) and its state of motion or rest. That is, retaining properties needs no explanation, changing or losing properties needs an explanation: a moving body does not come to rest of its own accord, "for rest

---

14 AT VII, 40 (*Meditations*, III).

is the contrary of motion and nothing moves by virtue of its own nature towards its contrary or its own destruction" (II, §37).

The *Second Law of Nature* further specifies the preservation of the state of motion. Each body in motion is determined to continue its own motion in a straight line, since motion is conserved exactly as it is in an instant without considering previous instants.[15] The magnitude involved here is called *determinatio*, and its conservation is asserted only for a *single* body in motion. Descartes does not maintain that determination is conserved in a body during or after interactions nor that the sum of the determinations of the interacting bodies after collision stands in any determinate quantitative relation to the sum before collision. Determination, as he elsewhere makes clear,[16] has a direction *and* a magnitude both of which are conserved only in the absence of interactions. Its measure turns out to be matter times directed motion (*mv* taken as a vector).

It is easily seen that the consistency of a theory containing the concepts quantity of motion (|*mv*|), which is added arithmetically, and determination (*mv*), which is added vectorially, is going to be problematical any time two determinations which do not coincide in direction are added together. The length of the diagonal of a parallelogram is always shorter than the sum of the lengths of two adjacent sides except in the limit where the angle is zero (cos $a = 1$). Thus it would seem that the use of the parallelogram rule, which as we have indicated is one of the essential elements of Descartes' system, must constantly lead to internal contradictions. However, as we shall show below, the two measures are only inconsistent under certain assumptions which hold in classical physics but do not hold in Descartes' theory. Specifically, it is not assumed in Descartes' theory that the two magnitudes, motion and determination, are independent, in fact this is explicitly denied: motion (|*mv*|) is a mode of matter, determination (*mv*) is a mode of this mode. We shall show in section 2.5 that when interpreted within Descartes' conceptual system, there is no contradiction. The vectorial composition and resolution of *motions* is not permitted, and neither the (vectorial) composition and resolution of determinations nor the treatment of the oblique impact

---

[15] *Principia*, II, §39. It should be noted, however, that Descartes' concept of "speed" (*celeritas, velocitas*) does not refer to an instantaneous quantity but rather denotes the space traversed in a *finite* time. Determination, on the other hand, is introduced explicitly as an instantaneous magnitude: "...that each part of matter, considered in itself, always tends to continue moving, not in any oblique lines but only in straight lines ... For [God] always conserves it precisely as it is at the very moment when he conserves it, without taking any account of the motion which was occurring a little while earlier. It is true that no motion takes place in an instant; but it is manifest that everything that moves is *determined* in the individual instants which can be specified as it moves, to continue its motion in a given direction along a straight line, and never along a curved line" (emphasis added). On the development and the systematic consequences of Descartes' concept of speed, see Chap. 1, above.

[16] In the *Dioptrics* and in a series of letters for Fermat and Hobbes; section 2.5 of this chapter discusses the concept of determination in great detail.

of two or three bodies results in any inconsistency – when carried out according to the rules Descartes presents.

This is not to say, however, that Descartes' theory is analogous to classical mechanics or that it differs from it only in the measure of the scalar magnitude: motion as opposed to kinetic energy. On the contrary, as indicated already in the statement that determinations may not be considered as *independent* magnitudes like momenta (in fact only the directional aspect of determination is independent while its magnitude depends directly on and changes with the quantity of motion), his theory differs from classical mechanics in many significant respects. This will become even more clear when we examine Descartes' rules of impact. Our comparison of the conservation laws of classical mechanics with those of Descartes serves only the purpose of examining whether Descartes' theory possesses the minimum of conservation laws necessary for a deterministic impact theory of the behavior of elastic particles that expresses causal relations as equations.

In conclusion we can say that Descartes recognized the importance of conservation principles for the foundations of natural science and that he also saw that the principles must apply both to scalar and directed magnitudes. Furthermore, the conservation laws and the physical theory erected on them are internally consistent. However, as we shall show in section 2.4, his theory remains underdetermined and is unable unequivocally to predict and explain the outcome of some two body interactions, since the directed magnitude conserved is not independent and is conserved only in the absence of interactions.

2.3 The Logic of Contraries and its Application in Physics

Although Descartes, as is well known, often spoke disparagingly about logic, what he rejected was not logic as such but what he considered an unproductive logic, i.e, a logic that merely administered the known instead of discovering the new. As an alternative to unproductive syllogistics he sought a productive logic that could function as an *ars inveniendi*. In the letter to his translator that served as a preface to the French edition of the *Principia*, he advised that one "must study that logic which teaches how to use one's reason correctly in order to discover the truths of which one is ignorant,"[17] and in the letter to Henry More quoted in the motto to this chapter he states unequivocally that he is looking in physics for truths as necessary as those in logic.

If the standard by which logic is to be judged is the ability to produce or guide the production of new knowledge about the world, it is not surprising that Aristotelian syllogistics or even immediate inference according to the law of non-

---

[17] AT IX, 13-14.

contradiction are found wanting. Even Aristotelian physics did not base physical explanation on this kind of logic. The basic explanatory principle for motions in Aristotelian physics was not the exclusion of contradiction but the opposition of logical contraries. Both in Aristotle and in the Aristotelian physics codified in Scholasticism the central principle of explanation was the actualization of potential in motion from one contrary to the other. Mutation between the contradictories affirmation and negation was confined to generation and corruption, while motion (in the general sense of change) was derived from contrariety.[18] We shall see that a logic of contraries also forms the basis of Cartesian physics as well as that of the systems of his contemporaries, whether mechanistic or scholastic.

In this section we shall first of all review some basic characteristics of logic in the works of Aristotle (2.3.1) and systematically explicate the logic of contraries (2.3.2) in order to be able to explain Descartes' program. Then, we shall briefly sketch the application of this kind of logic in one part of Aristotle's physics (2.3.3). Finally, we shall illustrate the prevalence of the logic of contraries in the 17th century (2.3.4), before we present Descartes' use of this logic to derive the basic categories of his physical system (2.3.5).

2.3.1   Contrariety and Contradiction

The law of noncontradiction is the fundamental principle of logical inference, although it itself, as we shall argue, is based on a more fundamental semantic principle of the exclusion of inconsistency. We shall claim (i) that in Aristotle the principle of noncontradiction is based on the principle of inconsistency of contrary predicates and (ii) that this contrariety is in turn based on an ontology of real opposition or real incompatibility of properties of things.

The classical formulation of the law of noncontradiction is to be found in Aristotle's *Metaphysics*, where three versions of the principle are presented, which later philosophers have classified as "logical," "ontological," and "psychological," although it seems that Aristotle did not distinguish rigorously among them.[19]

---

[18] See Aristotles *Physics* V, 1, 225a34ff and the commentaries of Aquinas (1963, §670) and Scotus (*Opera*, 1891, vol. 2, p. 326).

[19] Our discussion of Aristotle relies heavily on H. Maier, *Die Syllogistik des Aristoteles* and J. Lukasiewicz, "Aristotle on the Law of Contradiction." For a contemporary grounding of the principle of noncontradiction on the more general principle of inconsistency, see Strawson 1952.

[The *logical* formulation:] The most certain of all [principles] is that contradictory statements are not true at the same time (1011b13f).

[The *ontological* formulation:] It is impossible that the same thing should both belong and not belong to the same thing at the same time and in the same respect (1005b19f).

[The *psychological* formulation:] No one can believe that the same thing can [at the same time] be and not be (1005b23f).

Aristotle believed that the logical principle could be shown to be valid because it expresses the ontological: "To say of what is that it is and of what is not that it is not, is true" (1011b26f).[20] The psychological principle he reduced immediately to the ontological: Since belief in something is a property of the believer, if someone were to believe two inconsistent statements, he would have two contrary properties at the same time, which is (ontologically) impossible (1005b16-32).

While the psychological version bases the relation of contradiction directly on that of contrariety and thus on real incompatibility, the logical version does so only *indirectly*. Aristotle presents what can best be called a *semantic* argument for the validity of the law of noncontradiction: it is argued that the law is the prerequisite for meaningful discourse. A predicate that signifies something determinate or conveys information not only includes some things it also excludes some things. If I maintain that "X is a man" is true, then I say at the same time at least that he is not, for instance, a trireme and that statements like "X is not a man" or "X is a non-man" are false.

However, a negative statement cannot be warranted on the basis of direct experience alone but also depends on an inference. We cannot experience what is not the case, but only what is the case. From the fact that a subject possesses a property incompatible with the one we are interested in, we infer that it "lacks" the latter; and since "privation is the denial of a predicate to a determinate genus" (1011b19), we infer from the truth of the ascription of the privation the falsehood of the ascription of the positive property.[21] Thus a contradiction between

---

20 See. also *Categories*, Chap.10, 12b16f: "Nor is what underlies affirmation or negation itself an affirmation or negation. For an affirmation is an affirmative statement and a negation a negative statement whereas none of the things underlying an affirmation or negation is a statement ... For in the way an affirmation is opposed to a negation, for example, 'he is sitting' – 'he is not sitting', so are opposed the actual things underlying each, his sitting – his not sitting."

21 We believe that Lukasiewicz has the same argument in mind when he writes that "it is in general impossible to suppose that we might meet with a contradiction in perception; for negation, which is part of any contradiction, is not perceptible. Really existing contradictions could only be inferred" (Lukasiewicz 1979, §19b). The most prominent opponent of this point of view is Geach (1972, p. 79), who remarks: "What positive predication, we might well ask, justifies us in saying that pure water has no taste? Again, when I say there is no beer in an empty bottle, this is not because I know that the bottle is full of air, which is incompatible with its containing beer." While it is clear that one need not know what else is in the bottle in

two statements about empirical matters of fact presupposes that we have two positive statements ascribing incompatible predicates to the same subject, and that we then transform one of them into the formal negation of the other.[22]

Although knowledge about the incompatibility of properties is empirical, this does not mean that this knowledge depends each time on actual empirical experience. Aristotelian logic is a logic of natural language, and natural language embodies at any given time a great deal of knowledge about the incompatible properties of things; in fact, the meaning of many words is learned just by identifying them as the contraries of others. Their meanings thus depend on the meaning of the name of the genus, the meaning of the term for the property and the meaning of the relation of contrariety. Hence the application of the principle of the inconsistency of contrary predications depends on the knowledge conveyed by language of what properties are incompatible. This is the basis of Aristotle's opinion that to know a predicate is to know its genus and its contrary (see 1004a9; 105b33; 109b17; 155b30; 163a2).

To sum up: at any given time the concepts in a language incorporate a great deal of empirical knowledge about those properties of objects that are incompatible with each other.[23] Any application of the principle of noncontradiction to empirical statements depends on this knowledge about contraries.[24]

---

order to state that there is no beer in it, positive knowledge is nonetheless necessary to justify this statement. One cannot simply see or feel that there is or isn't beer in the bottle. *Mistaken* judgments which involve no optical illusions illustrate this clearly; and many of the standard tricks of magicians depend precisely on such mistaken inferences (e.g., that the hat is empty or that there is nothing up his sleeve). Although we may readily admit that a gestalt can be built up over time so that we begin to *perceive* immediately what we originally had to infer (e.g., the absence of something), this psychological process does not effect the inferential justification of the statement.

[22] This enables Aristotle then to turn around and draw inferences about the state of the world from the consistency of language: "If then it is impossible to affirm and deny truly at the same time, it is also impossible that contraries should belong to a subject at the same time ..." (1011b13f).

[23] See the following observations of Sigwart: "Which representations are incompatible cannot be derived from a general rule, but rather is given with the factual nature of the content of the representations and their relations to one another. We can conceive our sense of sight as so constituted that we could see the same surface shine in different colors, as it in fact emits light with different refractibility, just as we hear different overtones in a single tone and distinguish different tones in a chord; it is purely factual that the colors are incompatible as predicates of the same visible surface, but different tones as predicates of the same source of sound are not, no more than sensations of pressure and temperature, which can be ascribed to the same subject in very different combinations (cold and hard, cold and soft, etc.)" (Sigwart 1904, vol. I, p. 179).

[24] The relation between the real incompatibility of properties and the logical inconsistency of propositions attributing them to the same subject caused considerable difficulties in the early development of Logical Empiricism. The problem arose, why not all "elementary" or "atomic" propositions are compatible with one another. For instance, Ludwig Wittgenstein in his *Tractatus logico-philosophicus* (6.3751) wrote:

2.3.2 The Logic of Contraries

There are two aspects of Descartes' dissatisfaction with the use of traditional logic in science which we can distinguish analytically without maintaining that Descartes himself clearly distinguished between them. The first aspect is the inadequacy of traditional syllogistic logic to provide the *formal* techniques of inference necessary for modern science. This is connected to the introduction into science of more powerful mathematical techniques of inference which were able to expand the scope of scientific knowledge. The second aspect is the search for a logical calculus of *material* propositions, for a source of logically necessary statements which nonetheless have empirical content. This is a basic characteristic of the rationalist program from Descartes to Kant.

Descartes' alternative to syllogistic inferences and immediate inference based on the law of noncontradiction can be seen as the attempt to solve both the above mentioned problems through a logical calculus of exhaustively contrary predicates and through a calculus of contraries with quantifiable intermediates.

Logically *contrary* predicates are used to designate mutually incompatible physical or moral properties, i.e., properties that cannot be possessed and thus not be ascribed to the same subject in the same regard at the same time, such as hot and cold, just and unjust, red and green. In order to be logically incompatible such predicates must however share certain common features, they must be mutually exclusive species of a common genus, or as contemporary logicians put it, they must belong to the same "range of incompatibility": only colored things, such as apples but not symphonies, can be red or green; gasses but not molecules can be hot or cold, competent adults but not infants can be just or unjust.

"For example, the simultaneous presence of two colours at the same place in the visual field is impossible, in fact logically impossible, since it is ruled out by the logical structure of colour." In his "Some Remarks on Logical Form" (1929), Wittgenstein then introduced a distinction between contrariety and contradiction in his own terminology: "I here deliberately say 'exclude' and not 'contradict,' for there is a difference between these two notions, and atomic propositions, although they cannot contradict, may exclude each other" (p. 35). The process of exclusion is then described: "The propositions 'Brown now sits in this chair' and 'Jones now sits in this chair' each, in a sense, try to set their subject term on the chair. But the logical product of these propositions will put them both there at once, and this leads to a *collision*, a *mutual exclusion* of these terms" (p. 36; emphasis added). Wittgenstein's "collision" and "exclusion" correspond to the "struggle" and "expulsion" adduced by most scholastic logicians up to the 17th century to explain the incompatibility of contraries (see the quotation from Toletus in section 2.3.4 below). The difference between Wittgenstein's position and that of a scholastic logician like Toletus lies of course in Toletus' insight that the relevant struggle takes place between physical entities, not between propositions, and that their incompatibility is an empirical fact and therefore stated on the basis of empirical knowledge. For an analysis of the role of this problem in Wittgenstein's eventual rejection of the *Tractatus,* see Allaire 1966.

By *exhaustively* contrary predicates we mean two predicates which are incompatible and which together exhaust the possibilities of a given range of incompatibility, such as odd and even.[25] These can be distinguished from another kind of contrary predicates: those predicates which have *intermediates*, i.e., which constitute the extremes of a continuous scale, such as the temperatures between hot and cold.

However, any range of incompatibility can be divided into two areas, any genus divided into two species by a simple and purely formal process of negation. Red and nonred exhaust the colors, Europeans and non-Europeans exhaust *homo sapiens*, etc. Such "non"-predicates formed by negating a term were called infinite or indefinite terms in traditional logic. What Descartes is looking for, however, is not a definite and an indefinite term which exhaust a range of incompatibility, but for two definite predicates which either exhaust the possibilities or define the extremes of quantifiable intermediates. These two predicates should be independently determined, i.e., neither should be generated solely by the negation of the other species within the genus or defined primarily by this relation.

In contrast to contrariety, a contradiction is a special kind of logical incompatibility characterized since Aristotle by the opposition of affirmation and negation. Its importance lies in the automatic applicability of the so-called law of the excluded middle or law of the excluded third and hence in the possibility of obtaining a true statement by negating a false one. This also holds for the negation of one of two exhaustively contrary predicates, if their common genus is necessarily predicated of the subject. If two statements contradict each other in the strict sense, then both of them cannot be true at the same time, but since both of them also cannot be false, one of them *must* be true. From the falsity of the one statement one can infer the truth of the other and vice versa, thus making a two-valued calculus of propositions possible. However, the law of noncontradiction cannot guarantee that *both* statements have determinate meanings. This is because the negation involved in contradiction is indeterminate; we learn nothing in particular about anything by this kind of negation, which tells us only that a given predicate, whose meaning we understand, cannot be ascribed to a particular subject (or that an entity's membership in one particular class does *not* imply its membership in another particular class) but says nothing else. For instance, "X is not red" tells us neither what if any color X has, nor even whether X is visible; the sentence is true of salt, prime numbers, and moral virtues among other things, and in almost every imaginable context says nothing determinate about them.

In contrast, a calculus of contrary predicates would, if successful, allow the inference of statements with some determinate meaning from the negation of

---

[25] Such predicates have often been called *contradictory* predicates. This can lead to confusion since, strictly speaking, terms cannot be contradictories; only statements or propositions can be contradictories. Sigwart (1904, pp. 23-25) deals with some of the problems associated with the use of such predicates; and Wundt (1906, vol. II, pp. 62f. and 80f.) gives examples of their use in empirical sciences.

statements with determinate meanings. Even as a purely formal calculus, the contrary of a predicate is at least as determinate as the genus or range of incompatibility common to them both. Furthermore, whether two exhaustively contrary predicates both have meanings more determinate than that of the common genus depends on the semantics of the conceptual system to which they belong. To take a simple example: in an arctic environment being white means to blend into the background of snow and ice; being nonwhite, i.e., colored, means to contrast with the background. Thus, whatever the *linguistic form* of the two predicates white and nonwhite (colored), they would both have a quite determinate meaning, we suppose, in the conceptual system of arctic inhabitants. Consequently, knowing it to be false that a thing of a particular kind is white tells us something quite determinate about it, namely, that it contrasts with the background.[26] Thus, the calculus of contraries provides a formal technique for drawing inferences which always have as much material content as the common genus of the predicates and can under certain circumstances have as much content as the negated species: natural numbers are even or odd, motion in a vertical line is upwards or downward, etc.

It is evident that the ideal case combining both certitude and informativity would be possible if a genus comprising two and only two positively determined species were necessarily predicated of an object, such that the negation of the statement attributing one of them to an object yields the affirmation of the statement attributing its contrary to the object.[27] And it is this logic that Descartes wants to apply. Examples of such pairs of contraries actually used by Descartes are: *divine* and *created* (substance), *thinking* (active) and *extended* (passive) substance, *motion* and *rest*, and *determination to the right* and *determination to the left*.

Whereas in Aristotle's time a calculus was available in logic only for exhaustively contrary predicates, where the principle of *tertium non datur* applied, by the time of Descartes a calculus had been developed that was also capable under certain circumstances of yielding determinate results when dealing with contraries that admitted of *intermediates*. This new method of dealing with intermediates, introduced by medieval logicians in Oxford and Paris, was the calculus of "latitudo formarum" (discussed in section 1.2.3), which also allowed for the simultaneous representation of contrary predicates. If, say, the heat in a subject is uniformly difform terminated at no degree and is represented by a right triangle, then the quantity of the contrary quality, cold, could be represented by a complementary right triangle. Thus the heat and cold in a subject could be represented by a rectangle.[28]

---

26 See also Strawson 1952, p. 8.

27 The informativity of this logic is of course not confined to cases where *two* species exhaust the genus. If there are a limited number of determinate species, a limited number of negations are necessary to establish the sole remaining species.

28 Contrary qualities were conceived in general as incompatible. It seems that Buridan was the first to allow that some contrary qualities, for instance, hotness and

2.3.3   Aristotle's Physics of Contraries

We have seen in the preceding sections that the law of non-contradiction in Aristotle was based on the ontological principle that contraries cannot coexist in the same substance and that a logic of contrary predicates (both exhaustively contrary predicates and contraries with quantifiable intermediates) could in principle be productive under certain circumstances. We will now show that both these variants of the logic of contraries were applied by Aristotle to generate the basic doctrines of his physics.

A considerable part of Aristotle's scientific knowledge is derived from and justified by the logic of contraries sketched above. One central element of Aristotle's physics is the application of the logic of contraries to the explanation of motion, i.e., qualitative change, quantitative change, and locomotion, each of which involves a transition from one state to its contrary (*Physics* I, 6-10, 189a10ff, *Metaphysics* XII, 2; 1069b3ff). The elements, too, are defined by contrary predicates and each can be transformed into another, if one of its properties is changed into its contrary (cf. *De gen. et corr.* II, 2; 329b18ff). Natural motion along a radius of the world sphere allows for two contrary directions: *up* and *down*, which allow no intermediates but which correspond to the two opposite ends of the radius representing the qualities *light* and *heavy*. Four of Aristotle's five elements are based on contraries: Fire and Earth move up and down and being absolutely light or heavy respectively therefore occupy the periphery and the center. In between are two elements which are relatively light

---

coldness, may coexist in the same substance and that their sum is constant. See Maier 1952, pp. 304f. Oresme first represented this thesis by means of the method of configuration of qualities and motions:

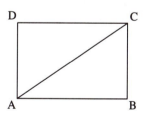

Fig. 2.1 (Oresme 1968, p. 214)

"And so let the hotness of subject AB be uniformly difform terminated at no degree in point A and at the highest degree in point B, or at the least degree with which the coldness cannot stand. Therefore the hotness will be imagined by Δ ABC, and so the coldness is to be imagined by Δ DCA. Then let the figure be inverted and let A be put in place of D and B in place of C. And then it is evident that the coldness is uniformly difform terminated at no degree in point B. And it is the same for all the contraries which exist together, so that however the figuration of the one is changed, so the figuration of the other will be equivalently changed in a contrary fashion enough to make up the uniformity of the whole aggregate. Whence it is evident that if one of the contraries is imagined by a convex figure, the other existing [with it] at the same time is to be imagined by a concave figure, and vice versa ..." (Oresme 1968, pp. 212-14). Similar figures applied to contraries can also be found in later treatises. See Clagett's Introduction to Oresme 1968, pp. 75 and 81ff. See, e.g., Paul of Venice 1503 p. 16v.

and heavy: Air and Water. Hence, Air moves up until it reaches Fire, and Water moves down until it reaches Earth.

Another way to define elements by contraries is derived from basic tactile qualities: heat, cold, dryness and wetness. These four qualities can be combined in four possible ways to define the elements: hot and dry (Fire); hot and wet (Air); cold and wet (water); cold and dry (Earth) (cf. *De gen. et corr.* II,2, 329b18ff; *Meteor.* IV,1, 378b10f). (The combinations hot/cold and wet/dry are excluded since none of the elements themselves may contain incompatible properties.)

Now, if all material bodies or substances contain all the elements (see *De gen. et corr.* II,8, 334b31), then a logical and physical problem arises: Each of the four adjacent pairs of the elements (Air and Fire, Fire and Earth, Earth and Water, Water and Air) contains a pair of shared properties and a pair of contraries. Each of the opposite pairs Air and Earth, Fire and Water has two pairs of contrary properties. Any time we are constrained to conceive of elements (or bodies) which have incompatible properties as constituting one body, i.e., any time we have a "true mixture"[29] not a mere aggregate, we will obtain a subject with inconsistent predicates: a logical contradiction. According to Aristotle, however, many combinations of elements are true mixtures. The solution to the problem lies in the fact that these fundamental contraries admit of intermediates, so that the new mixed substance can be conceived to have intermediate qualities, e.g., between wet and dry.

The logic of contraries thus allows the conclusion that the new substance will either have an *intermediate* quality (if the contraries are conceived as extremes of a continuous scale) or that one of the original qualities will *prevail* (if the contraries are conceived as exhausting the range of possibilities).

It might seem that all knowledge embedded in natural language is simply empirical. However, even though the knowledge that two given qualities are contraries with intermediates is acquired empirically, the knowledge that the mixture of two substances bearing the qualities will generate a substance with an intermediate property need not be empirical. Rather it can be inferred according to the deductive rules of the conceptual system from the definition of intermediates as composed of the extremes and from the definitions of continuous magnitudes. Hence, even if the concepts of the new qualities generated out of contrary ones arise as generalizations of empirical knowledge, nonetheless, once a conceptual system using and defining the contraries exists, the logic and semantic relations of the system can determine assertions for which no adequate empirical basis is available. This is why Aristotle and his successors could and did derive new information about the world from language.

It is a *logical* problem that an inconsistency arises in a conceptual system which characterizes the mixture of two materials by the attribution of two contrary predicates to the same subject; and it is similarly a *logical* consideration

---

[29] In a true mixture two or more materials generate a new one with determinate qualities out of the contrary qualities of the components (327b10-238b25). For a discussions of these topics in scholastic natural philosophy see Maier 1952, pp. 1-140.

that in order to avoid inconsistency, one has to demand a change in one or more of the contrary qualities so as to render them compatible and the description of their mixture consistent.

2.3.4   Contraries in 17th Century Logic and Physics

That the contrary predicates in language provide the basic principles of explanation in physics was part of the standard knowledge in science up to the time of Descartes. The four elements and the contrary qualities which allow one to be transformed into the other were basic to physics as codified in textbooks in the early 17th century. In many works[30] the elements and basic qualities were presented in "Pythagorean tetrads" or squares of oppositions analogous to the standard presentation in logic of the square of oppositions of judgments. Scholastic logic texts and commentaries on Aristotle's *Physics*, including those probably used by Descartes in his studies at La Flêche,[31] carried on the tradition of deriving physical knowledge from the logic and semantics of language. One of

---

[30]   See Heninger 1977, esp. pp. 103-108 for a number of such tetrads. A tetrad almost identical to the one given here from the *Cosmographia* of Oronce Fine can be found in Leibniz, *Dissertatio de arte combinatoria*, (1666) GP IV, 34. The Square of Opposition is taken from De Soto *Summulae* (1554), p. 52.

Oronce Fine, *Cosmographia*   De Soto, *Summulae*

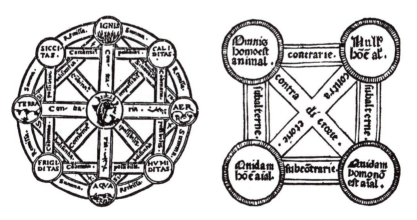

Fig. 2.2 (Heninger 1977, p. 106)   Fig. 2.3 (Soto 1554, p. 52)

[31]   See Cronin (1966, pp. 32-3) for a discussion of the texts used at the Jesuit school La Flêche.

the most prominent of these authors, Franciscus Toletus, analyzed contrariety as follows[32]:

> Three [conditions] are necessary for those which are called contraries. *First, that they belong to the same genus*: because those which belong to diverse genera or predicaments are not called contraries but diverse. *Second, that they be positive forms*. Insofar as one is a privation of something that is, they are not called contraries because blindness and vision are not contraries: these two [conditions] do not suffice since man and horse fulfill these two but are not contraries. A *third* [condition] is necessary, namely, *that they struggle with one another and act in the same subject*, from which by their mutual interaction one expels the other, as heat expels cold from the hand. This action takes place according to the forms themselves or according to those of which they consist, as heat struggles against cold according to itself, while white and black act by qualities of which they intrinsically consist ... *Let this suffice, the rest is Physics and Metaphysics.*

In the same text Toletus distinguishes between contraries that have intermediates (e.g., black and white) and those that do not (even and odd). Furthermore, by pointing out that authorities sometimes disagreed as to whether a particular pair of contraries (e.g., sickness and health) had intermediates, he makes it clear that empirical knowledge is involved in determining which is which. He also stresses that the terms must be predicated of a "proper subject," i.e., a subject that may receive one or the other predicate. For instance, a "mixed body" may be black or white; but an angel is neither sick nor healthy, nor is it even or odd.[33] The logic of contraries as a tool for acquiring knowledge about the material world and for constructing theories about it was not only an integral part of the philosophical tradition, it was also part of standard text book knowledge used in scholastic education in the 17th century and was thus immediately available to anyone with a certain amount of academic training.

The use of contraries as basic principles of physics in Descartes' time was not confined to the scholastics. Mechanistic science distinguished itself from scholasticism not by abandoning the logic of contraries but rather by determining different concepts as contraries, for instance by including circular motions (of the celestial bodies) which according to Aristotle had no contraries. Moreover, since dynamics was developed by means of concepts and techniques adapted from statics, the logic of contraries was also applied to these concepts. Such contemporary "mechanical philosophers" as Galileo, Hobbes, and Marci relied heavily on the use of the contrary predicates in language to enable them to represent causal necessity in the world by logical necessity in their systems.

---

[32] Toletus, *Logicam* (1589), p. 105*r* (emphasis added); almost verbatim the same discussion can also be found in the *Summa* of Eustace of St. Paul (Part I, pp. 38-39), which Descartes once recommended as a competent scholastic presentation (AT III, 251).
[33] Toletus 1589, p. 176.

The representation of physical oppositions by contrary concepts combined with an extension of reasoning from statics typical of the mechanists of the 17th century is well illustrated in the following passage from Hobbes[34]:

> Nothing can hinder motion but contrary motion: that the motion of the water when a stone falls into it, is point blanke contrary to the motion of the stone. For the stone by descent causeth so much water to ascend as the bignesse of the stone comes to ... and this rising upwards is contrary to the descent, and is no other operation than that we see in scales, when of two equal bullets in magnitude that which is of heavier metal maketh the other rise.

In his treatise, *De Proportione motus*, Marcus Marci not only uses the logic of contraries in argument, he bases his system explicitly on this logic. The first of the definitions with which his work begins is a definition of contrariety[35]:

> [Def.] 1, Contraries are said to be those which remove or impede their contraries.

He then distinguishes degrees of contrariety from absolute contrariety and applies this distinction to motion:

> [Def.] 4, Motions are absolutely contrary, which conduct the same mobile from the same point towards the opposite direction in the same right line.

> [Def.] 5, Motions are contrary *secundum quid*, which from that same point or origin of motion make an angle greater or smaller than a right angle but less than two right angles.

Marci's seventh definition points out that, if the two motions make an angle neither greater nor smaller than a right angle (i.e., exactly 90°), they are not opposed at all but constitute a "perfectly mixed" motion. In the second postulate, which follows the definitions, Marci then establishes the connection between the logic of contraries as applied to motions and the analysis of forces in statics.[36]

Although these three mechanist contemporaries of Descartes all apply a logic of contraries to physics, they all also differ from Descartes in taking motion as essentially directed. Thus the contrary of motion is not rest, but rather the contrary of motion (in one direction) is *motion* (in the opposite direction).

2.3.5   Descartes' Physics of Contraries

Descartes' theory of matter is constructed according to the logic of contraries discussed above. His first step is to divide substance as the bearer of properties

---

[34] Hobbes, letter to Cavendish, Feb. 8, 1641 (*Works* 7, pp. 458-459). For Galileo's use of contraries, see Chapter 3 below, sections 3.2.1, 3.2.3, 3.3.1, and 3.7.3.

[35] Marcus Marci, *De proportione Motus* (1639), no pagination, sigs. A3v, A4v, B1v-B2r.; see also Gabbey 1980, p. 245.

[36] Marci 1639, B1v-B2r.

and subject of predicates into two kinds: *contingent* (created, finite) and *necessary* (divine, infinite) substance. Created substance is then divided into *thinking* (active) and *extended* (passive) substance. Physics is the study of created, extended substance. On this basis Descartes then introduces the three pairs of contrary predicates from which he constructs his physics. These are: (1) *matter* and *vacuum* (something and nothing), (2) *motion* and *rest*, and (3) determination *towards one side* and determination *toward the opposite side*.

These pairs of predicates are introduced in the second book of the *Principia*. Using the first pair, matter and vacuum, Descartes argues that "space" and "material substance" refer to the same entity, for there cannot be empty space or a vacuum, "that is, a space in which there is absolutely no substance ... because it is entirely contradictory for that which is nothing to possess extension (*quia omnino repugnat ut nihili sit aliqua extensio*)" (II, §16). Since nothingness can have no predicates, what has extension cannot be a vacuum. Thus the subject of extension is substance and extended substance is matter. This pair of contraries (matter and vacuum) is used to analyze the concept of extended substance and to show that it cannot be further specified.

After introducing the conservation laws of the system of matter, Descartes presents three laws of nature. The first two laws define invariants in the behavior of single bodies, the third governs two-body interactions. Each of the first two laws of nature is based on one of the pairs of contrary predicates introduced above.

1. From the contrariety of *motion* and *rest* Descartes derives his *First Law of Nature*, the law of inertia. Rest and motion are both positive and contrary predicates,[37] and "nothing can move by its own nature towards its contrary or its own destruction." Thus, in general, if a substance is to lose a property and take on a contrary property, then there must be an *external* cause.

> The first of these [laws] is that each thing, insofar as it is simple and undivided, always remains in the same state as far as it can and never changes except as the result of external causes. Thus, if a particular part of matter is square, we can be sure without more ado that it will remain square forever, unless something coming from outside changes its shape. If it is at rest, we hold that it will never begin to move unless it is pushed into motion [*ad id*]by some cause. And if it moves, there is equally no reason for thinking that it will ever cease this motion of its own accord and without being checked by something else. Hence we must conclude that whatever moves, so far as it can, always moves. (II, §37; AT VIII, 62)

The physical striving of a body to remain in its state of motion is derived from a logical principle: the logical opposition of the predicates motion and rest. We can see that the concept state of motion (encompassing motion and rest) which

---

[37] Descartes criticized scholastic philosophers who "attribute to the least of these motions a being much more solid and real than they do to rest, which they say is nothing but the privation of motion. For my part I conceive that rest is just as much a quality, which must be attributed to matter while it remains on one place, as is motion, which is attributed to it while it is changing place" (*Le Monde*, AT XI, 40).

Descartes bequeathed to classical physics is developed to name the genus or range of incompatibility of the two contrary predicates, rest and motion.

2. Since motion as the opposite of rest is defined as a scalar magnitude without reference to direction, Descartes needs to introduce a *Second Law of Nature* to guarantee that the conserved motion keeps the same direction. Here he introduces for the first time (in this systematic exposition) the concept of determination. Determination is a *mode* of motion, which in turn is a mode of bodies.[38] As a specification of the concept of motion, which has a quantity, determination is supposed to have a magnitude as well as a direction. Its magnitude is entirely dependent on that of the motion it '"determines"; its direction is independent of the quantity of the motion. A thorough analysis of the concept of determination must be put off until section 2.5.

The *Third Law of Nature* containing the rules of impact will be analyzed in depth in the next section (2.4). To sum up, we have seen thus far that the basic concepts of Cartesian physics are derived by means of a logic of contrary predicates adopted from traditional logical and physical thought.

2.4 Descartes' Derivation of the Rules of Impact

2.4.1 The Means of Representation

We have argued that Descartes formulated conservation laws functionally similar to those laws used in classical physics for the deterministic description of a mechanical system, and we have already indicated that his concepts are very different from those of classical physics. This will become clear in the following discussion of his rules of impact which describe the basic interactions of bodies.

In classical mechanics the result of a central collision of two perfectly elastic bodies under extremely idealized conditions is calculated by simultaneously solving two equations based, respectively, on the conservation laws for (1) momentum and (2) kinetic energy. The two equations derived, (3) and (4), determine the velocities of each body after the collision.

---

[38] Although determination is defined as a mode of motion, it is also sometimes *used* as if it were a mode of the bodies themselves. For instance, in the section of the *Principia* in which Descartes first introduces the concept it seems that a body can have a determination in an instant ("in that instant at which it is at point A"), although it can only have motion during some finite length of time: "It is true that no motion takes place in an instant; but it is manifest that everything that moves is *determined* [*determinatus esse*] in the individual instants which can be specified as it moves, to continue its motion in a given direction along a straight line, and never along a curved line" (II, §39; AT VIII, 64; emphasis added).

$$m_1v_1 + m_2v_2 = m_1u_1 + m_2u_2 , \tag{1}$$

$$1/2 \, m_1v_1^2 + 1/2 \, m_2 \, v_2^2 = 1/2 \, m_1u_1^2 + 1/2 \, m_2u_2^2 , \tag{2}$$

$$u_1 = \frac{(m_1 - m_2)v_1 + 2 \, m_2v_2}{m_1 + m_2} , \tag{3}$$

$$u_2 = \frac{(m_2 - m_1) \, v_2 + 2 \, m_1v_1}{m_1 + m_2} , \tag{4}$$

Note that here the conservation of mass in the system need not be stated explicitly because the much stronger statement, that the mass of each body is conserved, is incorporated into the mathematical representation by using the same variable for the mass of each body before and after collision.

In early classical mechanics, although both quantity of motion (momentum) and *vis viva* (kinetic energy) were defined by the same basic magnitudes (mass and speed), they were nonetheless dealt with according to different algorithms (parallelogram rule for addition of *mv* and arithmetical addition of mv^2), and this difference required justification and led to controversies. In later (19th century) classical mechanics, on the other hand, no justification was needed and no controversies arose, since momentum and kinetic energy could already be defined on the basis of a presupposed mathematical theory as vector and scalar magnitudes, respectively, and the two different kinds of magnitudes could be represented by different kinds of symbols (mv^2 and *mv*). The equations give us both speed and direction (i.e., velocity) of the bodies after collision based on a knowledge of speed and direction before collision.

For Descartes the situation is quite different. His two fundamental magnitudes, quantity of motion and determination, have the same absolute value |*mv*|. The concept of vector had not yet been invented, and he compounded and resolved determinations by means of the geometrical parallelogram rule. In the geometrical representation (Fig. 2.4) the quantity of motion or (abstracting from the size of the body) the speed was represented by the radius of a circle, i.e., by the undirected distance between center (B) and circumference (AFD). The *particular* radius as actually drawn (e.g., AB) connects the positions of the body at the beginning and end of the time interval considered and represents, therefore, *both* the quantity of motion and the determination. But the distinction between quantity of motion and determination was of paramount importance since the determinations can be resolved and then recompounded according to the parallelogram rule, whereas the motions are added and subtracted arithmetically. For instance, if we resolve the determination according to the parallelogram rule,

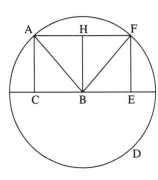

Fig. 2.4

the sides of the parallelogram (AC and AH) represent components of the determination (AB) but *do not represent motions or components of motions.* Consequently, the *sides* of the parallelogram may not be considered to represent actual motions, although the *diagonal* represents not only determination but also motion. Thus, it was in practice not always possible to distinguish between motions and the determinations of motions on the basis of the geometrical representation alone; and furthermore, it had constantly to be borne in mind that the radius actually drawn represented both. In order to refer in language to what the radius represents, the linguistic representation, e.g., of the quantity of motion (*motus*) had to be evoked and further qualified "in a particular direction" (e.g., *determinatus versus certam partem*). Thus one must keep in mind the *content of the physical argument,* part of which is represented only in language and its logical structure, in order to understand the meaning of the elements of the geometrical representation and to apply the appropriate calculus (arithmetical addition or vector addition). (This will be illustrated in detail in section 2.5, in which Descartes' derivation of the inverse sine law of refraction is analyzed.)

Furthermore, in the study of collision, which involves two bodies moving in a straight line in opposite directions, the motions (|mv|) of the two bodies are added to find the total quantity of motion conserved in the system, but since the *determinations* of the bodies are not conserved in collision, their behavior cannot be predicted by any purely computational algorithm. And, as we shall see, the impact rules are presented verbally with *pictures* of the bodies but without geometrical representation. Thus, the mathematical representation is not sufficient for deriving the outcome of physical interactions.

2.4.2   Contrary Predicates and Opposing Modes

Since the mathematical techniques for dealing with motions and determinations are insufficient to determine unequivocally the outcome of collisions, further qualifications of a non-mathematical nature must be introduced explicitly, and their introduction must also be argued for. The argument must appeal to some form of accepted knowledge in order to have any foundation; this knowledge, however, cannot consist simply of empirical experience. Descartes does not claim that his impact rules apply to empirical bodies; i.e., although the motions say of billiard balls are *ultimately* governed by these rules, billiard balls rolling on a smooth surface in an air medium are not considered an adequate realization of the theoretical ideal to which the rules apply. He also states explicitly that the motions of the bodies in fluid media (like air) is different from (though of course ultimately governed by) the ideal case;[39] but he offers no way of reducing a complex empirical interaction to the two-body ideal case governed by the theory.

---

[39] *Principia*, II, §§53 and 56; AT VIII, 70 and 71.

Thus there is *factually* no means of testing the theory. Furthermore Descartes cannot appeal to an accepted system of *theoretical* or *scientific* knowledge about material reality itself, since he was consciously rejecting the accepted Aristotelian system of knowledge and developing an alternative to it.[40] But since the physical phenomena with which he was dealing were conceptualized in natural language, he could and did appeal to the necessary relations embedded in the language between the concepts used, i.e., broadly speaking, to *logic*. The logic applied by Descartes to derive his impact rules is the logic of contrary predicates discussed in section 2.3. The logical contrariety of concepts is taken to express the physical opposition of properties of bodies; the contrary predicates correspond to incompatible modes of bodies and the inconsistency of the predicates is resolved when one or both incompatible modes undergoes change. Thus the collision of *bodies* is represented by the contrariety of *predicates* .[41]

Not every concept that has a conceptual contrary denotes an entity which can be opposed to another entity. The central concept, quantity of matter, for instance, refers to material substance, and substance is the contrary of nothing (II, §16); but a substance cannot be physically opposed to another substance or even to nothing. It is only the properties or modes of a substance or a body that can be opposed to one another.[42] The relevant modes of bodies that can be incompatible are *motion* and *rest*. Since motion as such is without direction, the possible opposition between directions, e.g., left and right is expressed as an opposition between further specifications of the concept motion, or in Descartes' terminology: *modes of the mode of motion*; i.e., the determinations of two motions can be opposed.[43] The title of §44 states, "That motion is not contrary to motion, but to rest; and that determination in one direction is contrary to determination in the opposite direction."

---

[40] The one area of scientific knowledge, to which Descartes could and did appeal, namely statics, had to be interpreted and its adaptation itself had to be justified.

[41] Leibniz clearly saw that Descartes was deriving physical theorems by applying the logic of contraries and presented one of them (concerning firmness) in syllogisms; in these syllogisms the common relation of the predicates is "maxime adversatur." Leibniz's criticism of Descartes is not directed against this procedure in general (in fact Leibniz applies it too), but against two specific points. On the one hand, Leibniz doubts that the characterization of certain concepts as most opposed is correct (thus he remarks that contrary motion is more opposed to a specific motion than is rest) and on the other hand he doubts the validity of Descartes' implicit axiom, that the cause of what is most opposed to something is also most opposed to the same thing. (see. Leibniz, "Critical Thoughts on the General Part of the Principles of Descartes," on Articles 54, 55, GP IV, 385-388, PPL, 403-407.

[42] See Aristotle, 225b10-11 and 3b24-27.

[43] There have been numerous studies of Descartes' physics, which we shall not be able to deal with here. For a discussion of various previous interpretations of Descartes' physics consult Gabbey 1980, which has set qualitatively new standards of analysis for the study of Cartesian science. Gabbey's interpretation emphasizes especially the opposition of modes as a key to understanding Descartes' system. Our debt to his work will be evident both in this and the next section.

Both oppositions admit of degrees and can be quantified. From the opposition between motion and rest, an opposition between "rapidity of motion and slowness of motion" can be derived "to the extent that this slowness partakes of the nature of rest." Thus not only motion and rest can be opposed but also to a certain extent *slow* and *swift* motion. Determinations, too, can also be more or less opposed, i.e., the opposition can be direct or more or less oblique, but since Descartes' impact rules deal explicitly only with direct oppositions of determinations, i.e., collisions in one dimension, we shall deal with oblique collisions only in a later section (2.5.3).

Descartes' general rule governing the interactions of bodies, i.e., governing the resolution of the incompatibility of their modes is stated as the *Third Law of Nature* :

> When a moving body encounters another, if it has less force to continue in a straight line than the other has to resist it, it is deflected in another direction, and retaining its quantity of motion it loses [*amittit*] only the determination of the motion. If, however, it has more force, it moves the other body with it and loses [*perdit*] as much of its motion as it gives to the other. (II, §40; AT VIII, 65)

This law of nature is then further qualified by the explanation of the measure of force:

> And this force should be measured not only by the size of the body in which it is, and by the surface which separates this body from another, but also by the speed of its motion and by the nature and [degree of] contrariety of the mode in which different bodies encounter one another. (II, §43; AT VIII, 65)

Thus, when two bodies with incompatible modes encounter one another (collide), the modes of one or both bodies must change in such a way that the incompatibility is removed. The major determinant of the relation of the changes in each body's properties or modes is the relation of the sizes of their *forces*. But the incompatibility of opposed modes in collision can be resolved in different ways, and more than one opposition of modes can be involved: motion versus rest and determination ($\rightarrow$) versus ($\leftarrow$) determination. The unequivocal calculation of the outcome of a collision in more complicated cases demands a rule specifying the relations between changes in the pairs of opposed modes. Since every change in the speed of a body entails a change in (the amount of) determination while not every change in determination (e.g., in direction) entails a change in the scalar speed of the body, determination being only a mode of motion, the relation of changes in both these quantities has to be introduced explicitly. Both the concept of physical change as the solution to an incompatibility of modes and the specification of the ratio of change of quantity of motion and determination are formulated by Descartes as one principle determining all collisions. This principle is used implicitly in the *Principia*

Philosophiae to derive some of the impact rules, although Descartes actually stated it explicitly only a year later in a letter of explanation to Clerselier[44]:

> That when two bodies collide and have in them incompatible modes, unquestionably there must occur some change of these modes to make them compatible, but this change is always the least possible. That is, if they can become compatible through the change of a certain quantity of these modes, a greater quantity will not undergo change. And it must be noted that there are in motion [*movement*] two different modes: one is the motion [*motion*] alone, or the speed, and the other is the determination of this motion [*motion*] in a certain direction. Of these two modes, one changes with as much difficulty as the other.

This principle, which Gabbey has called the "Principle of the Least Modal Mutation,"[45] guarantees that the result of interactions will be unequivocal by singling out one of the possible results (the smallest amount of change) as the one that must actually occur.[46]

Thus, to derive the rules of impact, Descartes uses (alongside the unstated but assumed conservation of matter) *one* conservation law, the conservation of the quantity of motion, as well as an extremal principle (the principle of minimal modal change), and the qualification, that force is transferred only from a stronger to a weaker body.

2.4.3 The Rules of Impact

There are three relevant possibilities of incompatible modes involving two bodies in one dimension, and these possibilities determine the structure of Descartes' presentation of the rules of impact. There can be an opposition: (1) between the *modes* of bodies, motion and rest, (2) between the *determination* of the motions of bodies in opposite directions, and (3) between the *intermediaries*, slowness and rapidity of the motion of bodies.[47] The order of the impact rules

---

[44] Descartes, letter to Clerselier, Feb. 17, 1645; AT IV, 185-186 (transl. adapted from Gabbey 1980, p. 236). Spinoza argued that this principle directly follows from the first law of nature (inertia); see *Renati Des Cartes Principiorum Philosophiae Pars I & II*, bk. II, §25.

[45] Gabbey 1980, p. 263f.

[46] However, when calculating the amount of change in the various possible results, it should be remembered that the quantity of determination changes whenever the quantity of motion changes. Thus, although each mode changes with equal difficulty, a change in determination involving only change of direction is in fact easier than a change in motion which also necessitates a change in the quantity of determination and thus counts double (i.e., reversing four units of determination without changing motion is equal to transfering two units of speed because two units of determination are attached to them).

[47] In fact there are four possibilities: two *kinds* of opposition, each of which can be *absolute* or admit of *degrees*. As Descartes says: "but strictly speaking, only a two-

that Descartes presents is determined by the increasing complexity of opposition between incompatible modes.

2.4.3.1 The Opposition between Determinations

The rules begin with the simplest case of the interaction of bodies, in which only the determinations of motion are opposed: two *equal* bodies, both *in motion* with the same speed (if both are at rest, there is no interaction), collide because their determinations are directly opposed – and they lie in the same line on the appropriate side of one another (see Fig. 2.5). This case is governed by *Rule 1*, which states that the two bodies rebound with unchanged speeds and exchanged or reversed determinations, the quantity of motion in the two-body system being conserved. Rule 1 is the only one of Descartes' impact rules which can actually be confirmed on the billiard table.[48]

Rules 2 and 3 introduce no new *oppositions*, but each introduces an additional *difference*: either in the *size* of two bodies that are equal in speed (Rule 2) or in the *speeds* of two bodies that are equal in size (Rule 3). In each case one of the bodies is said to be stronger than the other, and the stronger body compels the weaker to change its direction so that both move in the stronger body's direction at the same speed, conserving the quantity of motion in the two-body system.

It should be pointed out that although the stronger body is in fact the one with the greater quantity of motion, Descartes does not use the quantity of motion to define strength, nor does he consider in the discussion of these two rules any cases where the two bodies differ in *both size and speed*. It is also noteworthy that Descartes states *two* rules for what should actually be only *one* case, namely where one moving body has a larger quantity of motion than the other. This peculiarity recurs in the presentation of other impact rules and will prove to be significant for the status of the concept of quantity of motion (see section 2.4.4 below).

---

fold contrariety is found here. One is between motion and rest, or even between swiftness of motion and slowness of motion (that is, to the extent that this slowness partakes of the nature of rest); the other is between the determination of a body to move in a given direction and the encounter in this direction with a body which is either at rest or moving in a different manner; and this contrariety is greater or smaller according to the direction in which the body which encounters the other is moving" (II, §44; AT VIII, 67). The fourth possibility – oblique oppositions – will be dealt with in the next section.

[48] Even this, the simplest case and the only "empirically correct" one, is not strictly implied by Descartes' premises. Since the two bodies are equal in force, this case does not really fall under the Third Law of Nature, which holds for the relations between stronger and weaker bodies. See also Spinoza (1925, vol. 1, pp.211-212) *Renati Des Cartes Principiorum Philosophiae*, bk. II, prop. 24, who appeals implicitly to symmetry considerations when the two bodies are equal.

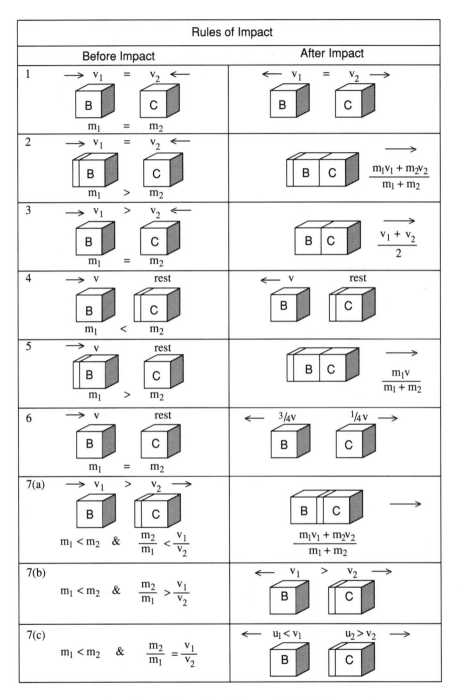

Fig. 2.5 (adapted from Aiton 1973, p. 36, with permission) In Descartes' own presentation the moving body B always comes from the right not from the left.

2.4.3.2 Double Opposition: Determination vs Determination and Motion vs Rest

Rules 4 to 6, which we shall consider in some detail below, introduce a *second opposition*, namely, the opposition between the rest of one body and the motion of the other. In these cases the bodies are doubly opposed, both in their modes of motion or rest and in the determinations (or modes) of these modes. Since a stationary body has no motion and therefore no determination, Descartes attributes to the resting body an ability to resist motion and determination in any direction. The second opposition is thus referred to as one between determination and the encounter with another body in its path (cf. II, §44). Descartes treats three possible variants: the body in motion is *smaller* than the body at rest (Rule 4); it is *larger* (Rule 5); they are *equal* (Rule 6).

2.4.3.3 Double Opposition: Determination vs Determination and Swiftness vs Slowness

The three-part *Rule 7* deals with the collisions of bodies moving with different speeds in the *same* direction, in which one body overtakes the other. This is a somewhat more complicated double opposition, in which neither states nor determinations of motion are *directly* opposed but rather are quantitatively incompatible: the swiftness of one is opposed to the slowness of the other and the determination of the faster motion is opposed to the encounter in its path with the slower body. Descartes considers three cases in which a *smaller and faster* body overtakes and collides with a *larger but slower* body.

Here again three possible differences are taken into account: the quantity of motion of the smaller faster body is (a) *greater than*, (b) *smaller than*, or (c) *equal to* that of the larger slower body. If the overtaking body is stronger than the larger but slower body, it transfers enough motion to the other so that both move with the same speed and determination; if weaker, it rebounds with unchanged speed and reversed determination. And if they are equal in strength, the overtaking body transfers some motion and rebounds with decreased speed and with reversed and decreased determination.[49]

Once again it should be pointed out that Descartes himself does not formulate the rule in terms of quantity of motion but rather in terms of the quantitative relations between the *ratio* of the sizes and the *ratio* of the speeds: for instance, "if C were larger than B but the excess of speed in B were greater than the excess in magnitude of C ..."(II, §52). (See section 2.4.4 below.)

---

[49] Descartes deals with this last possibility only in the French edition, but it can be derived in analogy to Rule 6.

2.4.4   Analysis of Difficulties

The three rules (4-6) governing straightforward double oppositions are particularly revealing of the difficulties of Descartes' system:

Rule 4 states that if a smaller body in motion hits a *larger* body at rest it will rebound with its original speed in the opposite direction. For "a resting body resists a great speed more than a small speed and does this in proportion to the excess of one over the other" (II, §49). This implies that no matter how fast the smaller body may be, it will never be able to move a resting body which is larger than it, no matter how small the difference in size. This follows from the application of the principle of minimal change or "least modal mutation," as Gabbey has called it.[50] Descartes explicitly draws this seemingly absurd conclusion, because for the smaller body to push the larger ahead of it, it would have to transfer *more than half* its own motion (and thus more than half its determination) to the other body; this is a larger change of modes than the *complete* reversal of its determination. Descartes gives no explanation in the *Principia Philosophiae*, but the reasoning behind the position based on considerations from statics can be found in his letters. For instance, in an exchange with Hobbes, who maintained that nothing that cannot be moved to some extent by the smallest force can be moved by any force at all (*nulla vi amoveri, quod non cedit levissimae*), Descartes replied that a *balance* loaded with 100 pounds on one side cannot be moved at all by 1 pound placed on the other. Nonetheless, it can easily be moved by 200.[51]

Rule 5 states that a larger body colliding with a smaller body at rest will push it ahead of itself in such a way that both move with the same speed and the quantity of motion is conserved. Since this rule is the only one for which Descartes elsewhere calculates a case of *oblique* collision (to which we shall return in section 2.5.3), we shall here give the formula for calculating the speed of the bodies after direct collision: If B is the large body moving with velocity v and C the small resting one and if the quantity of motion (and in this case its direction as well) is conserved in collision, then:

$$Bv_1 = u_1(C + B) \quad \text{or} \quad \frac{v_1}{u_1} = \frac{C + B}{B} = \frac{C}{B} + 1 \ .$$

---

[50]   See Gabbey 1980, p. 269. This rule is further complicated by the fact that a body *at rest* with respect to its contiguous bodies must actually be considered to be *part of* them, since "our reason certainly cannot discover any bond which could join the particles of solid bodies more firmly together than does their own rest ... for no other mode can be more opposed to the movement which would separate these particles than is their own rest" (II, §55; AT VIII, 71). It is thus hard to see how we can even estimate the *size* of a resting body without contradicting ourselves.

[51]   Hobbes' objection is cited by Descartes in a letter to Mersenne of Jan. 21, 1641, in which he answers the objection; AT III, 289. See document 5.2.9.

Rule 6 deals with the case where the body in motion is neither larger nor smaller than the body at rest. Both bodies are equally large and thus equally strong, the resting body having as much force to resist as the moving body has to drive it forward. Rules 4 and 5 solved this same double contrariety in favor of the larger body, but in this case an indeterminacy arises because both bodies are the same size and thus have equal force. Since there seems to be no logically necessary solution to the problem, Descartes calculates a mean result, as if both Rules 4 and 5 applied simultaneously. Calculated according to Rule 4, the moving body would retain its speed (say 4 units) and reverse its determination (also 4 units). On the other hand, according to Rule 5, the moving body would communicate half its speed and determination (say 2 units of each) to the resting body, retaining half its speed and half its determination in the same direction. As a compromise Rule 6 asserts that the moving body will both reverse its direction and transfer some motion and determination to the resting body. The result will be the mean between rules 4 and 5 and the two bodies will split the difference between 2 units and 4 units of speed. The moving body B will rebound in the opposite direction with a speed of 3 units (instead of 4), thus losing 1 unit of speed (and determination) and reversing 3 units of determination. The resting body C will move in the direction of B's original motion with a speed of 1 unit.

This last example shows that Descartes algorithm, even with the Principle of Least Modal Mutation, is not unequivocal. the first two possible solutions to the problem involve the same amount of modal change: following Rule 4, four units of determination, and following Rule 5, two units of speed and 2 of determination. But according to Rule 6, five units are changed: 1 unit of speed and 1 unit of determination are transferred and 3 units of determination are reversed. The actual answer that Descartes provides is determined not by the Principle of Least Modal Mutation, which is violated, but rather by an arbitrary decision to take a mean transfer of *motion*.

These three examples of double contrariety show that Descartes algorithms are neither unambiguous nor do they really employ the concepts quantity of motion and determination in the same way as kinetic energy and momentum are employed in classical mechanics. Far more important than the differences in the measures used is the fact that the concepts involved do not yet form the basis of a closed deductive structure. On the one hand there are still indeterminacies (e.g., whenever the forces of bodies are equal) and possibly also inconsistencies between the inferences drawn within this conceptual system, so that ad hoc considerations such as in Rule 6 have to be introduced to assimilate even relatively simple cases into the system.

On the other hand, some of the concepts introduced still bear the stigma of derived concepts, which depend for their interpretation on their constituent concepts. Thus, Rules 2 and 3 are treated as two different cases since the differences in the quantity of motion of the two bodies is due in one case to a difference in size and in the other case to a difference in speed. Rule 7, although it implicitly considers three different relations of the quantity of motion of the two interacting bodies, deals explicitly with different *ratios* of sizes and speeds. Hence, while quantity of motion is, on the one hand, a concept referring to an explanatory physical entity (it determines the sum of speeds of bodies after impact), it does not determine the direction and speed of each single body (not even together with determination) after collision. For this purpose the constituent concepts quantity of matter and speed must be invoked, as well as other considerations such as taking the mean of two possible but incompatible solutions.

Before we conclude our discussion of Descartes' impact rules, let us take a brief look at his illustrations. Many of Descartes' successors (and most historians of science) consider the impact of homogeneous spherical bodies taking these to be adequate models of the collisions of elastic bodies or point masses. But Descartes, who identifies the quantity of matter as volume, cannot use point masses, *and he does not use spheres.* A law governing the fundamental inter-actions of matter (i.e., collisions) cannot presuppose unexplained forces of cohesion in spheres. The parts of a body are held together not by forces of cohesion but only by their relative rest with regard to one another (II, §55). Hence, if two spheres were to collide, the parts around the point of contact would not be affected at all but proceed in their motion (the body would break up or be deformed) until they came into contact with the corresponding parts of the other body moving in the opposite direction. The *point* of contact would flatten out into a *surface* of contact. Descartes' own illustrations (see Fig. 2.6) picture *cubes* or *parallelopipeds* with the same rectangular surfaces on the side of collision. Differences in the sizes of bodies are represented by differences in the *lengths* of bodies with the same surface of contact. Thus in collision the whole body acts as a unit.[52]

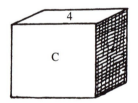
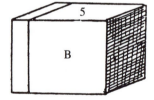

Fig. 2.6 (AT IV, 185)

---

[52] See especially the figures in the letter to Clerselier, Feb. 17, 1645, AT IV, 185; also *Princ. Phil.* II, §46, and AT III, 79; see documents 5.2.2, 5.2.3, and 5.1.12. William Neile (1637-1670), who like Descartes defined as one body all matter shar-ing the same motion, stated among the presuppositions of his discussion of impact that the colliding bodies are cubes and that "the whole square surface of the one meets in the same instant of time with the whole square surface of the other" (Neile, "Hypothesis of Motion," [May, 1669)], in: *The Correspondence of Henry Oldenburg*, vol. 5, pp. 519-524).

To sum up: A comparison between the concepts of classical mechanics and of Descartes shows two very different results. On the one hand, we see that Descartes introduced and used conservation laws as logical prerequisites for a deterministic causal theory of impact. On the other hand, the concepts introduced and the algorithms used are very different from those of classical physics and it turns out that even some of the simplest cases are underdetermined.

2.5 Determination

The concept of determination is one of the central concepts of Descartes' conceptual system, and it is especially in the use of this concept that inconsistencies and defects in the system come to the surface. Determination refers, on the one hand, to the supposedly *instantaneous* direction of a velocity conceived as *actual* space covered in a unit of time. It must furthermore guarantee the compatibility of the conservation of scalar motion with the application of the parallelogram rule for compounding motions. It is even called upon to explain the action of the mind on the body.[53] Most of Descartes' contemporaries had great difficulties in grasping what he meant by the term, and his disciples had no fewer problems than did his opponents.

The term determination itself is taken from logic, and it was used by Descartes in the sense in which it was used in logic. Determination refers to the further specification of a concept, for instance, the determination of a species within the genus. In particular, a determination is the modification of a predicate, the qualification of a quality.[54] It is so described in the major lexicons of philosophy of the 17th and 18th century; but more importantly, the term was simply *used* in logic with this sense, whether or not it was considered a *terminus technicus* that must be included in lexicon or index. Spinoza's famous "determinatio negatio est" exemplifies this use of the term, asserting that any specification of a concept excludes some other specifications.

Any change or motion was said in the Aristotelian tradition to be *determined* by its *terminus a quo* and its *terminus ad quem*. Descartes' innovation in the

---

[53] On instantaneous direction see the Second Law of Nature, *Principia*, II, §39; AT VIII, 63-64. The mind or will cannot increase or decrease the amount of motion in the world; it can only determine the motion of the body. Descartes uses the term "determine" on a number of occasions to describe the action of the mind on the body (see sepecially AT VII, 229 and XI, 225-226). This is an adaptation of a traditional way of speaking about freedom of the will: in decision the will determines *itself* to action. See Suarez, *Opera* vol. 10, pp. 459ff; Descartes (*Passions de l'âme*, §170; AT XI, 459) also uses the term in a similar sense.

[54] Gabbey (1980, p. 248) provides many examples of the logical use of "determination" and points out that it was derived from the Greek προσδιορισμός meaning division, distinction, definition.

application of the term to motions (for which he was criticized by Hobbes) consists in maintaining (i) that the determination is "*in*" the motion, i.e., it is a physical magnitude and not merely a specification of direction; (ii) it is a physical entity which has a quantity relevant to its causal efficiency; and (iii) the determination of a motion specifies it as directed towards a *line* or a *plane* perpendicular to the line of motion, never towards a *point*. The expression Descartes uses is always: "vers quelque costé" or "versus aliquam partem." It specifies not a particular line directed to a point but a set of parallel lines, so that parallel motions (which have different points as their *termini ad quem*) have the same determination.[55]

As a first approximation we can define determination as a *quantified* and *directional mode* of motion or motive force.[56] Pragmatically, we can say that determination designates the vector of motion (*mv*), i.e., the most important function of the concept is to give a justification for computing the results of physical interactions among bodies by mathematical means corresponding to vector addition.

Technically speaking, vectors were an invention of the 19th century,[57] but the *geometrical* technique embodied in the parallelogram of forces was well known in the 17th century under the title "composition of motions" and presented no technical difficulties.[58] It also presented no serious philosophical difficulties in either statics or kinematics. In statics the interpretation of the sides of the parallelogram as component forces, whose resultant is the diagonal, raised no philosophical problems, since it is easy to conceive of two forces pushing or pulling a body from different directions at the same time or to conceive that these two forces are at equilibrium with a third force equal to their resultant and acting in the opposite direction. The technique had been used by Stevin in the analysis of the inclined plane.[59] A purely kinematic interpretation also presented no insurmountable problems. At first it seems contradictory to speak of a body as moving simultaneously in two different directions, but there were two alternative interpretations which could avoid this formulation: the sides of the parallelogram could represent consecutive trajectories, so that the distance between the beginning and end of the resultant diagonal would represent only the *displacement*; or a point could be conceived to move along one side of the parallelogram while the side itself is moved sideways in some direction so that the compound motion of the point describes the diagonal. However, although the geometrical technique was easily mastered and could be applied without great difficulty in *statics*, the

---

[55] Hobbes objects that "towards a certain *side*" does not unequivocally *determine* a motion (letter to Mersenne, March 30, 1641; AT III, 344-5). See document 5.2.12.

[56] See Gabbey 1980, p. 258; Schuster 1977, p. 288

[57] See Crowe 1967, chap. 1.

[58] There were, however, serious technical difficulties involved in applying the compounding of motions to infinitesimals and accelerations, e.g., in calculating tangents to curves and ellipses. See Costabel 1960.

[59] See Dijksterhuis 1970, pp. 48-63.

science of motionless forces, as well as in *kinematics*, the science of forceless motions, it proved to be somewhat troublesome to apply the technique to the motions of forceful bodies or the forces of bodies in motion, i.e., in dynamics. The troubles involved in the dynamical application of the parallelogram rule and some of the solutions proposed by Descartes will be discussed in this section.

The established interpretations of the parallelogram rule in both kinematics and statics were presented by Fermat in patient detail and applied to dynamics in the second letter of his exchange with Descartes over the *Dioptrics*.[60] After distinguishing "two sorts of compounded motions," Fermat described the first sort as follows:

Let us imagine a heavy body [*un grave*] at point A [Fig. 2.7], which descends in the line ACD at the same time that the line advances along AN such that it always makes the same angle with AO ... (AT I, 468-9).

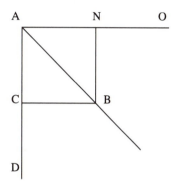

The second sort of compound motion deals with forces:

Let us suppose in the same figure a heavy body at point A, which is pushed at the same time by two forces, one of which pushes it along AO and the other along AD ... (AT I, 469-70).

Fig. 2.7 (AT I, 468)

Descartes was or course well aware that a body in motion can be part of a system in motion which in turn is part of another system in motion. He illustrates this state of affairs with the example of the motion of the hand of a watch worn by a (moving) sailor on a moving ship on a moving ocean current on the moving earth. However, while the hand of a watch takes part in various motions, it has (strictly speaking) only one proper motion, which is defined by the relation to contiguous bodies (II, §§30-32).

Descartes saw two major problems for a physical interpretation of the parallelogram rule: (1) It cannot be interpreted as a compounding of *motions* since a body cannot execute two different motions with respect to the same frame of reference at the same time. If "motion in a certain direction" has any determinate meaning, then it excludes motion in other directions. Properly speaking, "each individual body has only one motion which is peculiar to it" (II, §30). The sides of the parallelogram cannot be interpreted as two different motions of one and the same body at one and the same time, this being a contradiction in terms.[61] (2) If

---

60 Fermat, letter to Mersenne, Nov. 1637; AT I, 464-74. See document 5.2.6.

61 Hobbes later applied the same reasoning to determinations, which he took in the traditional sense of the *points* from which and to which the motion is directed:

the diagonal (or whatever it represents) is to be conceived of as *compounded* out of the sides of the parallelogram (or whatever the sides represent), i.e., if we are to conceive of the operation as genuine *addition*, then the component sides must be conceivable as genuine *parts* of the resultant.[62] But they cannot be conceived as parts if they are proportional to motions since in Descartes' system motions are scalar magnitudes whose sum is not equal to the resultant derived from the parallelogram rule. Descartes' solution is to introduce a new entity.

The physical magnitude whose parts are added and subtracted according to the parallelogram rule is called determination by Descartes. The concept was first introduced in Descartes' published works in the *Dioptrics*, where it is used to derive the law of refraction, and it is explicated in some detail in the subsequent debates with Fermat, Bourdin, and Hobbes, which Mersenne instigated and mediated.

In this section we shall analyze the use of the concept of determination to deal with the problem of directed quantities. We shall first of all present the derivation of the inverse sine law of refraction. Then we shall take up some of the difficulties which arise from the particular nature of the concept of determination and which were brought to light in the debates with Fermat and Hobbes. For the sake of intelligibility some of these later clarifications will be incorporated into the presentation of the derivation of the inverse sine law of refraction.

"Thirdly, it is to be objected that *one motion* cannot have *two determinations*; for in the figure drawn, let A be a body which begins to move towards C, having the straight path AC. If someone tells me that A is moved in a straight line to C, he *has determined* for me this motion; I can myself trace the same path as unique and certain. But if he says A is moved along a straight path toward the straight line DC he has not shown me the *determination* of the motion, since there are infinitely many such paths. Thus the motions from ABD to DC

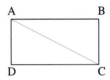

Fig. 2.8 (AT III, 344)

and from AD to BC are not determinations of *one single motion* of the body A towards C, but rather the determinations of two motion of two bodies, one of which goes from AB to DC, the other of which from AD to BC." (AT III, 344-5; see document 5.2.12.)

[62] The significance of the question, whether the entities represented by the sides of the parallelogram may be considered as *parts* of that represented by the diagonal, can be illustrated by the consequences drawn by Bertrand Russell. In his *A Critical Exposition of the Philosophy of Leibniz* (p. 98) he wrote: "It has not been generally perceived that a sum of motions, or forces, or vectors generally, is a sum in a quite peculiar sense – its constituents are not parts of it. This is a peculiarity of all addition of vectors, or even of quantities having sign. Thus no one of the constituent causes ever really produces its effect, the only effect is one compounded, in this special sense, of the effects which *would* have resulted if the causes had acted independently." In *The Principles of Mathematics* (p. 477) he formulated a "paradox of independent causal series": "The whole has no effect except what results from the effects of the parts, but the effects of the parts are non-existent."

2.5.1   The Argument of the *Dioptrics*

The inverse sine law of refraction is the showcase piece of Cartesian physical argument. Using his oppositions of contraries and his conservation laws as well as some at least plausible assumptions on the nature of light,[63] he inferred rigorously the "correct" law of refraction. Although many contemporaries sought and some seem to have stumbled upon the law of refraction,[64] only Descartes presented a conceptual system from which the empirical law could be inferred. It would be hard to exaggerate the historical and systematic significance of the discovery of the law of refraction for the foundations of Cartesian physics. The historical development of Descartes' theories will not concern us here, since this has been quite adequately discussed by recent research.[65] The systematic significance, which we shall be dealing with, can be seen by the following considerations.

From the definition of some general concepts and the application of simple logical and mathematical operations, Descartes derives a nontrivial (and long sought) functional relation between the states of a system before and after a physical interaction: the angle of incidence and the angle of refraction. This functional relation could be presented mathematically as a fixed proportion and needed only the empirical measurement of a constant (for the index of refraction between two media) in order to be formulated in an equation as an *empirical* law of nature. This law could be tested by experiment, and we know that Descartes did in fact try to have the necessary apparatus constructed to verify the law experimentally.[66] Thus Descartes not only proposed one of the first nontrivial empirical physical laws expressible as an equation (actually expressed as a proportion) and conceived one of the first fully scientific experiments to confirm a theoretical hypothesis expressible in such a form; he also derived the hypothesis

---

[63]   These assumptions are: (1) that light is transmitted instantaneously; (2) that it is an instantaneous action or inclination to move that can be taken to follow the same laws as an actual motion in time; (3) that the amount of impetus necessary to traverse a particular medium instantaneously with a particular intensity is analogous to the speed with which a body with a particular force traverses the medium, so that the *speed* of a ball is comparable to the *ease* of passage of a light ray.

[64]   Harriot, Snel, and Mydorge were all working on the problem. See Schuster 1977, pp. 308-321.

[65]   See Gabbey 1980; Sabra 1967, chaps. 3 and 4; and Schuster 1977, chap. 4. The standard work on Descartes' *Dioptrics* is the excellent analysis in Sabra 1967. While we follow in basic outline much of Sabra's presentation, there are important differences, especially in the analysis of Fermat's objections. The most recent monograph on Descartes's *Dioptrics* (Smith 1987), while somewhat unclear about the concept of determination, also presents some interesting material on the backgound in perspectivalist optics

[66]   See Ferrier's letter to Descartes, Oct. 26, 1629 (AT I, 38-52) and Descartes' letter to Ferrier, Nov. 13, 1629 (AT I, 53-59); also Descartes to Golius, Feb. 2, 1632 (AT I, 259); and Descartes to Huygens Dec., 1635 (AT I, 335-6).

from a comprehensive theory that was not primarily formulated to account for this particular phenomenon. The significance of the stated law is thus much greater than just the fact that it can explain a particular phenomenon, since it is presented as an instance of a comprehensive theory.

2.5.1.1 Reflection

Before deriving the law of refraction, Descartes first explicates his general procedure on the example of the law of reflection, a trivial case known since antiquity: angle of reflection equals angle of incidence. The law was illustrated at least since Alhazen (Ibn al-Haytham) with geometrical figures displaying the composition of motions. Descartes argument is in effect the demonstration that according to his system the *sines* of the two angles are equal and therefore that the angles must be equal.

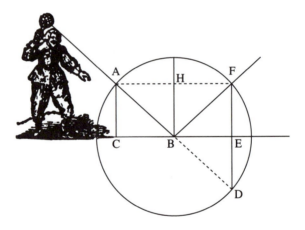

Fig. 2.9 (AT VI, 95)

Descartes takes up the traditional analogy between a light ray and a spherical body in motion, in this case a tennis ball, and illustrates the laws of reflection of light on the example of the reflection of a tennis ball struck by a racket onto the court (Fig. 2.9). He distinguishes first of all between the power (*puissance*) that makes the ball move and that which "determines it to move in one direction rather than another."[67] It is the force (*force*) by which it was struck by the racket that causes the ball to move in the first place and to continue in motion after hitting the ground. But motion as such is directionless, being merely the amount of displacement in a certain time. On the other hand, it is the position (*situation*) of the racket, and later of the ground, that determines in what direction the ball

---

[67] *Dioptrics*, AT VI, 94.

moves. A different position of the racket attached to the same force would determine the motion of the ball in a different direction, just as the position of the ground determines the same motion of the ball in a different direction. On the other hand, a different force could be attached to the same position of the racket giving the ball more motion in the same direction. But, according to Descartes, motion and its determination are not independent magnitudes: the determination, which is defined for a given body by its speed and direction, can change without affecting the quantity of motion (if only the direction changes); but if the scalar quantity of motion changes, the determination also changes. That is, determination has not only a direction which is independent of motion but also a quantity which is dependent on motion. Since determination is a directed magnitude, it can be divided into parts which are also directed magnitudes and which can vary in both *size and direction*. Descartes explains[68]:

> Moreover, it must be noted that the determination to move in a certain direction just as motion itself, and in general any sort of *quantity*, can be *divided* into all the parts of which we can imagine that it is composed. And we can easily imagine that the determination of the ball to move from A towards B is composed of two others, one making it descend from the line AF towards line CE and the other making it at the same time go from the left AC towards the right FE, so that these two determinations joined together direct it to B along the straight line AB.

Thus, while it is not yet completely clear what a determination is, it is sufficiently clear that the operation embodied in the parallelogram rule is *constitutive* of the meaning of the term: the diagonal represents the whole and the sides represent the parts.

In reflection the scalar motion of a tennis ball (or the ease of instantaneous passage of a light ray through the medium) is conserved so that the space traversed by the ball in a unit of time is the same before and after the collision with the reflecting surface. The scalar speed can be represented by any radius of the circle AFD (Fig. 2.9). The motion of the incident ball or ray has a (vector) determination represented by AB, which is by definition equal in absolute value to scalar speed, i.e., the directed distance AB is equal in length to the radius. It is important to remember that any particular, i.e., determinate, radius represents a determination; its length also represents the scalar speed. The surface of reflection CBE determines a coordinate system, in which opposition to the surface (interaction with the surface) is represented by perpendicular (vertical) vectors and non-opposition to the surface is represented by a parallel (horizontal) line pointed in either direction.

The actual determination AB of a ball's motion can be divided into two components, the one (AC) directly opposed (i.e., perpendicular) to the surface and the other (AH) not at all opposed (parallel) to the surface. Of course, as a purely

---

[68] AT VI, 94-5; emphasis added. This is the point that caused most of the misunderstandings which Descartes tried unsuccessfully to clarify in his letters. See AT II, 18-20; AT III, 163, 250-51; see documents 5.2.8, 5.2.15, and 5.2.16.

mathematical exercise there are infinitely many possible divisions of the determination AB which can be obtained either by choosing an angle different from 90° between the two sides of the parallelogram or by keeping the right angle and choosing a different coordinate system. However, the physical theory in this case determines that line AC is the line of opposition and that angle ACB is a right angle; this in turn determines that the parallelogram used for calculation must be a rectangle with one side on the reflecting surface. The actual surface of reflection fixes the coordinate system, and the need for a clear cut distinction between opposition and non-opposition fixes the right angle between the component determinations. That is, as Descartes puts it, the interaction with the *real* surface *really* divides the determination into parallel and perpendicular components[69]:

> ... the determination to move can be divided (I mean really divided and not at all in the imagination) into all the parts of which one can imagine it to be composed; there is no reason at all to conclude that the division of this determination, which is done by the surface CBE, which is a real surface, namely that of the smooth body CBE, is merely imaginary.

Although the motion of the ball is opposed to the rest of the immovable ground, the motion continues unabated, as we know from the conservation law. The perpendicular component determination is also opposed to the encounter with the reflecting surface and must therefore change in some unspecified way in order to resolve the double conflict. On the other hand, the parallel component determination is not at all opposed to the encounter with the surface and therefore does not interact with it, so that it is conserved unchanged. After the collision of the ball with the surface, this component determination can be represented by BE or any parallel line of equal length and in the same direction. Since the scalar motion is also conserved, the actual (compound) determination of the ball's motion must also have the same *absolute value* before and after collision. The perpendicular component determination involved in the interaction must change in such a way that it remains compatible with the conservation of scalar speed and the conservation of parallel determination; i.e., the vector sum of the parallel BE and whatever the new perpendicular determination turns out to be must be equal in absolute value to the radius of the circle, since the resultant actual determination is equal in absolute value to scalar speed. In geometrical terms, if the parallel determination (AH) is unchanged and the absolute value of the compound determination is also unchanged, then the ratio of the two must remain the same: AH/AB = BE/BX. This ratio expresses the *sine* of the angle made by the compound determination with the normal to the surface, i.e., what we now call the angle of incidence or reflection.[70] Since the sines of these angles are necessarily equal, the angles themselves are necessarily equal.

---

[69] Descartes, letter to Mersenne, Oct. 5, 1637; AT I, 452. See document 5.2.6. The italics indicating where Descartes is quoting himself have been removed.

[70] Descartes and his contemporaries sometimes used the term "angle of incidence" for the angle made by a ray with the *surface* and "angle of refraction" for the *deviation* of a refracted ray from its original line. See documents 5.2.4 and 5.4.3.

Returning to the figure, it can be seen that there are only two points where the conserved parallel determination and conserved scalar speed coincide: points F and D. Only one of these points lies *above* the surface. Here Descartes has used two conservation laws (motion and unopposed determination) and the opposition of determinations plus the qualitative consideration, that the ball or ray does not penetrate the surface, to derive unequivocally the angle of reflection. While it is clear that the result itself is nothing new, the conceptual and geometrical rigor of the inference is new; i.e., the law of reflection is shown to be a necessary consequence of the way the appropriate entities and their interactions were defined in Descartes' conceptual system. The same kind of argument is then used to derive the law of refraction, which was entirely new.

Before we turn to refraction, one peculiarity of this argument should be stressed. Unlike the quantities of motion and matter, determinations are not conserved in the interactions of bodies, but only *in the absence of interactions*. Or only that *part* of a determination that is not involved in an interaction is conserved. An interaction involves a conflict of determinations and some or all of the conflicting determinations must undergo change. Exactly how an interacting determination or that part of it that interacts must change is not immediately given. It is not necessarily conserved, but there is no algorithm directly governing its change, except for the dependence of the magnitude of the compound determination on scalar speed.[71]

2.5.1.2 Refraction

In refraction the surface separating two media fixes the coordinate system and the direction of opposition. We are to imagine that the refracting surface acts like a second tennis racket, that at the point of passage from one medium to the next increases or decreases the scalar speed of a tennis ball by a certain amount and also interferes with its downward determination. As in the case of the original tennis racket we must distinguish between the force imparted by the racket's power and the determination conferred by its position. The ball has a different speed in each medium, and the light ray requires a different amount of force or impetus to travel instantaneously through each different medium.[72] Whereas in reflection the scalar speed was conserved, in refraction a change in speed occurs that is specific to the relation between the two media. This change in speed at the point of transition is characterized by Descartes as an arithmetic compounding of motions in contradistinction to the compounding of determinations

---

[71] In the original formulation of the Third Law of Nature in *Principia philosophiae*, cited in section 2.4.2 above, Descartes even says merely that the colliding body *loses* (*amittit*) its determination and only later adds that it acquires a new one.

[72] In the *Dioptrics* Descartes says that light passes "more easily" in one medium than in another. In a later letter to Mersenne (Jan. 21, 1641; AT III, 291) he explains that less *impetus* is required. See document 5.2.9.

according to the parallelogram rule.[73] But nonetheless, in Descartes' example, if a ball moves with half its original air speed in water, we have a problem that is structurally similar to that of reflection.

To represent the *new* speed of the ball after entrance into the second medium, we can either construct a second circle with an appropriately altered radius,[74] or we can (with Descartes) use the same circle and radius but change the scale of measurement; the same length now represents, e.g., half or twice the motion and determination, and an unchanged determination is represented by a line of twice or half the length depending on whether the ball moves half or twice as fast in the second medium (Fig. 2.10).

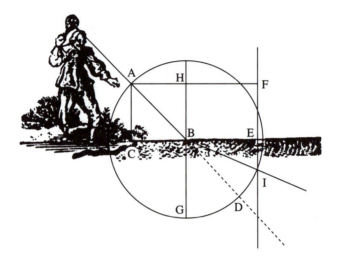

Fig. 2.10 (AT VI, 97)

Again, the horizontal component of the determination is not at all opposed to the encounter with the surface, and not being involved in any interaction, it is conserved without change. In Descartes' example, where the speed of the ball after refraction is half its speed before, the unchanged determination is represented by the line BE (which is *twice* as long as AH due to the change in scale). The vertical determination alone is opposed to the surface and must change in such a way as to be compatible with the new speed and the conserved parallel determination; i.e., the vector sum of changed perpendicular determination (whatever it

---

[73] Letter to Mydorge, March 1, 1638 (AT II, 19-20). See document 5.2.8.

[74] Historical evidence strongly suggests that Descartes, like Harriot, Snel, and Mydorge originally worked with an altered radius, formulating the law of refraction in terms of a constant ratio of the lengths of the radii (cosecant form). See Schuster 1977, pp. 268-368, for a highly plausible reconstruction of the original path of discovery.

turns out to be) and conserved parallel determination must be equal in absolute value to the new speed. There is only one point *below* the surface where this is true, namely I.

The new scalar speed and thus the absolute value of the new compound determination is dependent only on the medium; it is independent of the angle of incidence. Once again the ratio of the original parallel determination AH to the actual compound determination AB expresses the sine of the angle of incidence, and here the ratio of the unchanged parallel determination BE (differently represented) to the changed speed and compound determination BI (represented by the unchanged radius) in the new medium expresses the sine of the angle of refraction. The proportion between the two sines depends only on the factor of the change in scale of measure, which in turn depends on the ratio between the speeds in each medium; and this ratio is constant for any two given media. Thus, there is a constant relation between the sine of the angle of incidence (sin i = AH/AB) and the sine of the angle of refraction (sin r = BE/BI) for any two given media: AH/AB = k BE/BI. The constant (k = sin i/sin r) is an empirical magnitude that can in principle be ascertained by a single accurate measurement.

Having derived the inverse sine law for the refraction of a tennis ball, Descartes proposes an analogous argument for a ray of light. The inference from the behavior of the tennis ball to that of light is carried out in four steps. First, the surface of the tennis court in the example of reflection is replaced with a finely woven sheet which is punctured by the ball and in the process takes away half its speed. The sheet is then replaced by the surface of a body of water which is said to do much the same thing. Third, Descartes asks us to imagine that the surface of the body of water *increases* the speed of the ball instead of decreasing it, as if a second tennis racket acted at the surface to add (scalar) speed to the ball (in Descartes' example one-third more).[75] Finally, the refraction of light is compared to this process.

Whereas the ball that moved more slowly in the denser medium was supposed to break *away from* the perpendicular, light breaks *towards* the perpendicular. Descartes cannot argue that light moves faster in the denser medium, since he has already defined it as the *instantaneous* transfer of action; but he does make a similar kind of argument by asserting that light passes more easily (i.e., needs less force or impetus) through the denser medium than through the rarer. Thus, if light passes through water one-third more easily than through air, and if the *changed* ease of passage after interaction with the surface is represented by the *unchanged* radius of circle AFD (see Fig. 2.11), then the *unchanged* parallel component of the determination will be represented by the length BE (or GI),

---

[75] Descartes does not stipulate that the racket "hits the incident ball perpendicularly, thus increasing its perpendicular velocity," as Sabra (1967, p. 124) assumes; he says nothing about the slant of the racket, asserting merely that it should be thought to increase the scalar speed and not to affect the parallel component of the determination.

which is one-third shorter than AH. The intersection of line FEI below the surface with the circumference of circle AFD determines the path of the refracted ray BI. Since the relation between AH and GI is constant for any two given media and since AH and GI are proportional, respectively, to the sine of the angle of incidence and the sine of the angle of refraction, the ratio of the two sines is also constant.

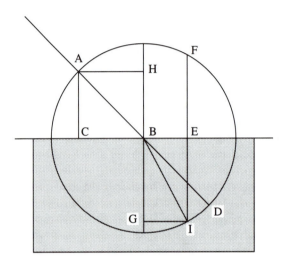

Fig. 2.11 (AT VI, 100)

Before turning to the two most important contemporary critics of Descartes' arguments in the *Dioptrics,* we should point out that although Descartes' results for light are correct from the point of view of classical mechanics, his analysis of the refraction of the *tennis ball* is incorrect. A material particle passing from air to water and continuing to move in the water with less than its former velocity breaks *towards* the perpendicular not away from it, just as light does. The reason for this is that the relevant magnitude is not simply the velocity of the particle, but rather its *momentum* in reference to the *density* of the medium; the energy needed to continue at half the speed in a denser medium can be considerably more than that associated with the full speed in the rarer medium.[76]

---

[76] For a detailed explanation and the relevant equations see Joyce and Joyce (1966), who also give a list of modern physics textbooks that argue (wrongly) along the lines of Descartes.

2.5.2   Criticisms of the *Dioptrics*

After Descartes had completed the *Dioptrics*, Mersenne arranged for a critical review by Pierre Fermat in 1637 (in fact prior to the actual publication), and three years later he organized a second debate with Thomas Hobbes on the explanation of reflection and refraction. Both the exchanges with Fermat and Hobbes served to clarify the concept of determination simply through the fact that Descartes was compelled to explicate the concept more than he had in the *Dioptrics*, and we have already incorporated many of these clarifications into our presentation. But there were also new elements in the two debates that deal with questions of the philosophical legitimacy of concepts and techniques of inference used in the *Dioptrics*. Fermat's objections point out problems in the relation of geometry to physical theory; Hobbes' critique forces Descartes to explicate the logical status of the concept of determination.

2.5.2.1 Pierre Fermat

Fermat's difficulties with Descartes' theory, aside from very real misunderstandings which were never cleared up,[77] centered on the Descartes' use of geometry in making physical arguments. To put his objection simply: A line drawn on a sheet of paper can be merely an instrument for geometrical construction, in which case it need not have a physical meaning. Or else it can represent directed distance covered in a given time, in which case it stands for a *motion*. Fermat maintained that Descartes' determinations were either simply projections of motions on an arbitrary line of direction having no physical meaning, or else they were actual motions. If they are simply projections, then they cannot "make" the tennis ball or the light ray *do* anything at all. If the lines drawn and called determinations are supposed to have physical meaning and are resolved and compounded according to the parallelogram rule for the compounding of motions, then they are in fact really motions and must be treated as such. Fermat's first letter stresses the point about projections; his second letter expounds at length on the second point, the parallelogram rule.

---

[77] The exchange with Fermat consisted of four letters: Fermat to Mersenne (for Descartes) April or May 1637 (AT I, 354-363); Descartes to Mersenne (for Fermat) Oct. 5, 1637 (AT I, 450-54); Fermat to Mersenne (for Descartes) Nov. 1637 (AT I, 463-74); and Descartes to Mydorge March 1, 1638 (AT II, 15-23). Fermat continued to consider motion and determination as independent magnitudes, specifically, he thought that motion can change without the determination's changing. He admitted (at least for the sake of argument) this misunderstanding 20 years later in his correspondence with Clerselier and Rohault. But even in a letter after this admission, he still interprets Descartes' determination as direction. See Fermat 1891, vol. 2, pp. 397-8, and 486; see also Sabra 1967, p. 129, fn. 77.

In his first letter Fermat denies that Descartes' argument is a genuine proof or demonstration. He sees that determination cannot mean *simply* direction since it has a quantity and can be divided into parts, but direction remains the primary aspect in his interpretation. He takes determination to be the projection of the motion onto a particular direction. He does not take it to be a component of the motion, *and he does not apply the parallelogram rule*. In fact in the second letter he even accuses Descartes of having confused determinations with motions because he applied the parallelogram rule to determinations.[78] He thus conceives determination as a measure of how far a body advances in a certain direction, and this measure is taken by dropping a perpendicular onto the line of direction (Fig. 2.12). Fermat then points out that the two directions onto which a body's motion may be projected need not be at right angles. Instead of projecting the motion (AB) onto the vertical and the horizontal directions, Fermat suggests that we project it onto the vertical (AC) – since he does not doubt that the vertical represents the line of

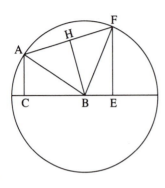

Fig. 2.12 (AT I, 359)

opposition – and onto a line (AF) that slants *away* from the surface and thus is also not opposed to it. This, he asserts, conforms to Descartes' specifications just as well as does the projection onto the horizontal. In fact there are an infinite number of directions which are unopposed to the surface and make an acute angle with the line of incidence AB.[79] If we extend the projection of the motion onto any one of these lines the same length (from H to F) in the second unit of time, we arrive at a point on the circumference of the circle, which must according to Descartes' specifications show the location of the ball or light ray. The actual determination of a motion can be divided into all the parts one can imagine it to be composed of; it can be projected onto infinitely many directions, and it is simply arbitrary to single out the one line of direction which

---

[78] Fermat, letter to Mersenne for Descartes, Nov. 1637; AT I, 466 and 468. See document 5.2.7.

[79] "Whereby it should be obvious that AF makes an acute angle with AB; otherwise, if it were obtuse, the ball would not advance along AF, as is easy to understand" (AT I, 359). Fermat's insistence that the angle BAF be acute makes it clear that he is talking about projections and not about the parallelogram rule. A line can only be projected on another line that makes an acute angle with it. This restriction does not apply to the side and the diagonal of a parallelogram; here, the angle made by the diagonal with either side may be obtuse as long as their *sum* is no greater than 180°. On this point we differ significantly with Sabra's interpretation. Sabra attempts to interpret Fermat (Fig. 2.12) as applying the parallelogram rule; this compels him to treat line HB as the line of opposition to the surface. Not only is there no textual basis for this, but it represents a position that would have been unique in the 17th century. See document 5.2.5.

gives us the law of reflection we already know. These geometrical constructions have no physical meaning. Fermat sums up his critique[80]:

> It is therefore evident that, of all the divisions of the determination to motion, which are infinite, the author has taken only that which can serve to give him his conclusion; and thus he has accommodated his *medium* [i.e., middle term] to his conclusion, and we know as little about it as before. And it certainly appears that an imaginary division that can be varied in an infinite number of ways never can be the cause of a real effect.

In his second letter Fermat takes up the use of the parallelogram rule in the explanation of refraction. He argues that if two independent forces act on a body, they will "interfere with and resist one another."[81] Thus the size of the resultant force or motion of the body will depend on the angles at which the forces impinge on it; the greater the angle, the smaller the velocity of the compound motion.

If the angle of refraction is to be found by applying the parallelogram rule, then it is actually forces or motions that are being resolved and compounded, not simply projections onto lines. If the surface of refraction (CBE), say, adds motion to the tennis ball at the point of transition from one medium to the next and only acts in the perpendicular, then the resultant force is to be calculated as the diagonal of the parallelogram whose sides are the (continued) original motion BD and the (added) perpendicular motion BG (Fig. 2.13). The diagonal of this parallelogram will be longer the more acute the angle GBD between the two motions; and the size of the resultant or compound motion of the body will depend on the angle of incidence. Thus the resultant speed of the body cannot be constant for the medium and independent of the angle of incidence as Descartes had maintained.[82]

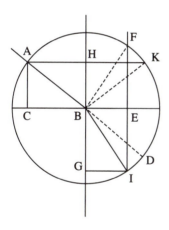

Fig. 2.13 (AT I, 473)

To sum up Fermat's critique: If the lines Descartes has drawn are simply projections onto directions, they have no physical meaning; but if they are to be compounded and resolved according to the parallelogram rule for the compounding of motions, they must represent actual motions, and operating with them leads to results quite different from those arrived at by Descartes. The geometrical techniques of inference used by Descartes may, according to Fermat, only legitimately be applied to actual motions.

---

[80] Fermat, letter to Mersenne, April or May 1637; AT I, 359. See document 5.2.5.

[81] Fermat, letter to Mersenne, Nov 1637; AT I, 470. See document 5.2.7.

[82] Fermat, letter to Mersenne, Nov 1637; AT I, 473-474.

2.5.2.2 Thomas Hobbes

The exchange of letters with Hobbes sheds light on the logical status of the concept of determination. We have already seen that determination is a mode of motion, which is itself a mode of bodies. One of the major points of discussion was the logical-ontological status of the entities resolved and compounded according to the parallelogram rule; and Hobbes also raises the problem of the compatibility of this rule with the conservation of motion.

To Hobbes' first letter (which is lost) Descartes replied that the sides of the parallelogram should not be called motions or even determinate motions as suggested by Hobbes, lest this lead to confusion, since motions are added arithmetically and not according to the parallelogram rule, which was to be reserved for determinations.[83] Hobbes responded that the confusion can easily be avoided and the diagonal sum reconciled with an arithmetical sum. Each of the sides of a parallelogram can be examined to see how much it actually *contributes* to the resultant diagonal (Fig. 2.14).

In effect Hobbes resolves each of the sides (BA and AC) of the parallelogram into two components (BF and FA; AD and DC), one parallel to the diagonal and one perpendicular to it. The components of the two sides that are parallel to the diagonal (FA and DC) add up as vector and as scalar magnitudes to the diagonal; these diagonal subcomponents can be seen as the *real contribution* of the component motions to the resultant motion. (Hobbes does not say what happens to the other subcomponents (BF and AD), which are perpendicular to the diagonal, but they are in fact equal and opposite and could have been seen to cancel each other out.) What Hobbes is actually saying is that the diagonal can be conceived as (vectorially) compounded of the sides because it is actually (vectorially and arithmetically) compounded of subcomponents of the sides. The sides compose the diagonal because each contributes a part of itself to it.[84]

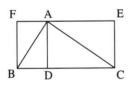

Fig. 2.14

---

[83] Descartes, letter to Mersenne for Hobbes, Jan. 21, 1641, AT III, 288. The exchange with Hobbes consisted of seven letters, starting with a letter from Hobbes which has been lost; Descartes to Mersenne (for Hobbes) Jan. 21, 1641 (AT III, 287-392); Hobbes to Mersenne (for Descartes) Feb. 7, 1641 (AT III, 300-313); Descartes to Mersenne (for Hobbes) Feb. 18, 1641 (AT III, 313-318) – this letter deals with Hobbes's own optical work (see Shapiro 1973); Descartes to Mersenne (for Hobbes) March 4, 1641 (AT III, 318-333); Hobbes to Mersenne (for Descartes) March 30, 1641 (AT III, 341-348); Descartes to Mersenne (in French; not for Hobbes), April 21, 1941 (AT III, 353-357). This exchange overlapped with Hobbes's "Objections" to Descartes' *Meditations*. Figure 2.14 is reconstructed from the diagram in Descartes' response; Hobbes's original has not been preserved. See documents 5.2.9-5.2.13

[84] Hobbes, letter to Mersenne, Feb. 7, 1641; AT III, 304-5: "In as much as the motion from A to B [i.e., B to A; see Fig. 2.14] is composed of the motions from F to A and from F to B [i.e. B to F], the compounded motion AB does not *contribute* more

Descartes rejects this argument as patently absurd: a subject must actually consist of what it is said to be compounded of. Hobbes is in effect asserting that a motion may be said to be compounded of two other motions even if it contains merely a *part* of (a contribution from) each of them. This is, as Descartes remarked, "just the same as to say that an axe is composed of the forest and the mountain because the forest contributes wood for the handle and the mountain [contributes] iron dug up from it."[85] If the entity represented by the diagonal is to be compounded out of the entities represented by the sides, then it must *contain* them, not just part of them but all of them.

This argument may well show that *motions* may not be resolved and compounded according to the parallelogram rule, but it certainly does not explain why anything else could be so compounded and resolved. It must still be explained why component *determinations* can be conceived of as genuine parts of the resultant determinations.

Descartes makes two comparisons which indicate how he wants the modes of modes to be conceived and how determinations can be considered to have *parts*. First of all, one can compare the motion of a body to the other fundamental mode of bodies, namely, shape and compare the determination of motion to that of shape.[86] Just as any real motion is a determinate motion, any real shape is a particular, determinate shape. But just as one can distinguish between a particularly shaped body, e.g., a flat or round body, and its flat or round *surface* (its "flatness" or "roundness"), so too can one distinguish between a determinate motion and the *determination* of the motion. A determination can be divided into parts just as a surface can be so divided. Let a body have a determinate shape, say it is a cube. Although the surface of the cube can be divided into six faces out of which it can be said to be compounded, this division of the surface does not divide the body itself. The parts (six square faces) of the surface of the cube are not the same as the surfaces of parts of the cube. Thus, too, the parts of the determination of a motion are not the determinations of the parts of the motion.[87] The general principle to which Descartes is implicitly appealing here is that *the parts of a mode of a subject are not the modes of the parts of the subject*. Therefore, the sides of the parallelogram, which represent the component parts of the determination of the body's actual motion cannot be taken indepen-

---

speed to the motion from B towards C than the components FA, FB can *contribute*; but the motion FB *contributes* nothing to the motion from B towards C: this motion is determined downwards and does not at all tend from B towards C. Therefore only the motion FA *gives* motion from B to C ..." (emphasis added). Hobbes makes a series of minor technical mistakes here (which Descartes harps on and corrects); they do not however affect the substance of his argument (to which Descartes also replies). See document 5.2.11.

85 Descartes, letter to Mersenne, March 4, 1641; AT III, 324.

86 Descartes, letter to Mersenne, March 4, 1641, AT III, 324-5.

87 This point is made more clearly in an earlier letter to Mersenne (July 29, 1640; AT III, 113), from which the details of the example are taken. See document 5.2.14.

dently as if they were determinations of component parts of a motion. They have meaning only in relation to a particular diagonal.

In order further to explicate the difference between a determinate motion and the determination of the motion, Descartes takes up an example from logic suggested by Hobbes which also gives some indication of how a determination can be divided into components.[88] The example involves proper names and singular terms, cases of what medieval logic sometimes called discrete or determinate supposition.[89] Although every man is a determinate man, for instance, Socrates, and thus "man" and "Socrates" can refer to the same entity or *suppositum*, the two terms have different meanings. The meaning of Socrates contains the "individual and particular differences" which individuate Socrates. Furthermore, one can distinguish conceptually between the individual denoted by Socrates and the determination of being Socrates, i.e., the individual properties ascribed to him. If one of the individual differences that make up what it is to be Socrates were to disappear, the man would still be a determinate man but would no longer have the determination of being Socrates. Descartes' argument implies that the determination of being Socrates is composed of component determinations such as, "for instance, the knowledge he had of philosophy." The parts of the determination of a motion can no more be ascribed to parts of the motion than can Socrates' knowledge of philosophy be ascribed to one of his parts, e.g., his arm or leg. One such component determination can change without affecting the other *component* determinations. The component parts of the determination of a motion are to be conceived as elements of the definite description of that particular motion.[90]

2.5.3   Oblique Impact

Descartes' comparison of a determinately moving body with a cubical body is systematically important because it shows that it is logically precluded that the component determinations be taken as determinations of parts of the motion itself. Thus, the sides of the parallelogram have no meaning independent of the resultant and may not be detached and operated upon separately. No more than Socrates' knowledge of philosophy can be predicated of his elbow, can the vertical component of the determination of a motion be ascribed to a part of that motion. No contradiction can occur between the scalar conservation law and the parallelogram rule, as long as the sides of the parallelogram are interpreted as components determinations, as parts of a mode of motion.

---

[88]   Letter to Mersenne, April 21, 1641, AT III, 354-6. See document 5.2.13.

[89]   See Kneale and Kneale 1969, p. 258ff.

[90]   Letter to Mersenne, April 21, 1641, AT III, 354-6. J.M. Keynes (1906, p. 469) still calls the components of a "complex term," e.g., "A and B and C ..." *determinants* of the term.

The only other case in which a direct conflict could arise between these two fundamental elements of Cartesian physics would be in the oblique collision of bodies. Here the sides of the parallelogram could represent the motions of *two different bodies* caused by the motion of a third body (represented by the diagonal), which collides with them at an angle and *causes* them to move as the sides of a parallelogram while itself ceasing to move. Such a case would indeed lead to a contradiction with the conservation of scalar motion, but only if such a collision were governed by the impact rules of classical mechanics. Let us conclude with a look at Descartes' handling of oblique impacts.

Descartes never deals explicitly and quantitatively with the kind of three-body collision sketched above. All his impact rules apply to two-body collisions. On a number of occasions, however, he did deal with a similar phenomenon in the transfer of the action of light.[91] He often uses the example of a body *pushing* against *two* others obliquely which in turn both push obliquely against a fourth body. It is clear that he considers this to be equivalent to having one body push a

second which pushes a third. But he is dealing with the transfer of action through bodies which are in some way constrained by their surroundings (see Fig. 2.15), not with the free motions of the bodies themselves. At one point, however, he does maintain that a three-body collision is in fact two successive two-body collisions,[92] so that it can at least in principle be reduced to the two-body collisions governed by the impact rules. It remains to be seen whether an interpretation of a two-body oblique collision can be given that is consistent with the scalar conservation law, the impact rules of the *Principia philosophiae*, and Descartes' use of the concept of determination.

Fig. 2.15

In one letter to Mersenne[93] Descartes does deal in a quantitatively precise way with a case of oblique impact. He considers the case of a large moving ball colliding obliquely with a small resting ball; the case should thus be subsumed under impact Rule 5 discussed above. Although a moving body cannot normally make another body move faster than it itself moves, this applies, Descartes tells us, only to motion in the same straight line. It does not necessarily apply when the motions are in different lines (see Fig. 2.16 and Fig. 2.17).[94]

> Thus, for instance, if the large ball B coming from A to D in a straight line encounters the small ball C from the side, which it makes move towards E, there is no doubt whatsoever that even if these balls are perfectly hard, the lit-

[91]   See AT II, 370; VIII, 187; and XI, 100.

[92]   "Since the continuous motion of these [balls] brings it about that this action is never, in any period of time, received simultaneously by two, and that it is transmitted sucessively, first by the one and then by the other" (AT VIII, 187; *Principia*, III, §135).

[93]   April 26, 1643, AT III, 648-655; see document 5.2.17.

[94]   AT III, 651-2.

tle one ought to leave more quickly than the large one moves after having encountered it; and constructing the right angles ADE and CFE, the proportion
which holds between the lines CF and CE is the same as holds between the
speed of the balls B and C.

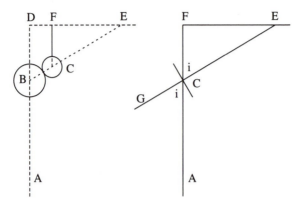

Fig. 2.16 (AT III, 652)   Fig. 2.17 (our figure)

This must of course be interpreted. We assume that Descartes meant to say that
the lines CF and CE also represent the actual paths along which the two balls
move after collision.[95] We also assume that the line CE is the line of direct
opposition between the two bodies, i.e., the normal through their centers of
gravity. Both assumptions are suggested by Descartes' illustration. Although
Descartes does not explicitly mention determinations in this passage, it can be
seen that they do indeed play a role; and the relation of determinations to motion
was directly thematized one page earlier in the letter. While the *length* of the line
CF represents the speed of the larger ball B after collision, the line CF itself can
be taken to represent not only the actual determination of B but also the vertical
component determination of C. Thus although the speed of C is greater than that
of B, their determinations in B's original direction are equal. Thus B transfers
enough motion to C to make it move with the same vertical determination as B
has but in the direction determined by the line of direct opposition, i.e., along
CE. The proportion between CF and CE, which Descartes says is also the proportion between the speeds of B and C, expresses the cosine of the angle FCE,
which is equal to angle ACG, the angle of incidence of ball B. Thus the ratio of
the two speeds after collision is a function of the angle of incidence. The speeds
can then be calculated.

Let v_1, u_1 be the speeds of ball B before and after collision; and let u_2 be the
speed of ball C after collision ($v_2 = 0$) and let i be the angle FCE, which is

---

[95] This condition implies that both bodies can be conceived as points (i.e., that Fig.
2.16 [Descartes'] and Fig. 2.17 [ours] are equivalent); it is a conclusion that is difficult to reconcile with Descartes' definition of material bodies.

equal to the angle of incidence ACG, then: $u_1/u_2 = CF/CE = \cos i$; i.e., the proportion between the speed of the two balls is determined by the angle of incidence, or $u_1 = u_2 \cos i$.

Assuming that the quantity of motion in the two-body system is conserved, then:

$$Bv_1 = Bu_1 + Cu_2 = u_1 \frac{B + C}{\cos i}$$

and:

$$\frac{v_1}{u_1} = \frac{B}{B} + \frac{C}{B \cos i} = \frac{C}{B \cos i} + 1.$$

The fifth impact rule, as we saw in section 2.4.4, can be formulated as

$$Bv_1 = u_1 (C+B) \text{ or } Bv_1 = u_1(C + B) \text{ or } \frac{v_1}{u_1} = \frac{C + B}{B} = \frac{C}{B} + 1.$$

Thus, impact Rule 5 can be seen as a special case of $\frac{C}{B \cos i} + 1$ where $i = 0$ and accordingly $\cos i = 1$.

Thus, at least in the one case where Descartes actually calculates an oblique collision, there need be no contradiction to the scalar conservation law. And, in fact, assuming the conservation law allows us to derive an even more general formula for the impact rule governing the collision. Since the conservation of scalar motion is one of the construction principles of any rule within the system for calculating impacts, this is not surprising. It merely shows that instead of a contradiction in the system between conservation of scalar momentum and empirically adequate rules of impact, we find a consistent system, from which impact rules are derived that are empirically inadequate.

2.6   Conclusion

Our discussion of Descartes' application of conservation laws, the logic of contraries, the parallelogram rule, and the concept of determination has shown that he successfully inferred by means of the conceptual system they constituted not only consistent laws of impact but also a nontrivial and empirically adequate law of refraction. And there can be little doubt that his system provided the basis for the later introduction of the concepts of (instantaneous) velocity and momentum and for the generalized application of the parallelogram rule in mechanics. However, we have also seen the great difficulties that Descartes' contemporaries had in applying some of his concepts.

Descartes is probably the only person who ever used his concept of determination productively in physics. His followers were unable to make any advances in science by applying the concept to explain new cases or to integrate disparate elements more coherently into a conceptual system. Although the *term* itself was retained and used on into the 18th century, due primarily to the popularity of Jacques Rohault's textbook, *Traité de Physique*, it came to mean simply *direction*. Rohault's presentation is faithful to Descartes' use of the concept – in fact it is simply a paraphrase –, but he applies the concept only to those cases already handled by Descartes. When Spinoza in his presentation of Descartes *Principia* attempted to deal with a somewhat more complex case of oblique collision (in which *both* bodies are in motion), he introduced a "*force* of determination" thus making determination independent of the "force of motion."[96] But perhaps the most illustrative case of the later use of the concept of determination is given by Clerselier, who a dozen years after Descartes' death renewed the debate with Fermat on reflection and refraction. When he tried to explain in his own way what happens in reflection, he did not derive the change in the perpendicular determination from the conservation of parallel determination and scalar motion. He simply asserts that the determination itself is *reversed* and thereupon derives the *compound* determination: when balls collide with the surface of reflection, "they will be constrained to *change the determination* which they have to go [downwards] ... *into that* to go or reflect [upwards]."[97] Clerselier is compounding and resolving independent entities according to the parallelogram rule; he does not need, nor does he in this case use, Descartes' circles representing scalar motion.

It proved almost impossible to apply the concept of determination to cases which Descartes had not already explained. It was not an everyday tool that could be used by any competent scientist; it was an idiosyncratic instrument that could be employed only by its inventor. One could not simply manipulate the representations of determinations and draw inferences: the slightest deviation from Descartes' terminology (the linguistic representation of the conceptual relations in the theoretical system) led to errors of reasoning. But these errors were not simply at random: they generally involve either taking determination as mere *direction* or as an *independent* magnitude. Both errors are allowed or even suggested by the mathematical means of representation. There is nothing in the geometrical representation itself of determination nor in the construction rules of geometry that prevents the sides of a parallelogram from being treated as independent magnitudes. The conceptual dependence of determination on motion is represented only in language not in the geometrical means of inference. Keeping the nonrepresented conceptual relations in mind while operating with the geometrical representations is a *tour de force,* not the application of an algorithm.

While such virtuosi as Hobbes and Fermat were able to argue with Descartes about the philosophical background of his theories, other scientists either accepted of rejected his explanations without real comprehension. Descartes'

---

[96] Spinoza 1925, vol. 1, pp. 213-216; *Renati Des Cartes...*, II, Prop. 27.
[97] Clerselier, letter to Fermat, May 13, 1662; *Oeuvres de Fermat*, vol. 2, pp. 478-9. See the Epilogue (section 4.2) and document 5.4.2.

exchange with Bourdin is a good case in point. Bourdin sees that Descartes' explanation of refraction makes use of right triangles for which the Pythagorean relations hold; a determination of "4 to the right" combines with one of "3 downwards" to make a resultant of 5. But instead of saying with Descartes: a *ball* covers 4 units to the right due to its determination, he says the *determination* covers 4 units or the determination carries the ball 4 units.[98] Descartes maintains that such seemingly minor changes totally misrepresent the theory; he points out that according to his theory it is not the determination that conducts but the force itself as determined (*sed virtus ipsa ut determinata*). He has, he thinks, explicitly excluded the possibility of manipulating the geometrical representation in the manner suggested by Bourdin. In fact Descartes' refrain every time Bourdin interprets the geometrical representation in his own words is: "quod ineptum est." However, Bourdin was not alone; even the Latin translator of the *Dioptrique* only a few years later translated similar mistakes into the Latin text. Where Descartes, using the language of indirect causality says the determination "makes" the ball move (*la fait aller, fait descendre, faisoit tendre*), the Latin simply says it moves the ball: *agebat, propellit, ferebatur*.[99]

But even Descartes was not virtuoso enough for his own system; for in one of his later comments on Bourdin he, too, makes a mistake: "But I believe that what perplexes him is the word *determination*, which he wants to consider without any motion, which is chimerical and impossible; in speaking of the determination to the right, I mean all that *part of the motion* that is determined towards the right ... [*qu'en parlant de la determination vers la droite, i'entens toute la partie du mouvement qui est determinee vers la droite*]."[100] It should by now have become clear that a component determination is neither a *part* of a motion nor even the determination of such a part.

Now even if Descartes' genius were taken as an explanation of his ability to invent and apply fruitfully the conceptual system of which determination was an essential component, it is the fact that genius was actually required to apply the concept fruitfully that explains why the system had no future in science. Whether or not genius is necessary to invent a new concept, the new concept must be integrated into a representational system whose use does not require genius, if it is to become part of a scientific tradition. The scientific adequacy of conceptual means is inversely proportional to the genius required for their application. The virtuosity needed to work with Descartes' system of explanations is a reliable indicator of the deficiency of the system.

---

[98] Descartes cites Bourdin's remarks in a letter to Mersenne (July 29, 1640; AT III, 105-119). For Descartes comments on Bourdin, see documents 5.2.14-5.2.16.

[99] See AT VI, 95-97 and 591-92. On the terminology of indirect causality in the 17th century, see Specht 1967, pp. 29-56.

[100] Letter to Mersenne, Dec. 3, 1640 (AT III, 251; second emphasis added). See document 5.2.16. Unaccountably, Gabbey (1980, p. 259) cites this passage as "the nearest Descartes came to a clear definition of the notion."

3
Proofs and Paradoxes:
Free Fall and Projectile Motion
in Galileo's Physics

> The treatise on motion, all new, is in order; but my restless brain cannot stop mulling it over, and with great expenditure of time, because this thought which lately occured to me concerning some novelty makes me overthrow all my previous findings.
> (Galileo to Fulgenzio Micanzio, Nov. 19, 1634; EN XVI, 163)

3.1 Introduction

According to a well established view the work of Galileo marks the beginning of classical mechanics.[1] His work does not yet represent the full fledged classical theory as it emerged in the contributions of Newton and others, but, following this widespread interpretation, Galileo did take the first decisive steps: he criticized and overcame the traditional Aristotelean world picture, he introduced the experimental method, he concentrated on a systematic and concise description of single phenomena rather than searching for their causes and elaborating an overarching philosophy of nature, and he succeeded in the mathematical analysis of some of the key problems of classical mechanics.[2]

Among the insights that Galileo attained in the field of terrestrial mechanics are such fundamental theorems of classical mechanics as:

---

[1] See, e.g., Clavelin 1974, p. 271: "While the defense of the Copernican doctrine enabled Galileo to outline some of the most significant ideas and methods of classical mechanics, his *Discourses* must be considered an integral part of that science."

[2] See, e.g., Drake's characterization of Galileo's pioneering contribution: "Physics began to depart from the Aristotelian conception with Galileo's investigations of motion in bodies free to descend, whether unrestrained, or supported from below, or suspended from above as with pendulums. ... The historical evolution of modern physics will not be thoroughly understood without the attention to the labors of men who put aside the ambition to explain everything in the universe and sought to find sure particular limited rules by careful measurements. So far as we know, Galileo was the first to do this for actual motions." (Drake 1979, p. VII)

(1) the proposition that all bodies fall equally fast in a vacuum, or more precisely, with the same constant acceleration;

(2) the law of fall, i.e., the proposition that the distances traversed by a falling body are proportional to the squares of times elapsed;

(3) the proposition that the trajectory of projectile motion is parabolic in shape.

In addition to such key insights Galileo's work comprises a number of more specific theorems which are valid in classical physics although they do not belong to its core today. One example is provided by the striking theorem that a body falling down differently inclined planes which correspond to the chords from the highest point of a circle always takes the same time in spite of the different lengths of these chords, a proposition which will be called in the following "Isochronism of Chords."[3] Another no less striking and also correct example is the theorem that the times a body needs to fall along different inclined planes of equal heights are simply proportional to the lengths of these planes ("Length-Time-Proportionality").[4]

The derivation of the fundamental laws of free fall and projectile motion in classical mechanics presupposes two basic principles which bluntly contradict medieval natural philosophy:

(1) The principle of inertia: In the Aristotelian tradition motion itself is conceived as caused by force. Classical physics is based on the assumption that only a change in motion, i.e., acceleration, needs to be explained by a specific cause.

(2) The principle of superposition of motions: In the medieval tradition of explaining motion by the impetus imparted to the moving body the composition of different motions in one body was conceived as struggle between opposing moving forces, usually ending with the domination of one by the other. In classical physics the trajectory of a projectile is constructed as superposition without any mutual interference of an inertial motion and accelerated motions caused by the moving forces.

Thus, according to the principles of classical physics free fall in a vacuum is simply a motion resulting from gravitational force which – in the proximity of the earth's surface – causes a constant acceleration independent of the properties of the falling body. The law of fall – the proportionality between the spaces traversed and the squares of the times elapsed – can be derived from the constancy of acceleration by integrating the constantly growing velocity over time, e.g., by using a simple geometric representation of this function. Finally, if air resistance is neglected, the trajectory of any projectile can be reconstructed as superposition of the uniform inertial motion along the direction of the initial projecting force and the vertical, accelerated motion of fall. The parabolic shape of the trajectory is an immediate consequence of this construction.

---

[3] Proposition VI. Theorem VI, *Discorsi*, p. 178; EN VIII, 221.

[4] Proposition III. Theorem III, *Discorsi*, p. 175; EN VIII, 215.

Galileo did not explicitly formulate such straightforward arguments. The underlying principles, however, appear to be so closely linked to his achievements, that it is usually assumed that they are part of his conceptual framework.[5] If this were correct, then he could in fact legitimately be called the first representative of classical mechanics, whereas his immediate predecessors and contemporaries such as Descartes appear to be still deeply rooted in premodern thinking.[6]

Thus it does not come as a surprise to those who hold this view that Descartes' reaction to Galileo's last and most comprehensive work on motion, the *Discorsi*, was most unfavorable.[7] Historical evidence of this reaction is provided by his correspondence with Marin Mersenne in 1638. After Mersenne had written him about Galileo's new book Descartes promised to look at it, as soon as it went on sale, but, he qualified, "only in order to be able to send you my copy, annotated if the book is worth it, or at least to send you my remarks on it."[8] It fits the picture of Descartes as a philosopher still deeply rooted in traditional conceptions of science that he had so little to say about Galileo's book, that he could choose the second option. On August 23, he again wrote to Mersenne saying, "I also have Galileo's book, and I have spent two hours leafing through it, but I find so little in it to fill the margins, that I believe I can put all that I have to say about it in one very short letter; and thus it is not worth it to send you the book."[9] Consistent with this condescending tone, the promised letter which Descartes sent to Mersenne on Oct. 11, 1638 is full of disparaging comments on Galileo's work. This letter seems to document Descartes' stunning refusal to acknowledge Galileo's major achievements. Historians of science are therefore almost unanimous in condemning Descartes' incapacity to recognize Galileo's contribution to the creation of classical mechanics.[10]

Since, according to the standard interpretation, Descartes was unable to overcome his traditional conceptions, even when confronted with the fundamental insights of classical physics, it should be somewhat puzzling that Descartes'

---

[5]   See, e.g., Drake 1964, pp. 602-603: "But if Galileo never stated the law [of inertia] in its general form, it was implicit in his derivation of the parabolic trajectory, and it was clearly stated in a restricted form for motion in the horizontal plane many times in his works."

[6]   This view is expressed drastically in Szabó 1977, p. 60: The value of Aristotle's and Descartes' natural philosophies for natural science may be seen at most in having stimulated Galileo and Newton to refute their errors.

[7]   According to Drake (1978, p. 387) Descartes' reaction "throws light on the reception of Galileo's science outside the circle of his friends and former pupils, illustrating the conception of science most opposed by Galileo and the ineffectiveness of his own work in altering traditional goals."

[8]   Letter to Mersenne, June 29, 1638; AT II, 194.

[9]   Letter to Mersenne, August 23, 1638; AT II, 336.

[10]   See the recent review of various interpretations given in Ariew 1986. Emil Wohlwill (1884, pp. 128-129) presents a remarkable exception claiming in fact that the essential points of Descartes' criticism are correct, even though Descartes failed to recognize Galileo's great achievements. For a similar view, see also Dijksterhuis 1924, pp. 301ff.

comments can also easily be understood as a critique from the viewpoint of the theoretical program of classical mechanics. While it is usually assumed that the fundamental principles of classical mechanics are implicitly part of Galileo's conceptual framework, Descartes begins his critique with the statement that exactly such principles are missing from the foundations of Galileo's mathematical deductions[11]:

> I shall commence this letter by my observations about Galileo's book. I find generally that he philosophizes much better than ordinary, in that he avoids as best he can the errors of the Scholastics and undertakes to examine physical matters by mathematical reasonings. In this I accord with him entirely, and I hold that there is no other way to find the truth. But he seems to me very faulty in continually making digressions and never stopping to explain completely any matter, which shows that he has not examined things in order, and that without having considered the first causes of nature he has only sought the reasons of some particular effects, and thus he has built without foundation. Now, in so far as his fashion of philosophizing is closer to the truth, one can the more easily know his faults; as one can better say when those who sometimes follow the right road go astray, than when those go astray who never enter on it.

What follows is a series of critical comments referring to various pages of the *Discorsi*, particularly to the foundations of and to purported gaps in Galileo's derivations. This critique was, as pointed out in Chapter 1, partly based on alternative assumptions about the principles of mechanics, some of which appear incorrect when interpreted in the theoretical framework of classical mechanics. But in Chapter 2 it was argued that these principles nevertheless were the first to fulfill basic requirements of a physical theory in the sense of classical physics. Moreover, it turns out that some of Descartes' specific remarks on the *Discorsi* can easily be interpreted as comments from the perspective of classical physics. With regard to the symmetry of projectile motion for instance, which is an obvious property of the parabolic trajectory according to classical mechanics, he objects that Galileo has only proved that the shape of the trajectory is parabolic for the downward part but not for the trajectory as a whole.[12]

---

[11] AT II, 380; translation adapted from Drake, 1978, pp. 387-388. See document 5.3.1.

[12] Letter to Mersenne, Oct 11, 1638. "It is to be noted that he takes the converse of his proposition without proving or explaining it, that is, if the shot fired horizontally from B toward E follows the parabola BD, the shot fired obliquely following the line DE must follow the same parabola DB, which indeed follows from his assumptions. But he seems not to have dared to explain it for fear that their falsity would appear too evident." (AT II, 387; translation, Drake 1978, p. 391) It has been claimed (e.g., Shea, 1978, pp. 150-151) that this criticism is insubstantial, but we shall see below that Descartes actually pointed to a serious problem in Galileo's treatment of projectile motion. (See document 5.3.1)

Fig. 3.1

If indeed it should turn out that Descartes' criticism of Galileo's theory of motion contains more substance than historians of science are generally willing to admit, then their interpretation of Descartes' reaction to Galileo's *Discorsi* must be incorrect. Moreover, if a contemporary philosopher and scientist such as Descartes had valid reasons for claiming that Galileo did not present proofs for some of his theorems but rather a collection of problematical and partly incompatible results, then the modern interpretation of Galileo as a first representative of the ideal of classical mechanics is also called into question.

Now, it is generally accepted that Galileo's early unpublished writings had a preclassical conceptual background.[13] In the case of Descartes, we had to reconstruct such a conceptual background from his terminology and from the structure of his arguments. In the case of Galileo, however, we are no longer forced to infer the relation between the medieval tradition and his initial studies of motion from such indirect evidence, because it is well known that Galileo was indeed familiar with this tradition: his notebooks on scholastic philosophy amply document his broad familiarity with medieval thinking.[14]

Another group of manuscripts document Galileo's attempts at a systematic study of motion on medieval and antique foundations.[15] Among these manuscripts there is a complete treatise, usually called *De Motu*, composed around the time Galileo was a lecturer at the university of Pisa from 1589-1592.[16] Apart from the *Discorsi* written about 40 years later, this treatise is the only comprehensive work on motion by Galileo that has been preserved. It contains theorems and proofs on falling bodies, fall along inclined planes, and projectile motion.

Although the manuscript of this treatise is coherent and complete, Galileo already very early seems to have decided not to publish it. However, he must have attributed considerable importance to this manuscript; at least we know that he kept it with him over 40 years, presumably in a folder labeled "De motu antiquiora scripta mea."[17] The reason for setting the manuscript aside was – as will become evident in this chapter – the fact that in his subsequent work he developed insights incompatible with the theory of motion he had originally intended to publish in *De Motu*. Several reworkings including two complete versions of the work provide evidence for his ongoing attempts to revise the theory

---

[13] For instance Drake (1978, p. 31) maintains that, in his early unpublished writings on motion, Galileo had exhausted "the traditional preoccupation with the search for causes as such."

[14] Galileo's scholastic notebooks on physical questions were published in EN I and are now also available in a translation by Wallace (1977).

[15] According to Wallace (1984, p. 230), Galileo's earliest writings on motion were the direct continuation of his scholastic studies of logical and physical questions.

[16] *De Motu* was first published in EN I, 251-340; it will be cited in Drabkin's translation (Galileo 1960b). For a thorough philological study of the treatise, including earlier versions and later reworkings, see Fredette 1969, partly summarized in Fredette 1972.

[17] See Fredette 1972, p. 327.

before he finally renounced the intention to publish this treatise.[18] Nevertheless, the contents of the *De Motu* manuscript remained important for Galileo, as is confirmed, for instance, by a later comment on this manuscript by Viviani[19]:

> All the things worth noticing, that are scattered in this manuscript, have been introduced, by the author himself, in their proper places, in the works he has published.

Abundant evidence is available of intensive efforts by Galileo to study motion further during the years he spent as a professor of mathematics in Padua. However, in 1609 the invention of the telescope and, shortly later, his move to Florence redirected his interests to other subjects, in particular to astronomical matters. Besides occasional returns to the study of local motion and the inclusion of some of his results in his final work on astronomy, the *Dialogo* ("Dialogue Concerning the Two Chief World Systems"), which led to his condemnation, he returned to the preparation of a systematic treatise on motion about 1630. This work finally resulted in the publication of his *Discorsi* ("Discourse on Two New Sciences"), which established the basis of his reputation as the founder of classical mechanics.

Whatever the nature and extent of Galileo's transition from the medieval conceptual background of his early work on motion to fundamental theorems of classical mechanics in his published works, it should be evident in the differences between the treatment of topics like free fall and projectile motion in *De Motu* and the *Discorsi*. However, since Galileo collected and preserved his attempts to revise his initial understanding of motion on separate folio pages containing memoranda, diagrams, calculations, proofs, and occasionally even results of experiments, as well as early versions of texts which were finally published in the *Discorsi*, we know not only the starting point and the outcome of Galileo's efforts but also the transitional phases. These folio sheets, some 160 pages altogether, have been preserved and are accessible.[20] These immediate

---

[18] Earlier versions and later reworkings of *De Motu*, including a dialogue version, are published in EN I, 341-408. For an English translation of some of these, see Drake and Drabkin 1969, pp. 115-131. Fredette (1972, p. 321) claims that Galileo first wrote the dialogue version and then two complete treatise versions of *De Motu* before he abandoned it. Fredette's ordering of the manuscript material was challenged by Drake (1976b), but Drake's criticism was rejected by Galluzzi (1979, p. 168, note 58). Wallace (1984, p. 231) follows Fredette's ordering, while Drake further develops his argument in Drake 1986. The most recent account of this debate is Wallace 1990.

[19] This comment, dating from 1674, is quoted from Fredette 1972, p. 326.

[20] They are preserved today at the Biblioteca Nazionale in Florence; many are transcribed in EN VIII, and many are also reproduced in facsimile in Drake 1979. Here, the manuscripts will be referred to by their folio numbers as in MS and, if transcribed in EN, by the page number of the transcription. For a more detailed description of this collection of manuscripts, see also Renn 1988, and the references given there. Comprehensive studies of Galileo's science incorporating an analysis of his unpublished manuscripts are Caverni 1895, vol. IV, Wohlwill 1883-1884 and 1909, Dijksterhuis

manifestations of Galileo's thinking, which accompanied the ongoing work on motion leading from *De Motu* to the *Discorsi* and thus document the transition, offer an invaluable opportunity to study the cognitive background of the origins of classical mechanics.

By analyzing some of these manuscripts we shall see that Descartes' criticism, though failing to recognize the importance of Galileo's discovery of key theorems of classical mechanics, had a sound basis in the conceptual and deductive structure of Galileo's final theory. It will become clear that some of the contradictory implications of preclassical arguments, which remained implicit in the work of Beeckman and Descartes, became explicit in Galileo's work and motivated continuous improvements of his theory. Although these improvements did give birth to new theorems and new proofs, they also gave rise to new paradoxes, and the contradictions were never actually overcome. Hence, in Galileo's work key theorems of classical physics were achieved within the conceptual framework of the original *De Motu* theory, and thus to most of his contemporaries, his solutions to problems were qualitatively indistinguishable from the numerous other deficient attempts to go beyond the limits of preclassical mechanics and to create a new science of nature.

3.2 Free Fall and Projectile Motion in *De Motu*

In this section we shall examine Galileo's first systematic treatise on motion with a view to the two fundamental problems of mechanics, which will be thematized in this chapter: the fall of bodies and projectile motion. We shall see that, while he severely critizes Aristotle's natural philosophy, he basically remains within an Aristotelian framework. Like most of his contemporaries, he used the notions of impressed force and impetus when dealing with the forces of bodies in motion, conceiving force as the cause of motion and not of acceleration.

3.2.1 The Laws of the Motion of Fall

Among those engaged in the creation of a new science of nature in 16th century Italy, the works of Archimedes on statics and hydrostatics became accessible and popular, in particular through the efforts of the self-taught mathematician and engineer Niccolò Tartaglia. Attempts were made by Tartaglia, and other engi-

---

1924, Koyré 1939, Settle 1966, Clavelin 1968, Wisan 1974, Drake 1978, Galluzzi 1979, Wallace 1984, as well as the numerous papers by Naylor.

neers, and mathematicians of this period to extend the techniques of statics and hydrostatics into natural philosophy.[21]

We will show that Galileo, as Benedetti had done before him,[22] in his unpublished treatise *De Motu,* severely criticized Aristotle's treatment of natural and violent motion and took up Archimedean hydrostatics to develop an alternative theory of motion. According to Aristotle's physics, the motion of heavy bodies toward the center of the earth is a natural motion since it is a motion towards their natural place. Similarly, the upward motion of light bodies is a natural motion towards their natural place, the sphere of fire. The speed of fall of heavy bodies increases in proportion to their weight and decreases in proportion to the density of the medium in which they move.[23] And vice versa, the rarer the medium the faster the motion; hence it follows that motion in a void would have to be instantaneous, which is impossible.

Galileo's unpublished treatise *De Motu* is not the first work to contain arguments which show that Aristotle's treatment of falling bodies leads to paradoxes, and which suggest that motion in the void would actually be possible and not instantaneous.[24] Galileo's first argument refers to the dependence of the speed of a falling body on its weight, and is based on the plausible assumption that the combination of two bodies, which separately would fall with different speeds, moves with a speed that is intermediary between that of its two constituents, since the slower part retards the faster one, while the faster will accelerate the slower[25]:

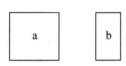

Fig. 3.2 (EN I, 265)

Suppose there are two bodies of the same material [Fig. 3.2], the larger *a*, and the smaller *b*, and suppose, if it is possible, as asserted by our opponent, that *a* moves (in natural motion) more swiftly than *b*. We have, then, two bodies of which one moves more swiftly. Therefore, according to our assumption, the combination of the two bodies will move more slowly than that part

---

[21] Some such attempts are documented in the English translations of 16th century works on motion and mechanics collected in Drake and Drabkin 1969. For a discussion of Tartaglia's role in making the works of Archimedes accessible, see the Introduction to this volume. The significance of Tartaglia's contribution, in particular, and that of the engineering tradition, in general, to early modern science is strongly emphasized in Olschki 1927. On their importance for Galileo's early science, see Drake 1986, pp. 438-439, and Settle 1987.

[22] See Benedetti 1585 (English translation in Drake and Drabkin 1969).

[23] This, at least, is the understanding of Aristotle's theory found in works of early modern science. For reconstructions of Aristotle's original thinking on falling bodies, see Young 1967 and Casper 1977; for a discussion of its later reception, see Sorabji 1988.

[24] For a discussion of a similar criticism of Aristotle by Benedetti and its relationship to Galileo's, see Drake 1986, pp. 438-439.

[25] DM, 29; EN I, 265.

which by itself moved more swiftly than the other. If, then, *a* and *b* are combined, the combination will move more slowly than *a* alone. But the combination of *a* and *b* is larger than *a* is alone. Therefore, contrary to the assertion of our opponents, the larger body will move more slowly than the smaller. But this would be self-contradictory.

According to Galileo's plausible assumption, the combined body should move with an intermediate speed, while according to Aristotle's rule it should move with a greater speed than any of its constituents. But, obviously, these two statements about one and the same body contradict each other.

In *De Motu*, not only Aristotle's claim that the speed of a falling body is proportional to its weight is refuted, but also his assumption of a proportionality between the speeds of a body and the rarenesses of those media in which it moves is shown to lead to a paradox when referred to a particular hydrostatic phenomenon[26]:

> For if the speeds have the same ratio as the [rarenesses of the] media, then, conversely, the [rarenesses of the] media will have the same ratio as the speeds. Hence, since wood falls in air but not in water, and, consequently, the speed in air has no ratio to the speed in water, it follows that the rareness of air will have no ratio to the rareness of water. What can be more absurd than this?

Although Galileo has thus explained why Aristotle's treatment of the motion of fall is inconsistent, he does not, however, draw the conclusion from this analysis that in a vacuum all bodies fall with the same speed. This is remarkable, since the first paradox, which derived from the dependency of the speed of a falling body on its weight presumed by Aristotle, can most easily be resolved if one concludes (as does classical mechanics) that the speed of fall does not depend on the body's weight. In *De Motu*, however, a different conclusion is drawn from this paradox, namely[27]

> ... that bodies of the same material but of unequal volume move [in natural motion] with the same speed.

The assumption that the speed of a falling body does not depend on its weight but only on its material indeed avoids the first paradox since it can no longer be inferred that the combination of the two falling bodies should fall faster than the single bodies.

But how can the second paradox, which results from relating the ratio of speeds of bodies falling in different media to the ratio of the densities of these media, be avoided? According to classical mechanics, the medium retards the motion of a falling body mainly by the effect of friction, which vanishes for fall in a vac-

---

[26] DM, 33; EN I, 269.
[27] DM, 29; EN I, 265.

uum, while in the case of fall in a medium no simple relationship exists between the speed of fall and the density of the medium.[28]

The resolution of the second paradox presented in *De Motu* relies on a very simple relationship between the weight of the falling body and the density of the respective medium, a relationship that is in fact incompatible with classical mechanics. According to *De Motu*, the speed of fall in a given medium depends on the difference between the weight of the body and the weight of an equal volume of the medium. This simple relationship, together with the conclusion that bodies of the same material fall with the same speed, is the basis of Galileo's theory of the motion of fall in *De Motu*.

It is no surprise that the presumed dependence of the speed of fall on the difference between the weight of the falling body and that of the medium is suitable to resolve the second paradox, since, according to Archimedean hydrostatics, this difference determines whether or not a given body will fall in a given medium. Hence, on the basis of this relationship the conflict, described by the second paradox, between the densities of different media and the respective velocities of fall can no longer arise.

Galileo's resolution of the second paradox is thus not an ad hoc assumption but a consequence of Archimedean hydrostatics. In fact, his resolution of the first paradox, too, the assumption that bodies of different sizes but of the same material fall with the same speed, appears as a plausible consequence of a hydrostatic theory of fall, since in hydrostatics it is not the absolute but the specific weight of a body, i.e., a property of its material, that accounts for its behavior in a medium. Thus the basic relationships between the speed of fall and the weights of the falling body and of the medium in *De Motu* follow directly from Archimedean hydrostatics.

In the case of the same body moving in different media, the following law is obtained[29]:

> For, clearly, in the case of the same body falling in different media, the ratio of the speeds of the motions is the same as the ratio of the amounts by which the weight of the body exceeds the weights [of equal volumes] of the respective media.

As a consequence of this law, fall in a vacuum is (contrary to Aristotle) possible and takes a finite time[30]:

> Therefore, the body will move in a void in the same way as in a plenum. For in a plenum the speed of motion of a body depends on the difference between its weight and the weight of the medium through which it moves. And likewise in a void [the speed of] its motion will depend on the difference between its own weight and that of the medium. But since the latter is zero, the difference

---

[28] For a study of the motion of fall in air according to classical mechanics, including a comparison with Galileo's analysis, see Coulter and Adler 1979.

[29] DM, 36; EN I, 272.

[30] DM, 45; EN I, 281.

between the weight of the body and the weight of the void will be the whole weight of the body.

In the case of different bodies of equal size moving in one and the same medium, Galileo states the following law[31]:

> And similarly we have a clear answer to our second problem – to find the ratio of the speeds of bodies equal in size, but unequal in weight, moving (with natural motion) in the same medium. For the speeds of these bodies have the same ratio as do the amounts by which the weights of the bodies exceed the weight of the medium.

Galileo claims that various other cases can be derived from these two laws.[32]

Following a 16th century practice of using Archimedean insights to solve problems of natural philosophy, Galileo thus achieved a systematic and deductive theory of the motion of fall in different media, a theory that is, however, obviously incompatible with classical mechanics. But how was he able to deal with problems of motion by using concepts from hydrostatics? The answer is that he, like his contemporaries, interpreted concepts of statics within the framework of Aristotelian notions such as that of "heaviness." At the beginning of *De Motu*, in agreement with Aristotelian natural philosophy, Galileo explains the foundational idea that the speed of the motion of fall is due to heaviness[33]:

> And since motion proceeds from heaviness and lightness, speed or slowness must necessarily proceed from the same source. That is, from the greater heaviness of the moving body there results a greater speed of the motion, namely downward motion, which comes about from the heaviness of that body; and from lesser heaviness (of the body), a slowness of that same motion.

Here, heaviness and lightness, as well as speed and slowness, are introduced as contrary qualities in the sense of the traditional logic of contraries. The downward motion of a heavy body is for Galileo, just as for Aristotle, the natural motion to its natural place. The doctrine of proper or natural places is discussed at the beginning of *De Motu* in a chapter with the title, "That Heavy Substances are by Nature Located in a Lower Place, and Light Substances in a Higher Place, and Why."[34] This chapter is followed by one in which Galileo explains that "natural motion is caused by heaviness or lightness."[35]

Also in agreement with traditional natural philosophy motion is conceived in *De Motu* as being due to a force which generates a velocity. This relationship between force and speed can be seen, for instance, in the smooth transition from

---

[31] DM, 37; EN I, 272.

[32] See, e.g., DM, 31; EN I, 267. The systematic treatment of all other cases actually presupposes a conceptualization of the notion of specific weight in terms of compounded proportions that was not yet available to Galileo at the time he composed *De Motu*. For a comprehensive study of the development of this notion in Galileo's later work and the role played by compounded proportions in it, see Napolitani 1988.

[33] DM, 25; EN I, 261.

[34] DM, 14; EN I, 252.

[35] DM, 16; EN I, 253.

the *force* driving an immersed body to the *speed* of its motion, a transition which, as we have seen, is crucial for Galileo's theory of motion[36]:

> ... when solids lighter than water are completely immersed in water, they are carried upward with a force measured by the difference between the weight of a volume of water equal to the volume of the submerged body and the weight of the body itself.[...].

> It is clear, then, that in all cases the speeds of upward motion are to each other as the excess of weight of one medium over the weight of the moving body is to the excess of weight of the other medium over the weight of the body.

Finally, the concept of velocity used in *De Motu* is clearly the Aristotelian concept of a distance traversed in a given time, as is confirmed, for instance, by Galileo's discussion of Aristotle's argument on the role of the medium in the motion of fall. In this discussion, Galileo makes implicit use of the Aristotelian proposition that the quicker of two bodies traverses the same space in less time in order to pass from a ratio of velocities to a ratio of times[37]:

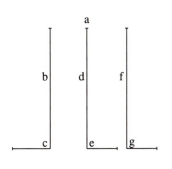

Fig. 3.3 (EN I, 277)

Thus, Aristotle's first assumption, when he saw that the same body moved more swiftly through the rarer than through the denser medium, was this: that the ratio of the speed of motion in one medium to the speed in the second medium is equal to the ratio of the rareness of the first medium to the rareness of the second. He then reasoned as follows. Suppose body *a* [Fig. 3.3] traverses medium *b* in time *c*, and that it traverses a medium rarer than *b*, namely *d*, in time *e*. Clearly, the ratio of time *c* to time *e* is equal to the ratio of the density of *b* to the density of *d*.

Although the diagram indeed suggests that the given spaces traversed by the two motions are equal, Galileo does not explicitly mention this condition, and simply treats velocity and time as corresponding to the contrary qualities rarity and density. Shortly after *De Motu,* in his treatise *Le Meccaniche*, which deals with the principles and uses of mechanical instruments, Galileo explicitly defines that motion as being "speedier than another which passes an equal distance in less time."[38]

Summing up, we have shown that the criticism of Aristotle as well as the Archimedean theory of fall contained in *De Motu* fundamentally depend on Aristotelian notions. The manuscripts documenting Galileo's reworking of parts of *De Motu* are indicative of his discovery of puzzles and inconsistencies within

[36] DM, 33; EN I, 269; and DM, 34-35; EN I, 270.

[37] DM, 41; EN I, 277.

[38] DM, 148; EN II, 156.

the theory of motion presented in this treatise.[39] Perhaps the most troubling puzzle concerns the laws of motion of fall themselves as they were derived in *De Motu*. At the end of the careful and painstaking arguments by which Galileo obtained these laws, he remarked briefly that they do not actually correspond to what is observed in nature[40]:

> ... those ratios will not be observable by one who makes the experiment. For if one takes two different bodies, which have such properties that the first should fall twice as fast as the second, and if one lets them fall from a tower, the first will not reach the ground appreciably faster or twice as fast.

The observation reported in this passage refers to what is a characteristic property of the motion of fall according to classical mechanics: the equality of the speeds of fall for different bodies. Others had made this observation before Galileo. But it led neither them nor, as we have seen, Galileo to the insight into this fundamental property of the motion of fall. In the next section, we will see that he was rather referring to the fact that the actual motion of fall is accelerated when he explained the discrepancy between his laws of motion and observation.

3.2.2   Acceleration as an Accidental Characteristic of Fall

In the rules on the motion of falling bodies presented in the preceding section, Galileo deals only with the global, extensive aspects of the motion. In his analysis the speed of the motion of fall is the speed of the overall motion, i.e., distance traversed by a body in an actually given time. After having thus determined the extensive characteristics of the motion of fall, he undertakes an analysis of its internal characteristics in a later chapter of *De Motu*. In agreement with antique and medieval natural philosophers Galileo characterizes the motion of fall as an accelerated motion. In *De Motu*, Galileo introduces his discussion of the

---

[39] For the evidence of these reworkings, see note 18. Several commentators on *De Motu* have pointed out the existence of internal contradictions in Galileo's treatise. Besides Fredette 1969, the most comprehensive study of *De Motu*, including an analysis of such contradictions, is Galluzzi 1979, pp. 166-197. Among the puzzles of Galileo's early theory of motion that become evident from his reworkings of *De Motu*, is the problem whether or not the upward motion of a light body can properly be classified as a natural motion since its "Archimedean" explanation by extrusion brings an external force into the play and hence is in conflict with the Aristotelian understanding of natural motion as an intrinsic property of the moving body. Another puzzle arises from Galileo's use of statics in his theory of motions along inclined planes in *De Motu*, which suggests a dependence of the speed of motion on the weight of the moving body, contrary to Galileo's conclusions discussed in the present section.

[40] DM, 37-38; EN I, 273. For a discussion of Galileo's early experimentation and its historical context, see Settle 1983.

acceleration of a body falling from the state of rest by stating that "the speed of natural motion is increased toward the end."[41]

On the basis of the relationship between the weight of a falling body and its speed, established in the first part of his analysis of natural downward motion and implying, for a constant weight, a uniform motion of the falling body, Galileo accounts for the actually changing speed of the falling body by a change of its weight[42]:

> Since, then, a heavy falling body moves more slowly at the beginning, it follows that the body is less heavy at the beginning of its motion than in the middle or at the end.

In order to establish the reason for this diminution of weight in the beginning of natural motion, Galileo first excludes the possibility of a loss of natural weight in the falling body[43]:

> But the natural and intrinsic weight of the body has surely not been diminished, since neither its volume nor its density has been diminished. We are left with the conclusion that that diminution of weight is contrary to nature and accidental. If, then, we have found that the weight of the body is diminished unnaturally and from without, we will then surely have found what we seek.

If it is not the natural weight that is diminished in the falling body, the cause of the loss of weight must be an external one, but this external loss of weight cannot be due to the medium[44]:

> But that weight is not diminished by the weight of the medium, for the medium is the same in the beginning of the motion as in the middle. The conclusion remains that the weight of the body is diminished by some external force coming to it from without – for only in these two ways does it happen that the body becomes accidentally light.

Galileo has now specified the cause of acceleration in natural fall in a way that suggests an analogy with the cause of projectile motion. In fact, the traditional problem of explaining projectile motion consisted precisely in explaining the continuation of a non-natural motion in the absence of a visible external mover. Galileo rejected a possible role of the medium as the cause of acceleration, just as the original Aristotelian explanation of projectile motion by the medium was generally rejected in preclassical mechanics and replaced by an explanation in terms of an impressed force. In *De Motu*, Galileo adapts this explanation using it to account for the acceleration of the motion of fall[45]:

---

[41] DM, 85; EN I, 315.

[42] DM, 88; EN I, 318.

[43] DM, 88; EN I, 318.

[44] DM, 88; EN I, 318.

[45] DM, 88-89; EN I, 318. Galileo believed his explanation of acceleration to be similar to that of Hipparchus, possibly on the basis of a study of *reportationes* from the Collegio Romano; see Wallace 1984, p. 244. For a discussion of the relationship

If, then, we find out how a body can be lightened by an external force, we will have, in that case, discovered the cause of the slowness. Now the force impressed by a projector not only at times diminishes the weight of a heavy body, but often even renders it so light that it flies up with great speed. Let us see, therefore, and carefully investigate whether possibly that very force is the cause of the diminishing of the weight of the body at the beginning of its motion. And in fact it definitely is that force impressed by a projector that makes natural motion slower at the beginning.

In agreement with his explanation of motion by heaviness, Galileo conceives the force impressed by the mover on a heavy body as a taking away of heaviness. Following a longstanding tradition, he compares the impressed force to the heat that fire communicates to iron and, by this analogy, explains its permanence in the projectile after its separation from the mover as well as its gradual disappearance from the moving body[46]:

But now, in order to explain our own view, let us first ask what is that motive force which is impressed by the projector upon the projectile. Our answer, then, is that it is a taking away of heaviness, when the body is hurled upward, and a taking away of lightness, when the body is hurled downward. But if a person is not surprised that fire can deprive iron of cold by introducing heat, he will not be surprised that the projector can, by hurling a heavy body upward, deprive it of heaviness and render it light.

The body, then, is moved upward by the projector so long as it is in his hand and is deprived of its weight; in the same way the iron is moved, in an alterative motion, towards heat, so long as the iron is in the fire and is deprived by it of its coldness. Motive force, that is to say lightness, is preserved in the stone, when the mover is no longer in contact; heat is preserved in the iron after the iron is removed from the fire. The impressed force gradually diminishes in the projectile when it is no longer in contact with the projector; the heat diminishes in the iron, when the fire is not present. The stone finally comes to rest; the iron similarly returns to its natural coldness.

From the comparison of the impressed force with heat it follows that vertical projection and subsequent fall are a single continuous motion[47]:

For the motion in which the stone changes from accidental lightness [i.e., upward motion] to heaviness [i.e., downward motion] is one and continuous, as when the iron moves [i.e., changes] from heat to coldness.

At the turning point of the motion, the impressed force impelling the body upward and the weight of the projectile are in equilibrium. The ensuing motion

---

between Galileo's explanation and its antique predecessors, see Drake 1989a; for a reconstruction of Hipparchus's theory of motion, see Wolff 1988.

[46] DM, 78-79. According to Duhem (1906-1913, Vol. 1, p. 110) it was common, in the Middle Ages as well as in early modern times, to conceive the impressed force generating the motion as a quality comparable to heat or cold and to make assertions about its varying degrees. For a discussion of the antique and medieval background of Galileo's understanding of impressed forces, see Moody 1951.

[47] DM, 99; EN I, 327.

of fall corresponds exactly to a motion of fall from the state of rest, at which a similar equilibrium holds between the weight of the body and the force necessary to prevent it from falling[48]:

> Thus, when a stone, which had been thrown up, begins to move down from that extreme point at which equilibrium occurs between impelling force and resisting weight (i.e., from rest), it begins to fall. This fall is the same as if the stone dropped from someone's hand. For even when no force impelling the stone upward has been impressed on it, and it falls from the hand, it leaves with an [upward] force impressed on it equal to its own weight.

The gradual disappearance of the force that has to be impressed on a heavy body in order to keep it from falling explains its acceleration during the motion of fall. Galileo illustrates his conclusion by example of a body c with a weight of 4 units[49]:

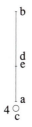

> Hence [Fig. 3.4] when c is at rest, there will be on it an impressed force of 4 tending to drive it upward. And if it is released by the agency that is impressing this force on it, it will begin to fall back possessing an impressed force of 4. Therefore, it ... will fall very slowly at the beginning and then move downward more swiftly according as the force opposing that motion becomes weaker.

Fig. 3.4 (EN I, 321)

Acceleration will stop once the inherent quality impelling the body upward and counteracting its weight ceases to exist[50]:

> For since the [motion of the] body is accelerated because the contrary [i.e., upward] force is continuously diminishing while [in consequence] the natural weight is being attained, it will stand to reason that the whole contrary force will finally be lost and the natural weight resumed, and, therefore, that acceleration will cease since its cause has been removed.

Galileo's causal explanation of acceleration by the difference of a constant force – the body's weight – and the gradually diminishing impressed force implies in fact the existence of an upper limit to the velocity of a falling body which corresponds to the velocity determined by its (specific) weight according to the laws of fall discussed in the previous section.

In order to explain the observation that light and heavy bodies fall in approximately the same time – in spite of the laws of fall derived above –, Galileo claims that lighter bodies move more swiftly at the beginning of their

---

[48] DM, 91; EN I, 320.
[49] DM, 92-93; EN I, 322.
[50] DM, 100; EN I, 328-329.

motion of fall than heavier ones. He explains how this behavior follows from a general property of contrary qualities[51]:

> all contrary qualities are preserved longer, the heavier, denser, and more opposed to these qualities is the material on which they have been impressed.

The motion impressed on a projectile or the heat communicated to iron provide examples of such contrary qualities[52]:

> Motion is more strongly impressed by the same given force in a body that is more resistant than in one that is less resistant, e.g., in the stone, more than in light pumice; and similarly, heat is more strongly impressed by the same fire upon very hard, cold iron, than upon weak and less cold wood.

The motion of fall is thus different in character for bodies of different material because they react in a different way to the moving forces impressed upon them. Galileo illustrates this behavior with an example[53]:

> Suppose that there are two bodies [Fig. 3.5], equal in size, one of wood, the other of lead, that the weight of the lead is 20, and of the wood 4, and that both are held up by the line *ab* ... But when they are released by the [removal of the] line, they still retain, at the first point of their departure, an impressed contrary quality that impels them upward; and this quality is lost not instantaneously but gradually. The lead has 20 units of this contrary quality to be used up, and the wood 4. Now if this quality were uniformly used up in each body, so that, when one unit of the quality in the lead departed, one unit in the wood did likewise, and, if, as a consequence, both of them had recovered one unit of weight, then, no doubt, both of them would move with equal speed.

Fig. 3.5 (EN I, 336)

But because the behavior of the impressed force depends on the body's material, the motion of fall is actually not uniform for bodies of different materials[54]:

> But in the time in which one unit of the quality departs from the lead, more than one unit has left the wood; and, as a consequence, while the lead has recovered only one unit of weight, the wood has recovered more than one. It is because of this that the wood moves more swiftly during that time. Again, in the time in which two units of the [contrary] quality depart from the wood, less than two depart from the lead. And it is because of this that the lead moves more slowly during that time. On the other hand, because the lead finally reacquires more weight than the wood, it follows that by that time the lead is moving much more swiftly.

---

[51] DM, 108; EN I, 335.
[52] DM, 79; EN I, 310.
[53] DM, 109-110; EN I, 336-337.
[54] DM, 110; EN I, 337.

In this passage, Galileo describes the acceleration of a falling body in terms of the units of the impressed quality that the body looses in a given time or, equivalently, in terms of the units of weight that it recovers in this time.[55]

Galileo explains that, due to the acceleration of the motion of fall, the speeds of falling bodies of different materials will not be observed as determined in the first part of *De Motu*. According to his determination of the ratios of speeds of falling bodies of the same size but consisting of different materials, the heavier body should have a speed greater than that of the lighter body in proportion to its greater weight; but[56]

> ... those ratios will not be observable by one who makes the experiment. For if one takes two different bodies, which have such properties that the first should fall twice as fast as the second, and if one lets them fall from a tower, the first will not reach the ground appreciably faster or twice as fast.

Galileo's explanation for this approximate equality of the speeds of falling bodies consisting of different materials is the different patterns of acceleration characterizing the light and the heavy body[57]:

> From what has thus far been written, anyone can readily discover the reason why heavy bodies do not, in their natural motions, adhere to the ratios which we assigned to them when we discussed the matter, i.e., the ratios of the weights of the bodies in the medium through which they are moving. For, since at the beginning of their motion, they do not move in accordance with their weights, being impeded by the contrary force, it will certainly not be surprising that the speeds do not adhere to the ratios of the weights.

Although acceleration thus provides an explanation for the discrepancy between his laws of motion and observation, Galileo does not include a law of acceleration in his early theory of motion, since he conceives acceleration as a merely accidental property of the motion of fall[58]:

> Rules cannot be given for these accidental factors since they can occur in countless ways.

In this way the discrepancy between the laws of fall and observation remains a puzzle for the *De Motu* theory.

---

[55] Galileo's description of the fall of different bodies in a medium is examined from the point of view of classical mechanics in Coulter and Adler 1979 and from the point of view of a 16th century experimentalist tradition in Settle 1983. Settle and Miklich (see Settle 1983) have repeated Galileo's experiment, confirming his description, and provide an explanation for the paradoxical results based on psychological and physiological differences in perception of simultaneity when dropping heavy and light bodies.

[56] DM, 37-38; EN I, 273.

[57] DM, 105; EN I, 333.

[58] DM, 69; EN I, 302. This statement refers specifically to motion along inclined planes.

3.2.3 The Trajectory of Projectile Motion

Galileo's analysis of oblique projection is contained in the last chapter of *De Motu*,[59] which bears the title:

Why Objects Projected by the Same Force Move Farther on a Straight Line the Less Acute are the Angles they Make with the Plane of the *Horizon*.

The way this question is posed presupposes that an obliquely projected body first follows a straight line before it starts to turn downwards, a characterization which corresponds to the diagram accompanying Galileo's text (see Fig. 3.6).It is furthermore presupposed as generally accepted that this first straight part of the trajectory is longer the closer the direction of the shot is to the vertical. Galileo does not claim that the shape of the trajectory is symmetrical. This brief characterization of the geometrical properties of the trajectory, to which little is added in the course of the chapter, corresponds to the generally accepted views on projectile motion dominated at the time by the ideas of Niccolò Tartaglia.

In 1537, Tartaglia had published the first comprehensive and systematic treatise on what he conceived as the new science of artillery.[60] In this treatise, he assumed that the trajectory consists of a straight part that is followed by a section of a circle and then ending in a straight vertical line (see Fig. 3.7).

This form of the trajectory roughly corresponds to Tartaglia's dynamical ideas because the first straight part of the trajectory reflects the initially dominant role of the violent motion, whereas the last straight part is in accord with the eventual dominance of the body's weight over the violent motion, which follows from a logic of contraries, as well as from the body's tendency to reach the center of the earth. The simple geometrical shape of Tartaglia's trajectory immediately allowed him to draw a number of conclusions about projectile motion by geometrical reasoning. The fact that Tartaglia's construction of the trajectory was particularly suited for drawing such conclusions was among the most important reasons why this construction was so influential throughout the 16th century, although it only roughly corresponds to the visual impression of the motion of projectiles and certainly could not be justified by precise observations of their trajectory.[61]

---

[59] DM, 110-114; EN I, 337-340.

[60] Tartaglia 1537, for an English translation, see Drake and Drabkin 1969. A later treatment by Tartaglia of projectile motion is found in Tartaglia 1546. For historical surveys of projectile motion in the time between Tartaglia and Galileo, see Hall 1952 and Barbin and Cholière 1987.

[61] Tartaglia's trajectory also became the basis for measuring instruments used by gunners; see Wunderlich 1977. Modern interpretations of the history of ballistics often overlook this practical context and judge progress in this field exclusively by the similarity of the trajectories found in the historical sources with that countenanced by classical mechanics; see, e.g., Barbin and Cholière 1987, pp. 63-66. As Tartaglia's explanation (Tartaglia 1564) and the description of the trajectory in Cardano (1550, Liber primus, p. 394) show, early modern engineers and natural philoso-

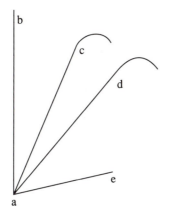

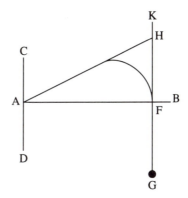

Fig. 3.6 (EN I, 340)[62]   Fig. 3.7 (Drake and Drabkin 1969, p. 87)

Simple geometrical assumptions about the shape of the trajectory were not the only tools that were available in the 16th century for a mathematical treatment of projectile motion. Concepts and methods from statics were also applied to ballistical problems. Tartaglia, for instance, compared the cannon to a balance, and attempted, in this way, to provide answers to such questions as the one stated in the title (cited above) of Galileo's last chapter of *De Motu*. But neither the use of geometry nor of statics allowed Tartaglia or any of his followers to clarify the problem of composition of motions, which was treated on the basis of a logic of contraries, or to derive the shape of the trajectory. Galileo provides two solutions to the question posed in the title. Both solutions refer to the assumption that the first part of the trajectory is a straight line, and both make use of a logic of contraries applied to the impressed force causing the motion of the projectile. His first explanation closely resembles an earlier attempt by Benedetti to solve a similar problem; it is based on an application of the statics of the inclined plane to the cannon. Hence, the treatment of projectile motion in *De Motu* makes use of the same conceptual tools employed by Tartaglia and his followers.[63]

---

phers were better aware of the actual shape of the trajectory than the widespread use of Tartaglia's construction of the trajectory, taken by itself, would suggest.

[62] Naylor (1974c, pp. 325ff) reports that the transcriptions of the diagram in EN I and DM do not correspond to the original *De Motu* manuscript; our diagram has been altered accordingly.

[63] See Tartaglia 1564. Benedetti criticized Tartaglia's approach but used the same conceptual tools as Tartaglia in his own treatment of problems of projectile motion. Trying to avoid problematical consequences resulting from Tartaglia's comparison of the cannon with a balance, he compared it instead with an inclined plane (see Drake and Drabkin 1969, pp. 224ff).

Galileo's first solution presupposes that the projectile offers more resistance to the same moving force, if it is shot by a more steeply inclined cannon. From this presupposition and from the principle "that all contrary qualities are preserved longer, the heavier, denser and more opposed to these qualities is the material on which they have been impressed," he draws the conclusion that the projectile shot by a more steeply inclined cannon receives more of the moving force than if it had been shot by a less steeply inclined cannon:[64]

> First I say that though the motive force remains the same in the cannon, yet more of the force is impressed on the iron ball, the more erect the cannon stands. And the reason for this is that the ball then resists the force more.

The presupposition that the projectile encounters more resistance the more erect the cannon stands is a direct consequence of the theory of motion along inclined planes, presented earlier in *De Motu*. According to this theory as well as to traditional statics, the resistance of a body to movement up a weakly inclined plane is weak; and resistance is stronger along a more strongly inclined plane. If the cannon is conceived as an inclined plane along which the cannon ball is projected, it follows that the resistance of the ball is greater the more inclined the cannon is with respect to the horizontal.

Galileo's first explanation refers exclusively to one component of the projectile motion, accounting for the variation of the strength of impressed force at different angles of projection; it does not take into account the simultaneous effect of the two forces responsible for the motion of the projectile. His second explanation, on the other hand, follows from an analysis of the compatibility of the terminal points of the two motions to be compounded.

According to Aristotle's physics, two motions are direct contraries if the *terminus a quo* of one motion is the *terminus ad quem* of the other, as is the case, for instance, for upward and downward motion which, for this reason, cannot be compounded. The problem is more difficult for two motions that are not contraries in this exclusive sense, such as the two motions to be compounded in projectile motion. In his second explanation, Galileo makes use of the traditional logic of contraries to argue that, in this more complicated case, the terminal points of the motions may be more or less compatible with one another depending on their directions. As a consequence, in the case of nonvertical projection, in which the two motions are less opposed, the downturn starts earlier than in the case of vertical projection, in which the two motions are directly contrary[65]:

> But if the motion is along the perpendicular *ab*, the body is in no way able to turn down away from that path, unless, by moving back along the same line, it moves toward the terminus from which it originally started.

---

[64] DM, 108; EN I, 335; DM, 112; EN I, 338.
[65] DM, 113; EN I, 339.

Whereas vertical projection is one extreme case in which the downturn cannot begin until the motion carrying the projectile away from the starting point has completely ceased, horizontal projection represents the other extreme case in which the downturn can begin immediately because it does not interfere at all with the motion increasing the distance from the starting point[66]:

> But when the body moves along *ae*, which is almost parallel to the horizon, the body can begin to turn downward almost immediately. For this turning down does not interfere with distance [measured along the ground] from the starting point.

What does Galileo's second explanation, focusing on the terminal points of the two motions to be compounded, state about the moving forces? The Aristotelian characterization of contrary motions by their terminal points, implies, as we have seen, that upward and downward motion are directly opposed to one another. If this characterization is made the basis of an analysis of projectile motion, it follows that the moving forces, too, must be directly contrary in this particular case, and hence, in upward projection, the moving force must have completely disappeared before the downturn can begin[67]:

> When a ball is sent up perpendicularly to the horizon, it cannot turn from that course and make its way back over the same straight line, as it must, unless the quality that impels it upward has first disappeared entirely.

But this conclusion bluntly contradicts Galileo's earlier statements about upward projection, according to which, at the turning point of the motion, an equilibrium between the weight of the body and the impressed force holds. While the analysis of oblique projection is based on conceiving upward and downward motion as direct contraries, Galileo had, in his explanation of acceleration, denied this property[68]:

> So far, then, are these motions from being contraries, that they are actually only one, continuous, and coterminous.

The use of the logic of contraries in the second part of Galileo's analysis of projectile motion is hence incompatible with his explanation of acceleration.

3.2.4 Summary

In *De Motu*, Galileo develops an analysis of the motion of fall according to which, contrary to Aristotle, bodies of the same material fall with the same speed whatever their size, and motion in a void is possible. Like other contem-

---

[66] DM, 114; EN I, 340.

[67] DM, 113; EN I, 339.

[68] DM, 94; EN I, 323. This contradiction was noticed by Drabkin, see DM, 113, note 7.

porary treatments of problems of motion, Galileo's theory makes use of arguments taken from Archimedean statics and hydrostatics. According to the laws of motion presented in *De Motu*, which are incompatible with classical mechanics, the speeds of bodies of different materials but of equal size moving in the same medium are proportional to the differences between their weights and the weights of equal volumes of the medium, if their acceleration is not taken into account. In spite of his criticism of Aristotle, Galileo conceives heaviness as the cause of speed, and speed in the Aristotelian sense as a distance traversed in a certain time.

While *De Motu* contains simple laws for the speeds of fall of bodies in media, no such simple rules are provided for the acceleration of the motion of fall. In natural motion acceleration is conceived as an accidental phenomenon by which discrepancies between the observed behavior of bodies and Galileo's laws of motion are explained. As in classical mechanics, acceleration is conceived as continuous, but contrary to classical mechanics it is accounted for by the difference between the falling body's weight and a gradually decreasing impressed force. As a consequence of this explanation, there exists a limit velocity which the falling body cannot surpass. The composition of the impressed force and the natural tendency of a heavy body to move downward is treated by a logic of contraries, from which it is inferred that rarer bodies accelerate more quickly than denser bodies but do not accelerate as long or as much. In this way Galileo attempts to explain the observation that heavy and light bodies tend to reach the ground at approximately the same time – at least for short distances of fall. He conceives acceleration in terms of the units of the impressed quality contrary to weight which the falling body loses in a given time, but he does not enter into a more detailed quantitative analysis of the acceleration actually occuring in falling bodies.

Galileo's treatment of projectile motion implies that in the case of horizontal projection the two motions of the projectile do not disturb one another, a conclusion that is compatible with classical mechanics. But just like the treatments of projectile motion by many Aristotelians of the time, Galileo's analysis, too, is based on explaining the violent motion of a projectile by an impressed force. In essential agreement with Tartaglia, he describes the trajectory of oblique projection as a line consisting of a straight part along the direction of the shot followed by a curved part downward.

Galileo claims that the straight part is longer the closer the direction of the shot is to the vertical. He attempts to explain this property of projectile motions by applying the statics of the inclined plane to the cannon and the logic of contraries to the impressed force of the projectile. In order to deal with the composition of the two moving causes of the projectile – the impressed force and the natural tendency downward – he characterizes their compatibility by the compatibility of the respective goals of the moving causes as given by the terminal points of the two motions. It follows from this characterization that the moving causes are mutually exclusive in the case of vertical projection in contradiction to Galileo's explanation of acceleration in the motion of fall by a mixture of two causes.

Galileo's early treatise *De Motu* is thus seen to be still rooted in a medieval conceptual framework, employing basic notions of Aristotle as well as of his medieval and early modern successors. Several puzzles and inconsistencies in its argumentation gave Galileo sufficient cause not to publish it.

3.3   Free Fall and Projectile Motion in Galileo's Manuscripts between ca. 1600 and ca. 1604

There is general agreement among scholars that by 1604 Galileo, who was then a professor at the University of Padua, had arrived at the times squared law of falling bodies. It is in fact explicitly formulated in a letter to Paolo Sarpi, dated October 16 of that year.[69] Based on this letter along with other indications, a number of manuscripts dealing with attempts to prove this relationship can be dated to approximately this period. There is strong evidence that in the course of experimentation guided by the *De Motu* theory, probably around 1601-1602, Galileo was confronted with results that indicated the times squared law of free fall and the parabolic trajectory of projectile motion. Such results are incompatible with the conception of acceleration explicated in *De Motu*, for acceleration now appears as an essential property of the motion of fall and no longer as a transitory phenomenon. In this section we shall examine some manuscripts from this period which document Galileos attempt to provide a theoretical explanation of these results. We shall see that Galileo, in his attempt to integrate these results into his theory of motion, shifted the focus of attention to the conceptualization of acceleration, applying medieval techniques which he had studied a decade earlier. In the course of his studies, he derived the conservation of horizontal motion and the parabolic trajectory for the case of horizontal projection on the background of the *De Motu* theory. He was, however, unsuccessful in *proving* the time squared relation of free fall itself. We shall examine in detail his attempts to prove the law of fall from the erroneous assumption that velocities increase in proportion to spaces traversed.

3.3.1   The Discovery of the Form of the Trajectory and of the Characteristics of Acceleration in Free Fall

From contemporary and later correspondence it is clear that Galileo performed experiments in connection with some of the problems studied in *De Motu*. In a

---

[69] EN X, 115-116. (See document 5.3.3.)

letter to Guidobaldo del Monte of November 29, 1602,[70] Galileo discussed experiments involving pendulums and motion along curved surfaces and related them to a theorem on motion along inclined planes which directly follows from his theory in *De Motu*, the Isochronism of Chords.[71] A manuscript page, fol. 107v, that can be dated approximately to the same period preserves numbers which Galileo probably obtained from an experiment measuring the distances traversed in equal times along an inclined plane.[72] On the same manuscript page Galileo confronted the successive distances with the sequence of square numbers starting from unity. The fact that he also entered several other simple sequences of integers in this manuscript suggests that he initially took a number of simple rules for the acceleration along the plane into consideration. On the other side of the folio page (fol. 107r; see Fig. 3.8) a chain line is related to various sequences of numbers: a sequence of squares, a sequence corresponding to the successive sums of the numbers from one, and a sequence of cubes.[73]

There also exists a manuscript page written by Guidobaldo del Monte probably in the same period, in which he describes an experiment on projectile motion tracing the trajectory of a ball on an inclined plane. From later correspondence as well as from a description of the same experiment in the *Discorsi*, it can be concluded that Galileo was familiar with this experiment. We shall therefore present the description Guidobaldo gives of his experiment as well as the conclusions he drew from it[74]:

---

[70] EN X, 97-100.

[71] Galileo's derivation of the Isochronism of Chords within the conceptual framework of the *De Motu* theory is documented by several of his manuscripts, see, e.g., MS, fol. 151r (EN VIII, 378 and document 5.3.2). The relevant manuscript evidence is extensively documented and analyzed in Wisan 1974, pp. 162-171. For an analysis of the conceptual background of Galileo's proof, see Galluzzi 1979, pp. 266-267, Souffrin 1988, and Souffrin and Gautero 1989.

[72] MS, fol. 107; both sides of this folio contain diagrams, numbers, and calculations but no text. For a plausible reconstruction of Galileo's experiment see Drake 1978, pp. 86-90. Drake's reconstruction was first published in 1974 and followed an earlier, successful attempt to reconstruct Galileo's inclined plane experiment by Settle (an attempt which was not, however, based on manuscript evidence); see Settle 1961. The reconstruction of the fol. 107v experiment, as well as its role for the development for Galileo's understanding of motion, has been extensively discussed by historians. Drake's reconstruction has been challenged by Naylor who reconstructs the figures in this manuscript on the basis of an experiment on projectile motion; see Naylor 1977a, pp. 373-377, and Naylor 1980a, p. 554.

[73] For the reconstruction of Galileo's diagram as a catenary, see Naylor 1980a, p. 554.

[74] del Monte (ms). The manuscript is quoted in Libri 1838-1841, Vol. IV (1841), pp. 397-398. It is now preserved in the Bibliothèque Nationale, Paris (supplément latin 10246, p. 236). We should like to thank Gad Freudenthal for arranging the reproduction of the manuscript and updating Libri's reference. Guidobaldo's manuscript was discussed by Fredette (1969, pp. 154-155), who quotes most of this section of it. An English translation was made by Naylor (1974c, p. 327). See document 5.3.4 for transcription and translation of the text. Guidobaldo's experiment is

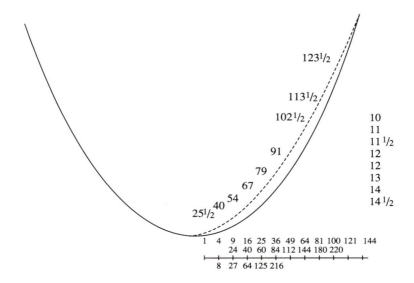

Fig. 3.8 (fol. 107r)

Guidobaldo's manuscript starts by describing some observations on the trajectory of projectile motion, without any reference to the particular experiment in question[75]:

> If one throws a ball with a catapult or with artillery or by hand or by some other instrument above the horizontal line, it will take the same path in falling as in rising, and the shape is that which, when inverted under the horizon, a rope makes which is not pulled, both being composed of the

discussed extensively in Fredette 1969 and Naylor 1974c. Galileo describes the experiment in the *Discorsi* (EN VIII, 185f/142). Fredette (1969, pp. 148-163) provides a convincing argument that Galileo must have been familiar with Guidobaldo's experiment before 1601. Here are the main points of his argument: In a letter to Cesare Marsili of Sept. 11, 1632 (EN XIV, 386), Galileo claims to have found the shape of the trajectory 40 years earlier, and also, that this discovery had been a primary goal of his studies of motion. In a letter to Galileo of Sept. 21, 1632 (EN XIV, 395), also concerning the discovery of the shape of the trajectory, Cavalieri reports that he had learned from Muzio Oddi 10 years before that Galileo had performed experiments with Guidobaldo del Monte precisely on this subject. Oddi had received his education in Pesaro from Guidobaldo, who was likely to have told him about his collaboration with Galileo. But since Oddi went to prison from 1601 to 1610, he could not have obtained – at least from Guidobaldo – any information about experiments jointly performed by Guidobaldo and Galileo during that time, nor after he was freed since Guidobaldo died in 1607. Hence, Oddi must have heard from Guidobaldo about the experiments with Galileo before 1601, which is – *cum grano salis* – in agreement with Galileo's own dating of his discovery of the shape of the trajectory.

[75] The translation is adapted from Naylor 1974c, p. 327; 1980a, p. 551.

natural and the forced, and it is a line which in appearance is similar to a parabola and hyperbola ...

Guidobaldo's description disagrees with Galileo's analysis of projectile motion in *De Motu*. Whereas Galileo had treated the first part of the trajectory as a straight line that is dominated by the impressed force, Guidobaldo claims that the entire trajectory is a symmetrical curved line. In fact, he compares the trajectory with the chain line and notes its similarity to a parabola or a hyperbola. Guidobaldo confirms his observations by an experiment that allows him to produce a trace of the trajectory of projectile motion:

The experiment of this movement can be made by taking a ball colored with ink, and throwing it over a plane of a table which is almost perpendicular to the horizontal. Although the ball bounces along, yet it makes points as it goes, from which one can clearly see that as it rises so it descends, and it is reasonable this way, since the violence it has acquired in its ascent operates so that in falling it overcomes, in the same way, the natural movement in coming down so that the violence that overcame [the path] from *b* to *c*, conserving itself, operates so that from *c* to *d* [the path] is equal to *cb*, and the violence which is gradually lessening when descending operates so that from *d* to *e* [the path] is equal to *ba*, since there is no reason from *c* towards *de* that shows that the violence is lost at all, which, although it lessens continually towards *e*, yet there remains a sufficient amount of it, which is the cause that the weight never travels in a straight line towards *e*.

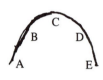

Fig. 3.9 (del Monte ms, p. 236)

In his protocol, Guidobaldo goes beyond the description of the trajectory and attempts to give an explanation of the projectile motion in terms of the forces by which it is generated. He tries to explain the symmetry of the motion and the curvature of the trajectory by a gradual disappearance of a moving force contrary to the projectile's weight.

When Galileo became acquainted with Guidobaldo's experiment and its result, he must immediately have recognized that it contradicted his treatment of the projectile trajectory in *De Motu*. Moreover, the experiment opens up an understanding of acceleration in free fall different from the view presented in *De Motu* that it is a transitory phenomenon. In fact, if Galileo tentatively accepted the parabolic shape of the trajectory he could have achieved at least an intuitive idea of the quadratic relation between times and distances in free fall.

Galileo was familiar with the mechanical generation of curves from the works of Archimedes. If the parabolic trajectory is conceived as the result of the mechanical superposition of two simple motions, one of them being a uniform motion along the horizontal, the other one representing the motion of fall, the latter motion must necessarily exhibit a quadratic relation between the times elapsed and the spaces traversed. Since, as we have seen in section 3.2.3, the composition of a horizontal and a vertical motion did not present a problem in the dynamical framework of the *De Motu* theory, and since Galileo was sufficiently

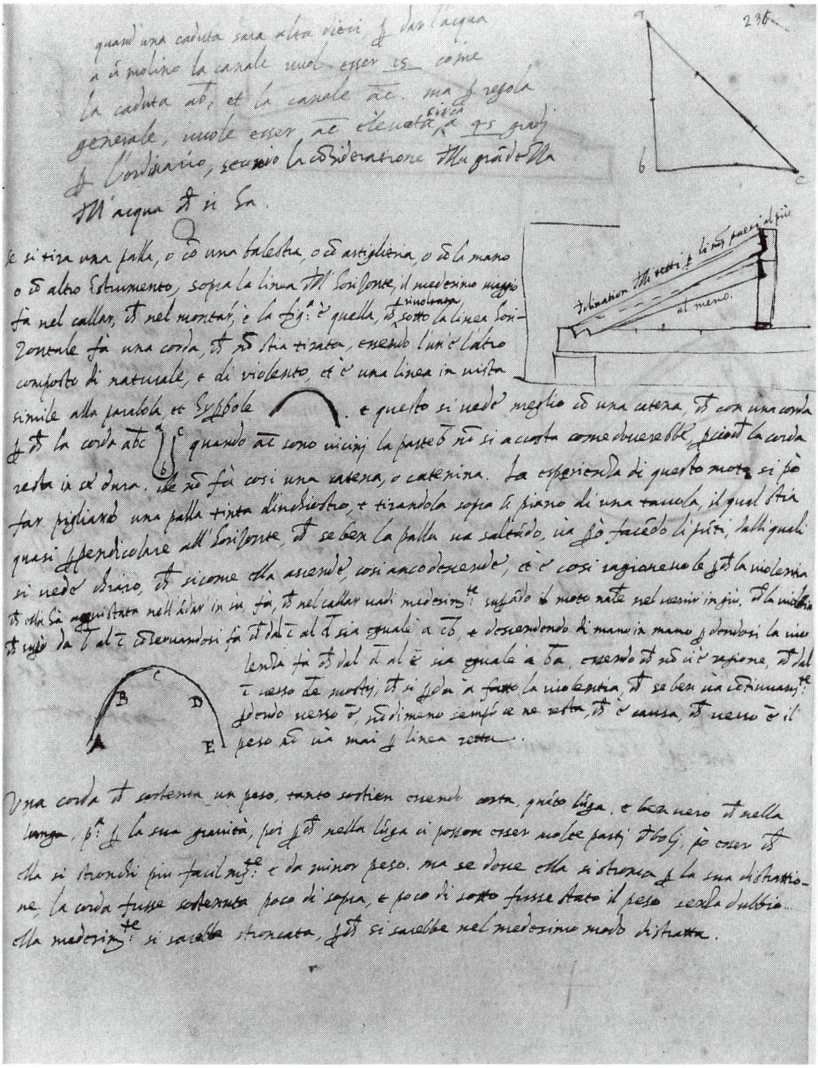

Plate I. Guidobaldo's experiment (del Monte MS, p. 236). Published with permission of the Bibliothèque Nationale, Paris.

familiar with Euclidean geometry, he could thus have immediately recognized this implication.[76]

It may have been precisely the recognition of this possible inference which inspired him to perform his experiments of measuring distances traversed in equal times by a body gliding down an inclined plane as documented by fol. 107v and to relate the shape of the chain line drawn on the other side of this manuscript (and compared to the trajectory in Guidobaldo's description of his experiment) to a sequence of square numbers. But it is also possible that Galileo pursued inclined plane experiments in connection with the rules for the motion along inclined planes derived in *De Motu*, and that he thus directly discovered the times squared relationship for the motion of fall along an inclined plane.[77] In any case, in Galileo's conceptual framework, the recognition of the parabolic shape of the trajectory suggested the recognition of the times squared relationship for the motion of fall and vice versa.[78]

In fact, as we shall see in the following, Galileo's recognition of the times squared relationship as a characteristic of the motion of fall made the assumption of a uniform horizontal motion as one of the components of projectile motion – and hence the derivation of the parabolic shape of the trajectory from a uniform

---

[76] This interpretation differs from that of Naylor (1974c, p. 333), who claims that Galileo would have required evidence less ambiguous than that provided by the Guidobaldo experiment before embarking on an explanation of either a parabolic or a hyperbolic trajectory. But it is quite conceivable that Galileo engaged in a speculative reconstruction of the trajectory on the basis of evidence that was even less conclusive. In fact, as Hill (1988, p. 667) convincingly argues, Galileo's familiarity with Apollonius and Archimedes equipped him to analyze a number of ordinary events in terms of mathematical curves. On the other hand, his later, more sophisticated experiments on projectile motion seem to presuppose a knowledge of the parabolic form of the trajectory.

[77] This reconstruction was suggested by Settle (1967, pp. 320-335).

[78] It is precisely the close relationship between these two elementary properties of projectile motion and the motion of fall that has made it so difficult for historians to establish which was discovered first, in spite of the relatively rich manuscript material documenting Galileo's thinking on motion. Drake has recently suggested another speculative reconstruction of the discovery of the times squared relationship, based on manuscript evidence; see Drake 1989b, pp. 35-49, and Drake 1990, Chapter 1. According to Drake, Galileo first measured the speeds of a motion of fall along an inclined plane to grow as the series of odd numbers, but he did not recognize the times squared relationship. Galileo then set out to relate fall to pendulum oscillations, first discovering the pendulum law and then the law of fall, both in the form of a mean proportional relationship (on proportions see section 1.2.1 above). Only then did he return to his inclined plane measurements, finally discovering the times squared law. According to Drake, Galileo discovered the parabolic shape of the trajectory only several years later as the result of a new set of careful measurements. While Drake's reconstruction makes Galileo a sophisticated experimentalist, it presupposes that he is an incompetent mathematician, unable immediately to recognize the elementary relationships between the sequence of odd numbers and that of the square numbers, between the expression of a relationship in terms of mean proportionals and in terms of double proportions or between a sequence of square numbers and a parabola.

horizontal and a vertical accelerated motion – plausible for him, given the background of his earlier analysis of a body moving along a horizontal plane. In *De Motu*, he had characterized motion along a horizontal plane as being neither natural nor forced[79]:

> A body subject to no external resistance on a plane sloping no matter how little below the horizon will move down (the plane) in natural motion, without the application of any external force. This can be seen in the case of water. And the same body on a plane sloping upward, no matter how little, above the horizon, does not move up (the plane) except by force. And so the conclusion remains that on the horizontal plane itself the motion is neither natural nor forced. But if the motion is not forced motion, then it can be made to move by the smallest of all possible forces.

In the same treatise, Galileo had posed the question whether or not a body moving with a motion that is neither natural nor forced would continue this motion perpetually[80]:

> [If] at the center of the universe there were a sphere that rotated neither naturally nor by force, the question is asked whether, after receiving a start of motion from an external mover, it would move perpetually or not. For if its motion is not contrary to nature, it seems that it should move perpetually; but if its motion is not according to nature, it seems that it should finally come to rest.

Galileo had left this question undecided in *De Motu* because he did not know how to escape from the dilemma described by the last sentence of this quotation resulting from the logic of contraries for two mutually exclusive properties of motion.[81] But his discovery of a simple rule for acceleration suggested that acceleration is not just an accidental but rather an essential property of natural motion and, correspondingly, deceleration an essential property of forced motion. As a consequence of this characterization of natural and forced motion, it follows

---

[79] DM, 66; EN I, 299.

[80] DM, 73; EN I, 305.

[81] Most historians of science have failed to realize that this question was a serious puzzle in Galileo's early theory of motion. In Wolff's comprehensive analysis of Galileo's conception of neutral motions and its historical background, for instance, he identifies Galileo's assertion that a "neutral" motion can be caused by the smallest force with a disposition in favor of neutral motions, and ultimately with Galileo's statements about the continuation of uniform motion along a horizontal (or spherical) plane; see Wolff 1987, pp. 247-248. Drake explains Galileo's hesitation in a similar context to make a clear pronouncement on inertial motion by the scruples of a careful experimentalist; with reference to a later comment on the continuation of motion by Galileo's disciple Castelli, Drake writes: "In a way it is a pity that Galileo never published his inertia idea in as general a form as his pupil thus ascribed it to him, though in another sense it is a great credit to Galileo as a physicist that he refused to go so far beyond his data to no purpose. Descartes did, being a less cautious physicist than Galileo; and being a more ingenious theologian than Castelli, he managed to derive the general law of inertia from the immovability of God" (Galileo 1969a, p. 171, note 26).

that a motion which is neither natural nor forced is a motion which is neither accelerated nor decelerated and hence a uniform motion. In other words, the contrariety between natural and forced motion now admits of an intermediary.[82]

Galileo's studies of acceleration using inclined planes made it therefore equally plausible for him to accept the quadratic relationship between times and distances as a characteristic property of natural downward motion as to accept the uniformity of horizontal motion also in the case of projectile motion. On the background of the *De Motu* analysis of projectile motion, which allows a composition of vertical and horizontal motion without mutual interference, the parabolic shape of the trajectory in the case of horizontal motion is then a strict consequence of these two properties of motion.

3.3.2 The New Law of Fall on the Background of the *De Motu* Theory

3.3.2.1 A First Derivation of the Law of Fall

While observations and experiments in connection with heuristic reasoning may have suggested to Galileo a quadratic relation between times and distances as a characterization of acceleration in free fall, they did not provide a proof. This quadratic relation between times and distances could, however, have been strictly inferred by Galileo from theorems about motion along inclined planes which resulted from an elaboration of his studies in *De Motu*. An argument by which the law of fall is obtained along these lines is indeed found in one of his manuscripts, fol. 147r.[83]

This argument depends on two presuppositions, both of them correct in classical mechanics. While Galileo had obtained the first presupposition, the Isochronism of Chords, as a direct consequence of the theory of motion along inclined planes in *De Motu*,[84] the second presupposition, the Length-Time-

---

[82] Except for the emphasis on the role of the logic of contraries as the traditional conceptual background of Galileo's argument, a similar reconstruction of his discovery of the continuation of uniform motion along the horizontal was given in Wertheimer (1945, pp. 160-167) and by Chalmers and Nicholas (1983, p. 335). Chalmers and Nicholas stress in their concise historical reconstruction that Galileo's argument neither presupposes nor, contrary to Wertheimer, implies the principle of inertia.

[83] MS, fol. 147r (EN VIII, 380). See document 5.3.5. Without referring to this particular manuscript, a similar derivation has been suggested as a possible root of the *discovery* of the law of fall; see Humphreys 1967. In view of the more direct ways to discover this relationship described in the previous section, this suggestion is not very plausible.

[84] See note 71.

Proportionality, was merely a plausible assumption, which Galileo at first also believed to be a consequence of this theory.[85]

Galileo's first presupposition states, in this case, that the time of fall along *ab* is equal to the time along *ac*, while the second presupposition consists in the assumption that the time through *da* is to the time through *ac* as the line *da* is to the line *ac*. At the beginning of his proof Galileo formulates the Isochronism of Chords as a theorem which he had already shown, and he states the proposition to be proved in what we shall call the "Mean Proportional Form of the Law of Fall"[86]:

> After it has been demonstrated that the times through *ab* and *ac* are equals [Fig. 3.10], it will be shown that the time through *ad* is to the time through *ae* as *da* is to the mean [proportional] between *da* and *ae*.

This formulation of the proposition is in fact equivalent to stating the law of fall as a quadratic relation or – using the terminology of the time – as a "double proportion."[87]

Galileo begins his proof by applying the Length-Time-Proportionality to the inclined plane *ac* and the vertical *ad* as well as to the inclined plane *ab* and the corresponding vertical *ae*:

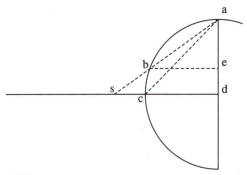

Fig. 3.10 (fol. 147r)

> For the time through *da* is to the time through *ac* as line *da* is to *ac*; but the time through *ac* (which is that through *ab*) is to the time *ae* as line *ba* is to *ae*, which is as *sa* is to *ad*.

---

[85] This is suggested by an argument in MS, fol. 177r (EN VIII, 386), translated in Wisan 1974, p. 200. Our interpretation follows Wisan 1974, pp. 188, and 200-201.

[86] MS, fol. 147r (EN VIII, 380). Later, in the *Discorsi*, this form of the law of fall is expressed as follows: "... if at the beginning of motion there are taken any two spaces whatever, run through in any [two] times, the times will be to each other as either of these two spaces is to the mean proportional space between the two given spaces." (*Discorsi*, pp. 170-171; EN VIII, 214). The following translation has been adapted from Drake 1978, p. 94.

[87] On the doctrine of proportions see section 1.2.1.

The equality of the times through *ac* and through *ab* is a consequence of the Isochronism of Chords.

> Therefore, by equidistance of ratios in perturbed proportionality [*ex aequali in analogia perturbata*], the time through *ad* is to the time through *ae* as line *sa* is to line *ac*.

Galileo's conclusion is an application of Proposition 23 of the Fifth Book of the *Elements* of Euclid in connection with the 18th Definition. This proposition states that (in modern terminology) if $a : b = b' : c'$, and $b : c = a' : b'$, then it follows that

$$a : c = a' : c'.$$

> And since *ac*, as has been demonstrated, is the mean [proportional] between *sa* and *ab*, while as *sa* is to *ab*, so *da* is to *ae*, therefore the time through *ad* is to the time through *ae* as *da* is to the mean [proportional] between *da* and *ae*, which was to be proved.

The conclusion that *ac* is the mean proportional between *sa* and *ab* follows from a geometrical proposition which Galileo had demonstrated in another manuscript.[88] Since $sa : ab = da : ae$, *sa* has the same proportion to the mean proportional between *sa* and *ab*, i.e., to *ac*, as *da* has to the mean proportional between *da* and *ae*. But since the proportion between *sa* and *ac* was found to be the proportion between the times through *ad* and *ae*, the proportion of these times is the same as that between the distance *ad* and the mean proportional between the two distances *ad* and *ae*, which was to be demonstrated.

The argument presented above shows that Galileo could derive a quantitative law of acceleration from two premises which by themselves do not presuppose an analysis of the internal structure of the motion of fall. The two statements on motion along inclined planes from which the law of fall is here derived are in fact related to that part of *De Motu* in which motion is considered only according to its extensive aspects, neglecting the question of whether it is accelerated or not. Nevertheless, this argument for the law of fall cannot be considered a proof on the basis of the *De Motu* theory, since one of the two premises on which it is based, the Length-Time-Proportionality, actually cannot be justified on the basis of this theory.

3.3.2.2 The Hollow Paradox

Not only did Galileo *not* succeed in integrating the law of fall into the *De Motu* theory, but he also encountered an inconsistency in his laws for the determination of velocities of bodies falling in media. As documented in section 3.2.1, these laws represent the core of his theory of the motion of fall in *De Motu*. In the same manuscript which contains the derivation of the quadratic rela-

---

[88] MS, fol. 58r, as pointed out by Drake 1978, p. 94.

tionship between distances and times in free fall presented above, Galileo shows that from this theory a paradox can be derived, which we may call the "Hollow Paradox"

According to the theory in *De Motu*, the speeds of bodies of the same material but of different weights fall in the same medium with the same speed. The derivation of the law of fall from two statements about motion along inclined planes in fol. 147r thus makes it plausible that this independence of the speed of fall from the weight of the falling body also holds for the motion of fall along inclined planes. In a sort of memorandum in fol. 147r,[89] Galileo makes a note to himself, indicating that this independence of the speed from the weight is also to be shown for motion of fall along inclined planes by the same kind of argument used earlier in *De Motu* against Aristotle's theory. In the course of the same memorandum, an argument of this kind is indeed developed, but, while the first part of the memorandum suggests a generalization of the *De Motu* argument to motion along inclined planes, the second part actually pursues a different generalization: to bodies of different materials. In the second part of the memorandum, it is shown that the *De Motu* theory of fall, from which it follows that bodies of different specific weights fall with different speeds, leads to a paradox[90]:

> It is to be considered that just as all heavy bodies rest in the horizontal [plane], the greater [weights] as well as the smaller, so they should move with the same speed [whether heavy or light] along inclined lines just as [they do] in the perpendicular itself. It would be good to demonstrate this, saying that if the heavier were faster, it would follow that the heavier would be slower, unequal heavy bodies having been joined, etc.

Fig. 3.11 (EN I, 265)

Moreover, not only homogeneous and unequal heavy bodies would move at the same speed, but also heterogeneous ones such as wood and lead. Since as it was shown before that large and small homogeneous bodies move equally, you argue: Let *b* be a wooden sphere [Fig. 3.11] and *a* be one of lead so big that, although it has a hollow for *b* in the middle, it is nevertheless heavier than a solid wood sphere equal [in volume] to *a*, so that for the adversary it should move faster than *b*; therefore if *b* were to be put into the hollow *i*, *a* would move slower than when it was lighter; which is absurd.

---

[89] MS, fol. 147r (EN VIII, 371-372). See document 5.3.5.

[90] The translation is adapted from Drake 1978, p. 95. In his interpretation of the memorandum (Drake 1978, pp. 95-96), Drake overlooks the fact that the second part does not specifically refer to motion along inclined planes but rather refers to motion of fall in general. Drake calls Galileo's argument an "unsuccessful gambit." But he fails to note the crucial difference between the argument given in the memorandum and the original *De Motu* argument, as is clear from his assertion that the equal speed of fall of bodies differing in weight already follows from the *De Motu* argument. But, in fact, this conclusion is in flat contradiction to the *De Motu* theory, as we have seen in the previous section.

Galileo's argument concerns a comparison of the sphere a, consisting of an exterior part of lead as well as of a hollow in the middle, and a wooden sphere (not pictured) having the same volume as the entire sphere a (including its hollow). If the lead part of the body a is sufficiently great, the weight of this body will be greater than that of the wooden sphere. It thus follows from Galileo's theory of fall in *De Motu* that the sphere a will fall more quickly than the wooden sphere having the same volume. But it will also move more quickly than the smaller wooden sphere b, which moves just as fast as the larger wooden sphere, since according to this theory bodies of the same material attain the same speed in their motion of fall.

In the last part of his argument, Galileo unites the two (pictured) bodies by putting b into the hollow of a. The resulting body should have a speed of fall which is intermediate between the speeds of the single bodies a and b, in particular, it should fall more slowly than a. But the weight of the body resulting from the union of a and b is the sum of the weights of the bodies a and b. Therefore, according to Galileo's theory of fall in *De Motu*, this body should move more quickly than body a which has the same volume but lesser weight. But since the union of a and b cannot move at the same time more slowly and more quickly than a, Galileo's theory of fall in *De Motu* leads to a paradox.[91]

---

[91]   Underneath this memorandum, Galileo wrote, at a later time, the word "Paralogismus" indicating that he had discovered a flaw in his argument. One could indeed object that, in his argument, Galileo had not properly determined the volume of the composite body because, when considering the specific weight of the composite body, he did not take into account the volume of the air (or whatever other medium the two bodies are moving in) which originally filled the hollow of the leaden sphere. One could argue that, if the wooden body is placed into the hollow of the leaden sphere so as to form the composite body, the volume of the air originally filling the hollow has to be counted as part of the volume of the composite body, because otherwise the composite body would have a volume smaller than the sum of the volumes of the single bodies. In other words, a certain volume of the contiguous air, equal to the volume of the hollow, has to be considered as part of the composite body. Although it may seem plausible to take into account this volume, from the point of view of hydrostatics - on which the whole argument is based - an arbitrary volume of the contiguous medium can in fact be added to the composite body without changing anything. Hence, this objection to Galileo's argument, based as it is on a simple addition of volumes, by no means refutes the argument, but rather serves to strengthen it by pointing to the crucial weakness of a theory of fall based on hydrostatic conceptions, i.e., the determination of the volume of the body falling in a medium, which is, to a certain extent, arbitrary. Nevertheless, in view of Galileo's later work on hydrostatics, it seems possible that the objection sketched above could have appeared convincing to him. Galileo's discussion of the floating of thin lamina of materials having greater specific weight than water is indeed based on an erroneous argument according to which the floating object has to be conceived as a composition of the contiguous air and of the floating thin lamina, whereas the modern understanding of this phenomenon is based on the surface tension of the water. (See Galilei 1612, Theorem VI, and the discussion preceding it.) It is in fact conspicuous that Galileo's argument in fol. 147r does not appear in his published works and that, in the above mentioned work

3.3.3 Attempts at a Deductive Treatment of Acceleration around 1604

3.3.3.1 Derivation of the Law of Fall from a Principle of Acceleration

In the preceding section it was shown that Galileo discovered serious problems in his theory of motion as presented in *De Motu*, while the quadratic relation between times and distances emerged as an assertion about the motion of fall which was central to his understanding not only of falling bodies but also of projectile motion and of motion along inclined planes. The argument for this assertion discussed in section 3.3.2.1 depends on an *unproven* assumption, the Length-Time-Proportionality and thus does not represent an independent proof. A date as early as the period immediately before 1604 for this argument seems likely but remains a conjecture. Fortunately, however, from a letter to Paolo Sarpi, written in 1604, we know for sure that already in this year Galileo was well acquainted with the correct law of fall, and we can classify a group of related manuscripts originating from his work at that time. Furthermore, this letter provides a deep insight into Galileo's attempts to find an independent proof of the law fall, based on a principle which is incorrect in classical mechanics.

In his letter Galileo claims to have found an indubitable principle from which he could derive the law of fall[92]:

Fig. 3.12
(EN X, 115)

> Thinking again about the matters of motion, in which, to demonstrate the phenomena [*accidenti*] observed by me, I lacked a completely indubitable principle to put as an axiom, I am reduced to a proposition which has much of the natural and the evident: and with this assumed, I then demonstrate the rest; that is, that the spaces passed by natural motion are in double proportion to the times, and consequently the spaces passed in equal times are as the odd numbers from one, and the other things. And the principle is this: that the natural movable goes increasing in velocity with that proportion with which it departs from the beginning of its motion; as, for example, the heavy body falling [Fig. 3.12] from the terminus *a* along the line *abcd*, I assume that the degree of velocity that it has at *c*, to the degree it had at *b*, is as the distance *ca* to the distance *ba*, and thus consequently, at *d* it has a degree of velocity greater than at *c* according as the distance *da* is greater than *ca*.

---

on hydrostatics, he still refers to the dependency of the velocity of fall on the specific weight; see Galileo 1663, p. 70.

[92] Galileo Galilei to Paolo Sarpi, October 16, 1604; EN X, 115-116. (Translation, adapted from Drake 1969.) See document 5.3.3.

When Galileo wrote this letter he obviously already took the law of fall for granted, formulating it in terms of a double proportion as well as in terms of the sequence of odd numbers which represent the distances traversed in equal successive times ("Odd Number Rule"). Both formulations are equivalent to the assertion that the distances are in the same proportion as the squares of the times which in fact is the correct law of fall of classical physics. What Galileo had been seeking (and what he assumed he had already found at the time he wrote this letter) was a solid ground for its derivation, because the *De Motu* theory did not provide a foundation for this law.

The principle Galileo intended to put as an axiom is an assertion about the increase of velocity in the motion of fall. The precise meaning of this principle can be inferred from the example subsequently given. In this more explicit formulation of his principle, Galileo specifies the increase of velocity in terms of its increasing degrees. The relationship between the increase of the velocity and the "proportion with which [the body] departs from the beginning of its motion" is now expressed as a proportionality between the degree of velocity and the distance from the starting point of the motion.

Galileo's use of the terminology degree of velocity relates his principle to the medieval doctrine of intension and remission, which is treated in his early scholastic notebooks, and on which the configuration of qualities discussed in section 1.2.3, is based. In fact, in one of his early notebooks, Galileo introduces the notion of degree as follows[93]:

> Second, one can consider degrees through which it [i.e., a quality] is constituted hot or cold, etc., and so can be more or less perfect; on this account the

---

[93] EN I, 119; Wallace 1977, p. 172. It has been claimed that Galileo's work on free fall is independent of the medieval traditions going back to the Mertonians and to Oresme, because one particular application of the conceptual tools developed in this tradition, the Merton Rule, is not documented, at least in its original form, by Galileo's working papers; see, e.g., Drake 1969, p. 350. But this argument is inconclusive because it focuses on one particular application of the conceptual tools in question, while, as we shall see, other applications can in fact be identified among Galileo's manuscripts. It has, on the other hand, been suggested that concepts and arguments associated with these medieval traditions found their way in several stages into Galileo's work, and that Galileo supposedly turned directly to medieval sources precisely when confronted with particular problems of motion. See Wisan 1974, p. 288, note 18, pp. 296-297, and also Galluzzi 1979, pp. 273-274, note 36, for a brief review of the relevant arguments by various other authors. However, although, as argued in section 1.2.3 above, the early modern tradition of the configuration of qualities did not include its use as a calculational tool, it cannot be doubted that some of its basic concepts were part of common knowledge, and that hence no particular medieval text has to be assumed to be the priviledged source of Galileo's familiarity with this tradition. In fact, the notion of "degree of velocity," for instance, was generally adopted in lectures at the Collegio Romano, from which Galileo drew much of his knowledge about scholastic traditions; see Wallace 1984, p. 268. For a recent study of the role of the Mertonian tradition in Galileo's work, also emphasizing conceptual similarity rather than dependence on technical results, see Sylla 1986.

quality has a certain latitude over which there are a number of degrees that do not vary the essence.

In his analysis of acceleration in *De Motu*, Galileo had conceived of the motion of fall and of vertical projection in terms of a continuous alterative motion, to which he could in principle have applied the medieval doctrine of intension and remission. Probably because he then considered acceleration as just an accidental characteristic of motion for which no firm rules can be given, he did not elaborate his analysis in these terms. But the manuscript fol. 107v [see Fig. 3.13], documenting an experiment on accelerated motion and discussed in the previous section, contains two diagrams which suggest that Galileo was not only familiar with the graphical representation pertaining to this medieval conceptual tradition, but that he also immediately associated it with accelerated motion when he did discover a rule for acceleration[94]:

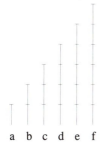

a   b   c   d   e   f

Fig. 3.13 (fol. 107v)

---

[94] MS, fol. 107v. One of the two diagrams may suggest that Galileo took the possibility of a discontinuous change of motion into consideration. It seems to be misleading, however, to assert, as Drake (1990, p. 39) does, a supposed "passage from a quantum-concept of speeds to the concept of continuous change" as an essentially new element in Galileo's mechanics. In fact, according to Wallace (1984, p. 267), Galileo adopted a scholastic view when claiming that velocity varies continuously. There is, however, manuscript evidence showing that Galileo did consider discontinuous changes of speeds; see MS, fol. 182r (EN VIII, 425-426), which Drake (1979, p. LII) dates as late as 1618 (see document 5.3.6). But this manuscript evidence does not support Drake's suggestion of a deep conceptual gap between the consideration of a gradual and a "quantum" change of speeds, but rather suggests that the choice between these considerations seems to be linked to the mathematical technique applied to a given problem. It shows that an attempt by Galileo to use arithmetic techniques in the study of accelerated motion led him into difficulties. In this manuscript, he attempts to construct the motion of upward projection by subtracting the naturally accelerated motion downward from the violent motion upward. In this construction, he divides the motion of upward projection into a certain number of equal parts and then uses simple arithmetic to determine the result of a step-wise combination of the violent and the natural motion. But his result remains inconclusive because it depends on the number of parts into which the motion was originally divided. For a critique of Drake's understanding of the role of discontinuity in the medieval and early modern analysis of motion, see also Franklin 1977.

In his letter to Sarpi, Galileo conceived the change of velocity precisely in terms of this medieval tradition with the distance from the starting point being identified with the latitude of the subject in the traditional terminology. Thus he attempted to solve the problem in exactly same way as Descartes did independently several years later in his answer to Beeckman's question, except that he seems to have been much more committed to a spatial instead of a temporal interpretation of the latitude than Descartes originally was. From the viewpoint of classical mechanics the assertion is therefore clearly wrong and does not imply the law of fall. Even more disappointing is that Galileo in his letter to Sarpi does not give any proof for his assertion that the law of fall could be derived from the given principle. No hint is given in this letter about the alleged derivation of the law which Galileo had in mind.

Fortunately, this derivation can clearly be identified in Galileo's manuscripts dating back to the period around 1604.[95] In a number of these manuscripts Galileo uses the traditional geometrical representation of uniformly difform qualities in attempts to derive the law of fall from the principle mentioned in the letter to Sarpi. They all have in common that Galileo inferred from the assumed linear increase of the quality the increase of the area of its geometrical representation in "double proportion." However, they differ considerably in their terminology, thus showing that it was not the technique of the derivation but rather the conceptualization of the problem and its solution that presented the real problem.

One of Galileo's notes relating to the letter to Sarpi can be found on fol. 179v.[96] It is an unfinished attempt to prove the law of free fall:

Fig. 3.14 (fol. 179v)

If in the line of natural descent two unequal distances from the starting point of the motion are taken, the moments of velocity with which the moving body traverses these distances are to one another in double proportion of those distances.

Let *ab* [Fig. 3.14] be the line of natural descent, in which from the starting point *a* of the motion two distances *ac* and *ad* are taken: I say, that the moments of velocity with which the moving body traverses *ad* are to the moments of velocity with which it traverses *ac* in double proportion of the distances *ad* and *ac*. Draw line *ae* in an arbitrary angle with respect to *ab*. [The note ends here abruptly.]

[95] These manuscripts, collected in MS, have been the object of numerous attempts of interpretation by historians of science; for a review of recent debates, see Romo Feito 1985. The precise chronological order of the manuscripts in MS is unclear; for an analysis of their physical features, including an attempt to use watermarks for a rough dating, see Drake 1972b and Drake 1979.

[96] MS, fol. 179v (EN VIII, 380). The translation is adapted from Drake 1978, p. 115.

The proposition Galileo intends to prove here refers only to the relationship between distances and the moments of velocity;[97] the derivation of the law of fall, i.e., of a relationship between distances and times, may have been intended as a subsequent step in Galileo's argument.

The diagram suggests that the moments of velocity are represented by the lines parallel to fg, not only the ones drawn but by all the infinite parallel lines that can be imagined between the starting point and the endpoint of the motion. Galileo does not explicitly state that the increase of the moments of velocity is in proportion to the distance traversed, but this assumption is implicit in his geometrical representation. The lengths of these lines grow in proportion to the distances from the starting point a. Hence the geometrical representation which Galileo employs suggests that he assimilated the meaning of moment of velocity to what, in his letter to Sarpi, he called the degree of velocity.

Galileo compares two sets of moments of velocities in the accelerated motion extended over two distances, representing the total motion of the body traversing these distances. He was probably going to conclude that the proportion between all the moments of velocity along the line ad to all the moments of velocity along the line ac equals the proportion of the area adl to the area ack, which is in double proportion to the distances. While it seems clear what Galileo wanted to prove, it is not clear what further consequences he wanted to draw from the propositions of his proof.[98]

In contrast to fol. 179v, another early fragment, fol. 85v, contains a complete proof of the law of fall from the principle Galileo formulated in his letter to Sarpi. The proof consists of two parts. The first part once again contains the derivation of the increase of the quantity represented by the area of the diagram "in double proportion" of the quality which increases proportionally to the space traversed. Again the most important aspect of the argument is the difference in terminology used by Galileo[99]:

---

[97] For a comprehensive study of Galileo's use of the concept of moment of velocity, see Galluzzi 1979, in particular Chapter III; for his discussion of fol. 179v, see pp. 285-286. The significance of the relationship between the infinitesimal meaning of the concept of moment and its meaning in Galileo's statics, thoroughly analyzed by Galluzzi, lies beyond the scope of this chapter.

[98] Drake claims that Galileo may have left the argument in fol. 179v unfinished because he discovered a problem with the erroneous principle mentioned in the letter to Sarpi, but his argument is based on an entirely fictitious mathematical difficulty which Galileo is supposed to have discovered in his argument in fol. 179v; see Drake 1970a, p. 23, and Drake 1978, p. 116. Wisan (1974, pp. 220-221) suggests that the argument is left unfinished because Galileo discovered an "elementary error" in his attempt to prove the law of fall from the erroneous principle mentioned in the letter to Sarpi. But, as we shall see in the following, this attempted derivation did not involve an elementary error, at least not from the point of view of Galileo's knowledge.

[99] MS, fol. 85v (EN VIII, 383). According to Favaro this manuscript is in the hand of Galileo's assistant Mario Guiducci. The passage is crossed out in the manuscript. We follow Wisan in interpreting this argument as an early version of the somewhat more polished argument in fol. 128, to be discussed in the following; see Wisan 1974,

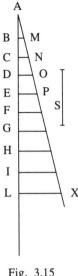

Fig. 3.15
(fol. 85v)

I assume that the acceleration of the falling body along the line *al* is such that the velocity increases in the ratio of the space traversed so that the velocity in *c* is to the velocity in *b* as the space *ca* is to the space *ba*, etc.

Matters standing thus, let the line *ax* be drawn [Fig. 3.15] at some angle to *al*, and, taking the parts *ab*, *bc*, *cd*, *de*, etc., to be equal, draw *bm*, *cn*, *do*, *ep*, etc. If therefore the velocities of the body falling along *al* in the places *b*, *c*, *d*, *e* are as the distances *ab*, *ac*, *ad*, *ae*, etc., then they will also be as the lines *bm*, *cn*, *do*, *ep*.

But because the velocity is successively increased in all points of the line *ae*, and not only in *b*, *c*, and *d*, which are drawn, therefore all these velocities are to one another as the lines from all the said points of the line *ae* which are generated equidistantly from the same *bm*, *cn*, *do*.

But those are infinite and constitute the triangle *aep*: therefore the velocities in all points of the line *ab* are to the velocities in all points of the line *ac* as the triangle *abm* to the triangle *acn*, and so for the remaining, i.e., in double proportion of the lines *ab*, *ac*.

In formulating his premise, Galileo refers to the aspect of velocity which he denoted in his letter to Sarpi by "degree," and which in fol. 179 was called "moment of velocity," as "the velocity in" a given point of the line of fall. In this way the parallel lines are interpreted as representing the velocities of the body in the places represented by the points of the line AL.

The core of Galileo's argument in this first part of the proof is an infinitesimal consideration in order to identify the meaning of the area. Galileo argues that the infinite number of parallel lines constitutes the triangle, which therefore represents "the velocities in all points of the line," i.e., the summation of the velocities.

Thus far, no relationship between the distances traversed and the times elapsed has been established. The short second part of Galileo's argument is devoted to this problem:

> But because in the ratio of the increase of [velocity due to] acceleration the times in which the motions themselves occur must decrease, therefore the time in which the moving body traverses *ab* will be to the time in which it traverses *ac* as the line *ab* is to that line which is the mean proportional between *ab* and *ac*.

Galileo's argument is somewhat problematical. Like Descartes in his diary note (see section 1.3.3), he starts with the observation that the ratios of the times decrease with the increase of the velocities. Like Descartes, he further assumes that

---

pp. 207-209. Contrary to Wisan, we interpret the line marked "S" next to the triangular diagram as representing space and not as being a geometrical mnemonic. The translation is adapted from Drake 1978, pp. 98-99. See document 5.3.9.

the decrease of times should be the same as the increase of velocities: "in the ratio of the increase of [velocity due to] acceleration the times in which the motions themselves occur must decrease." Unlike Descartes, however, he does not conclude that the ratios of the times must therefore be in inverse proportion to the ratios of the velocities. He inverts instead the double proportionality of the velocities by a transition to a "half" or "mean" proportionality of the times thus obtaining the law of fall in its mean proportional form.

Mathematically, the passage from double to mean proportion corresponds to the inversion of a "proportion of proportions" in the medieval sense of the term.[100] Dividing a proportion or finding its mean or its half, is the inverse operation to doubling a proportion in the following sense: To double a proportion $a : b$ means to continue the proportion $a : b$ by $b : c$ so that a proportion $a : b : c$ results, where $a : c$ is in double proportion of $a : b$, that is, $(a : c) = (a : b)(a : b)$. On the other hand, to divide a proportion $a : c$ means to find a mean proportional b such that the proportion $a : b$ results, which is in half proportion of $a : c$, so that $a : b : c$ holds. Galileo's problematical transition from the "double proportionality" between velocities and distances to "mean proportionality" between times and distances does not provide a rigorous conclusion – as for instance the application of the appropriate Aristotelian proportion by Descartes did provide. Though not rigorous, it does have the advantage of yielding the correct result, which – as we know – Galileo already possessed.

This interpretation of the proof on fol. 85v is supported by fol. 128 which contains a more explicit version of the same proof. The proof in fol. 128 begins with a general formulation of the principle from which the law of fall is to be derived and which is precisely the one Galileo proposed to Sarpi[101]:

> I suppose (and perhaps I shall be able to demonstrate this) that the naturally falling heavy body goes continually increasing its velocity according as the distance increases from the terminus from which it parted, as, for example, the heavy body departing from the point a [Fig. 3.16] and falling through the line ab. I suppose that the degree of velocity at point d is as much greater than the degree of velocity at the point c as the distance da is greater than ca; and so the degree of velocity at e is to the degree of velocity at d as ea to da, and thus at every point of the line ab it [the body] is to be found with degrees of velocity proportional to the distances of these points from the terminus a. This principle appears to me very natural, and one that corresponds to all the experiences that we see in the instruments and machines that work by striking, in which the percussent works so much the greater effect, the greater the height from which it falls; and this principle assumed I shall demonstrate the rest. (fol. 128)

---

[100] This interpretation follows Sylla (1986, pp. 73-74), who solves a much debated puzzle. Sylla's reconstruction is supported by the treatment of this mathematical technique in a standard textbook of the time, Tartaglia 1556b, pp. 111, 123, and 128. There is evidence that Galileo actually learned mathematics from this textbook; see Renn 1988, Appendix B.

[101] MS, fol. 128 (EN VIII, 373-374). The following English translation is adapted from Drake 1978, pp. 102-103. See document 5.3.7.

Again, the following proof consists of two parts the first of which serves to justify the identification of elements of the diagram with their meanings as well as to infer the double proportionality between the spaces traversed and the (summated) velocities:

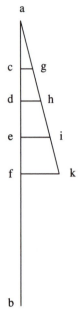

Draw line *ak* at any angle with *af*, and through points *c*, *d*, *e*, and *f* draw the parallels *cg*, *dh*, *ei*, *fk*: And since lines *fk*, *ei*, *dh*, and *cg* are to one another as *fa*, *ea*, *da*, *ca*, therefore the velocities at points *f*, *e*, *d*, and *c* are as lines *fk*, *ei*, *dh*, and *cg*. So the degrees of velocity go continually increasing at all points of line *af* according to the increase of parallels drawn from all those same points. Moreover, since the velocity with which the moving body has come from *a* to *d* is compounded from all the degrees of velocity it had at all the points of line *ad*, and the velocity with which it has passed through line *ac* is compounded from all the degrees of velocity that it has had at all the points of line *ac*, therefore the velocity with which it has passed line *ad* has that proportion to the velocity with which it has passed the line *ac* which all the parallel lines drawn from all the points of the line *ad* over to *ah* have to all the parallels drawn from all the points of line *ac* over to *ag*; and this proportion is that which the triangle *adh* has to the triangle *acg*, that is the square of *ad* to the square of *ac*. Then the velocity with which the line *ad* is traversed to the velocity with which the line *ac* is traversed has the double proportion that *da* has to *ca*. (fol. 128)

Fig. 3.16 (fol. 128)

Galileo uses the expression velocities at points familiar from the proof in fol. 85v and the expression degrees of velocity introduced in the letter to Sarpi in the sense of the term moment of velocity as it is used in the proof in fol. 179. The infinitesimal consideration closely follows the related argument in the first part of the proof on fol. 85v. The degrees of velocity are the infinitesimal constituents of the velocity understood as an overall property of motion represented by the area. His designation for this latter velocity is velocity with which a body has passed a given line.

In the subsequent second part of Galileo's proof we also find exactly the same argument as in the second part of fol. 85v:

And since velocity to velocity has contrary proportion of that which time has to time (for it is the same thing to increase the velocity as to decrease the time), therefore the time of the motion along *ad* to the time of the motion along *ac* has half the proportion that the distance *ad* has to the distance *ac*. The distances, then, from the beginning of the motion are as the squares of the times, and, dividing, the spaces passed in equal times are as the odd numbers from unity. Which corresponds to what I have always said and to experiences observed; and thus all the truths are in accord. (fol. 128)

In this way, Galileo obtains the same conclusion as in the proof on fol. 85v: the law of fall in its mean proportional form, which is subsequently expressed as the times squared relationship and as the Odd Number Rule. As in fol. 85v, his argument depends on the inversion of a "proportion of proportions" seemingly justified by the statement that "velocity to velocity has the contrary proportion of that which time has to time."

Hence there are in fact no substantial differences at all among the attempts to prove the law of fall on fol. 179, fol. 85v, and fol. 128. The version of the proof on fol. 128 is clearly the most elaborated. It is written in Italian and was surely the proof which Galileo was speaking of when he wrote his letter to Sarpi. The similarity of the language to the Sarpi letter and the state of elaboration of the proof in fol. 128 even suggest that this was the very manuscript Galileo had before his eyes when writing the letter.

The close relationship between the letter to Sarpi and the manuscript fol. 128 is confirmed by the passages following the parts on the law of fall in the two documents. Their purpose is to show the symmetry between projectile motion and the motion of fall. In both documents Galileo uses the principle underlying his derivation of the law of fall in order to show that

the naturally falling body and the violent projectile pass through the same proportions of velocity. (letter to Sarpi)	the velocity in the violent motion goes decreasing in the same proportion with which, along the same straight line, natural motion increases. (fol. 128)

Besides the principle proposed to Sarpi, Galileo's proof of the symmetry between vertical projection and free fall is based on a causal explanation of vertical projection:

For if the projectile is thrown from the point d to the point a, it is manifest that at the point d it has a degree of impetus able to drive it to the point a, and not beyond; and if the same projectile is in c it is clear that it is linked with a degree of impetus able to drive it to the same terminus a; and likewise the degree of impetus at b suffices to drive it to a. (letter to Sarpi)	For let the starting point of the violent motion be the point b, and the end point the terminus a. And since the projectile does not pass the terminus a, therefore the impetus it had at b was such as to be able to drive it to the terminus a; and the impetus that the same projectile has in f is sufficient to drive it to the same terminus a; and when the same projectile is in e, d, c, it finds itself linked with impetuses capable of pushing it until the same terminus a, not more and not less. (fol. 128)

As he had done in *De Motu*, Galileo gives the traditional explanation of vertical projection by a gradually decreasing impetus. In the above quotations the expressions "impetus" and "degree of impetus" are used in the same sense of a quality, which, at any given instant, is present in the projectile as the cause of its future motion. The effect of this cause is the motion until the terminal point of the violent motion. From the proportionality between cause and effect, Galileo concludes that the impetus decreases in the same proportion as does the distance to the terminal point of the motion.

Concluding his argument, he compares the violent upward motion with the natural motion downward and shows that the increase of the degrees of velocity in the motion of fall follows the same pattern as the decrease of impetus in vertical projection:

whence it is manifest that the impetus at the points *d, c, b* goes decreasing in the proportions of the lines *da, ca, ba*; whence, if it goes acquiring degrees of velocity in the same (proportions) in natural fall, what I have said and believed up to now is true.
(letter to Sarpi)

thus the impetus goes evenly decreasing as the distance of the moving body from the terminus *a* diminishes. But according to the same proportion of the distances from the terminus *a* the velocity increases when the same heavy body will fall from the point *a*, as assumed above and compared with our other previous observations and demonstrations: thus what we wanted to prove is manifest.
(fol. 128)

Galileo thus inferred the symmetry between vertical projectile motion and the motion of fall by means of a causal explanation of projectile motion and by making use of the principle underlying his derivation of the law of fall.

From his comments on this principle in the opening paragraphs of fol. 128, it becomes clear that he also directly supported this principle by a causal analysis of the motion of fall. In this analysis, he made the assumption that the effect of a percussent is proportional to the height of fall, interpreted as the cause of this effect[102]:

> This principle appears to me very natural, and one that corresponds to all the experiences we see in the instruments and machines that work by striking, where the percussent works so much the greater effect, the greater the height from which it falls.

This assumption of a proportionality between the effect of a machine and the vertical distance traversed was, for Galileo, an immediate consequence of a general principle underlying his analysis of simple machines.[103] According to this principle, which derives from traditional statics, the effect of the motion of a heavy body is always to be measured according to the vertical distance it traverses. However, in order to justify the principle proposed to Sarpi by the proportionality between the effect of a percussent and the height of fall, Galileo has to make the problematical assumption that the effect of the percussent is to be measured by its degree of velocity.

---

[102] MS, fol. 128; see above text to Fig. 3.16.
[103] See his early unpublished work on mechanics (EN II, 147-191; translated in Galileo 1960c). A close relationship between the principle of acceleration mentioned in the letter to Sarpi, and Galileo's earlier work on statics has been convincingly argued by Galluzzi (1979, p. 272).

Neither in the letter to Sarpi nor in any of the manuscripts related to this letter does Galileo propose a causal explanation of the motion of fall by an impressed force as he had presented it in *De Motu*. As noted earlier, this explanation is in fact incompatible with the law of fall, since it accounts for acceleration by a difference of two forces, one constant, the other gradually diminishing, thus implying a limit velocity.

3.3.3.2 Derivation of the Double Distance Rule

A number of manuscripts which can be dated to the period around 1604 document the fact that Galileo used the principle proposed to Sarpi as a basic assumption in constructing a theory of motion. Indeed, on the basis of this principle, he not only could give an argument for the law of fall and for the symmetry between vertical projection and free fall, but he now also possessed a premise from which he could attempt to derive the assumption about motion along inclined planes mentioned in section 3.3.2.1, the Length-Time-Proportionality. As is documented by one of his manuscripts, Galileo could do so by turning around the proof of the law of fall discussed in that section; in this way he obtained the Length-Time-Proportionality as a consequence of the law of fall and of the Isochronism of Chords.[104] But he also discovered that he could derive the Length-Time-Proportionality directly from the principle proposed to Sarpi so that he could now demonstrate the Isochronism of Chords on the basis of the law of fall and of the Length-Time-Proportionality.[105]

In order directly to prove the Length-Time-Proportionality from his principle, Galileo reformulated it so as to apply to motion along inclined planes as well; he assumed that[106]

> in natural descent, the moments of velocity are always found to be the same in points equally distant from the horizon[tal] according to their perpendicular distance.

Several of Galileo's manuscripts document his use of the erroneous principle proposed to Sarpi as well as the medieval conceptual tools related to it in a study of the relationship between the motion of fall and the uniform motion which results from deflecting the accelerated motion of fall into the horizontal. As we shall see in section 3.5.1, this relationship plays a crucial role in Galileo's

---

[104] MS, fol. 163v (EN VIII, 384-385), discussed in Wisan 1974, pp. 215-216. See Plate II..

[105] See MS, fol. 172r (EN VIII, 392-93); discussed in Wisan 1974, pp. 190-191.

[106] MS, fol. 179r (EN VIII, 388). As we are not primarily concerned with Galileo's theory of motion along inclined planes, we shall not discuss the proof of the Length-Time-Proportionality contained in the same manuscript. For studies of this proof, see Wisan 1974, pp. 217-220, Drake 1978, pp. 113-115, Galluzzi 1979, pp. 294-297, and, for a recent discussion, Souffrin 1986.

derivation of the size of the parabolic trajectory.[107] The following argument has been called the "Double Distance Rule"[108]:

> Let the motion from *a* to *b* [Fig. 3.17] be made in natural acceleration: I say, if the velocity in all points *ab* were the same as that found in the point *b*, the space *ab* would be traversed twice as fast; because all velocities in the single points of the line *ab* have the same ratio to all the velocities each of which is equal to the velocity *bc* as the triangle *abc* has to the rectangle *abcd*.

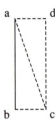

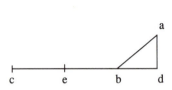

Fig. 3.17 (fol, 163v)   Fig. 3.18 (fol. 163v)

> From this it follows, that if there were a plane *ba* [Fig. 3.18] inclined to the horizontal line *cd*, and *bc* being double *ba*, then the moving body would come from *a* to *b* and successively from *b* to *c* in equal times: for, after it was in *b*, it will be moved along the remaining *bc* with uniform velocity and with the same with which [it is moved] in this very terminal point *b* after fall through *ab*. Furthermore, it is obvious that the whole time through *abe* is 1^1/$_2$ the time [*sesquialterum*] for *ab*.

The derivation of the Double Distance Rule consists of two parts. In the first, a relationship between a uniform and an accelerated velocity is established. This part is similar to the comparison between a uniform and a uniformly difform quality according to the traditional Merton Rule discussed in section 1.2.3. In Galileo's case, however, both of these motions are physically realizable: the uniform motion is a real motion along a horizontal plane, whose uniformity was guaranteed, as we have see above, by an argument based on the logic of contraries; the accelerated motion is either the motion of free fall or the motion of fall along an inclined plane. The brief second part of Galileo's argument represents an application of the comparison established in the first part to an accelerated motion along an inclined plane which is subsequently deflected into the horizontal so as to yield a uniform motion along a horizontal plane. Galileo claims that the same space is traversed twice as fast in the uniform motion as in the accelerated motion.

The first part of Galileo's argument establishes his first diagram as a representation of velocity along the line of reasoning familiar from the different versions

---

[107] Wisan has suggested that this derivation may have actually been the motive for Galileo's search for such a relationship; see Wisan 1974, p. 206.

[108] MS, fol. 163v (EN VIII, 383-384); discussed in Wisan 1974, pp. 204-207.

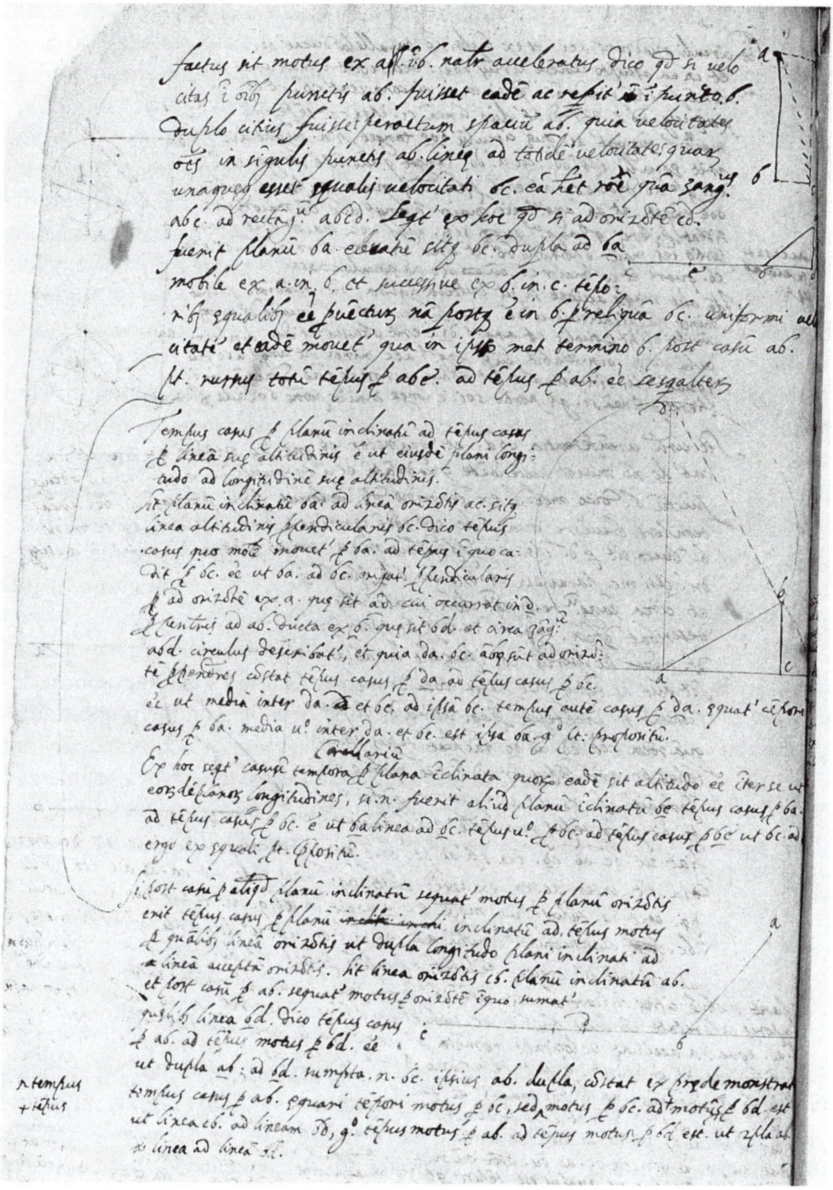

Plate II. Galileo's Double Distance Rule (fol. 163v). Published with permission of the Biblioteca Nazionale, Florence

of his derivation of the law of fall. The triangular shape of the area *abc* shows that Galileo again assumes the velocities, i.e., the degrees of velocity, to grow in proportion to the distance from the starting point of the accelerated motion. As in his derivation of the law of fall, he conceives of the overall velocity of the motion as being constituted by "all velocities in the single points of the line *ab*," since it is by determining the proportion of all the velocities in the accelerated motion to all the velocities in the uniform motion that he concludes the uniform motion to be twice as fast. Galileo interprets the proportion between the triangular area *abc* and the rectangular area *abcd* as representing the proportion between these two infinite sets of velocities, obviously supposing, as he did in his derivation of the law of fall, that the infinite parallel lines representing the velocities in the single points of the line *ab* constitute these areas.

The brief second part of Galileo's argument represents a straightforward application of the medieval kinematic proportions. First, since the distance *bc*, traversed along the horizontal, is double the distance *ba*, traversed along the inclined plane, the distances are in the same proportion as the corresponding overall velocities, and thus the times are equal.[109] Second, since the distance *be* is equal to the distance *ba*, the times through these distances are in the inverse proportion of the overall velocities, i.e., they are as 1 : 2. Summing up, it follows "that the whole time through *abe* is $1^1/_2$ the time for *ab*."

Although the Double Distance Rule is a correct statement about the motion of fall from the point of view of classical mechanics, Galileo's derivation of it depends on the principle mentioned in the letter to Sarpi, i.e., on the proportionality between velocities and distances in accelerated motion, which is actually not correct in classical mechanics. Nevertheless, Galileo's argument documented above is rigorously correct within the medieval conceptual system underlying the proof, since it contains neither an inconsistency nor an illegitimate conclusion.[110]

3.3.4 Summary

Galileo elaborated and extended the *De Motu* theory and performed experiments that were guided by it. In this way, he arrived at results that are incompatible with this theory: In addition to a number of internal inconsistencies in the *De Motu* theory, such as a paradox resulting from the conclusion that bodies of

---

[109] The same conclusion is the core of Galileo's derivation of the Isochronism of Chords, see note 71. The interpretation given here essentially follows Wisan 1974, pp. 204-207.

[110] This is disputed by some interpreters of Galileo's argument, who reinterpret it in the conceptual framework of classical mechanics; see, e.g., Sylla 1986, p. 84, note 90, where the concept of "average velocity" is used in the sense it has in classical mechanics in order to criticize Galileo's argument.

different materials fall with different speeds, he found, probably by experiments, that the trajectory of projectile motion is a symmetric curve, and that acceleration is permanent and not transitory. His discoveries made it plausible that the motion of fall is characterized by a quadratic relation between distances and times and that the trajectory is parabolic. As a consequence of the recognition that acceleration is not an accidental characteristic of the motion of fall but actually governed by a simple rule, Galileo shifted the emphasis of his reworkings of the *De Motu* theory, and attempted to provide a derivation of the law of fall.

By 1604, he was convinced that he had found a fundamental principle from which he could derive this and other propositions on the motion of falling bodies; he made the assumption, which in fact is not correct, that in a motion of fall the degree of velocity increases in proportion to the distance traversed. His derivation of the law of free fall makes essential use of concepts taken from the medieval tradition of the configuration of qualities. In particular, the increasing overall velocity of the motion of fall is represented by an area in the traditional triangular diagram. Galileo used the same concepts and the same representational technique to derive the so-called Double Distance Rule. According to this rule, which is correct in classical mechanics, a body that is deflected into the horizontal after a motion of fall traverses along the horizontal twice the distance of the preceding motion of fall in the same time.

Besides making use of the medieval configurations of qualities doctrine, Galileo's derivations of the law of free fall and of the Double Distance Rule also invoke relationships between velocities, distances and times. In the case of the Double Distance Rule only the Aristotelian kinematical proportions enter the argument, whereas the derivation of the law of fall contains a transition from velocities to times that rests on a problematical but plausible conclusion which cannot be justified by these proportions.

3.4 Free Fall in Galileo's Manuscripts after ca. 1604

After 1604 Galileo continued his studies of motion and by the time of his move from Padua to Florence in 1610 could claim to have all but completed three books on local motion.[111] Around this time, his interests turned to other fields, in particular astronomy, but in the course of the next 30 years he occasionally returned to problems of mechanics, especially when it was prudent to be occupied with matters without obvious ideological implications. Besides the final synthesis of his theory of motion, presented in the three books entitled *De motu locali* within the *Discorsi* of 1638, there are a number of manuscripts documenting intermediate stages which Galileo considered important enough to

---

[111] See Galileo Galilei to Belisario Vinta, May 7, 1610 (EN X, 348-353).

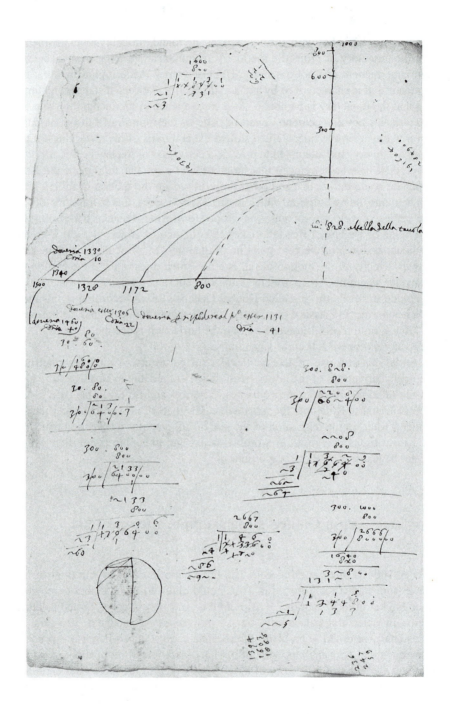

Plate III. An experiment by Galileo on horizontal projection, (fol. 116v). Published with permission of the Biblioteca Nazionale, Florence.

· preserve or have copied. In this section, we shall concentrate on two of these manuscripts. We shall see that, on the basis of a previously derived result, the Double Distance Rule, Galileo discovered a way to apply the traditional kinematic proportions to the motion of fall by equating a degree of velocity in accelerated motion to that of a uniform motion, a way that is different from the reasoning applied in his previous attempts to derive the law of fall. As a consequence he discovered that the degree of velocity in accelerated motion cannot be proportional to space but must be proportional to time, if the law of fall is supposed to hold. But since, up to this point, Galileo had attempted to derive the law of fall from the proportionality between velocities and spaces, this consequence leads to an open contradiction. Consequently, he is driven to new attempts at reconciling his premises, which in turn result in new paradoxes.

3.4.1 The Discovery of the Proportionality between the Degree of Velocity and Time

Two manuscripts, fol. 91v and fol. 152r,[112] document Galileo's discovery of a proportionality of the degrees of velocity in a motion of fall and the times of fall. Folio 91v represents a much more explicit and thus probably later version of an argument that is adumbrated in fol. 152r. In this argument, Galileo derives from the law of fall and the Double Distance Rule the paradoxical result that the degree of velocity increases not in proportion to distance, as he had assumed in his derivation of the law of fall, but in proportion to time.[113] In the course of a

---

[112] MS, fol. 91v. EN VIII, 280, 281-282, and 427. The manuscript contains three texts (see document 5.3.13, and Plate IV). For a corrected transcription of the second of these texts, which will be discussed in this section, see Wisan 1974, p. 227. The first and the third text will be discussed in section 3.5.3. MS, fol. 152r (partly transcribed in EN VIII, 426-427; see also Fig. 3.21).

[113] The argument in the second passage of fol. 91v was first identified as a derivation of the proportionality between degrees of velocity and times of fall by Wisan (1974, pp. 227-229). The relationship between this text and the notes in fol. 152r which document the same line of reasoning has not been recognized before. Galileo's discovery of the proportionality between the degrees of velocity and the times of fall has been the subject of numerous studies. Naylor (1980a, pp. 562-566) claims that Galileo encountered a conflict with the principle mentioned in the letter to Sarpi when studying the composition of velocities in projectile motion. This reconstruction cannot be correct, since, as we shall see in section 3.5.3, Galileo had not yet mastered the composition of impetuses in horizontal projection at a time when he already used the correct proportionality between degrees of velocity and times of fall. Wisan (1984, in particular pp. 276ff) has claimed that the discovery of this proportionality is due to a crucial experiment by which Galileo supposedly tested the principle mentioned in the letter to Sarpi. Her reconstruction of the supposed crucial experiment on the basis of a manuscript (MS, fol. 116v) has been convincingly refuted by Hill (1986, pp. 284-288). In his own reconstruction of Galileo's discovery Hill

rather complex analysis, documented by scattered notes in fol. 152r, Galileo elaborated the consequences of this discovery, with the result that further inconsistencies in the conceptual foundations of his derivation are disclosed. However, the analysis in fol. 152r also hints at a new derivation of the law of fall, a sketch of which is contained in a manuscript to be presented in section 3.4.2.

3.4.1.1 Double Distance Rule and Time Proportionality

Let us first turn to fol. 91v in which Galileo provides a detailed proof of the following assertion[114]:

---

argues that it depends on the prior discovery of the proportionality between the velocities and the square roots of the distances traversed in a motion of fall, a proportionality that is in fact equivalent to the proportionality between degrees and times, if the law of fall is taken for granted and if velocities are understood in the sense of degrees of velocity. Hill claims that the first proportionality is necessarily entailed by the Isochronism of Chords, but this is problematical. In fact, in his speculative reconstruction of Galileo's discovery, which is not supported by direct manuscript evidence, Hill argues that the Isochronism of Chords implies that the velocities along the chords are in the same proportion as the square roots of the vertical descents, obviously referring to velocity as an overall characteristic of the motion. He then assumes that the velocities along the chords are the same as the velocities along the corresponding vertical descents, a highly problematical assumption, as will be shown in section 3.4.2. On the basis of this problematical assumption, Hill finally draws the conclusion that the velocities of the motion of fall along the verticals must be in the same proportion as the square roots of the lengths of these verticals, now apparently referring to velocities in the sense of degrees of velocity, since he claims that Galileo tested this conclusion in the experiment on projectile motion also referred to by Wisan, an experiment which, in fact, involves instantaneous velocities. Hill's failure properly to distinguish between overall velocity and degree of velocity, together with the use of a problematical assumption, makes his reconstruction unconvincing; see Hill 1986. The interpretation of Galileo's refutation of the proportionality between the degrees of velocity and the distances traversed given in Sylla 1986, pp. 79-82, is a speculative reconstruction not based on manuscript evidence, in which use is made of an argument similar to the one later published in the *Discorsi* (see section 3.6.1). In his most recent account of Galileo's rejection of the Sarpi principle, Drake (1989b, p. 55, and 1990, pp. 104-105) claims that Galileo abandoned this principle after recognizing a paradox concerning motion along inclined planes, but he does not provide a detailed reconstruction. For a discussion of this paradox, see the following section.

[114] MS, fol 91v (EN VIII, 281-282 and Wisan 1974, p. 227). This passage consists of one long paragraph which has been divided up here for better understanding. See document 5.3.13 and Plate IV.

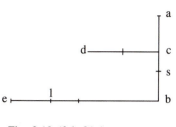

Fig. 3.19 (fol. 91v)

In motion from rest the moment of velocity and the time of this motion grow in the same ratio. For let there be a motion through *ab* [Fig. 3.19] from rest in *a*, and let an arbitrary point *c* be assumed; and let it be posited that *ac* is the time of fall through *ac*, and the moment of the acquired speed in *c* is also as *ac*, and assume again any point *b*: *I say that the time of fall through* ab *to the time through* ac *will be as the moment of velocity in* b *to the moment in* c.

As in fol. 179v, discussed in section 3.3.3.1, Galileo describes the increase of velocity during the vertical fall through the points *c* and *b* in terms of the moments of velocity acquired at these points; he claims that they are in the same proportion as the corresponding times of fall, in contrast to the proportionality between degrees and distances assumed in the letter to Sarpi.

In order to represent the times of fall in his diagram, Galileo uses the law of fall in its Mean Proportional Form; he then reformulates his claim with respect to the lines representing the times of fall:

Let *as* be the mean [proportional] between *ba* and *ac*; and since the time of fall through *ac* was set to be *ac*, *as* will be the time through *ab*: it thus has to be shown that the moment of speed in *c* to the moment of speed in *b* is as *ac* to *as*.

The proof proceeds in three steps. In the first step, Galileo considers deflections into the horizontal of the motion of fall at point *c* and at point *b*. He applies the Double Distance Rule in order to determine the distances traversed in uniform motions along the horizontal lines through *c* and through *b* in the times of fall through *ac* and *ab*, respectively; these distances are represented by the lines *dc* and *be*. The two uniform motions thus obtained are characterized by unequal distances traversed in unequal times and can therefore not be directly compared by means of the Aristotelian proportions. In the second decisive step, however, Galileo uses the proportionality between distances and times in uniform motion following from an Archimedean proposition[115] in order to find the distance along the horizontal through *b* that is traversed in the time of fall through *ac*; this distance is given by the line *bl* in Galileo's diagram. In this way, a third motion is introduced which can be compared to each of the two uniform motions resulting from the Double Distance Rule by simply applying the traditional kinematic proportions. Since, as Galileo states in his proof, "the moments of speed are to one another as the spaces, which according to these moments are traversed in the same time,"[116] a statement which corresponds to one of the two

---

[115] See sections 1.2.1 and 1.2.2.

[116] Galileo's use of the notion of moment as referring to a velocity to which the Aristotelian proportions can be applied is documented in MS, fol. 151r (EN VIII, 378); see also note 71.

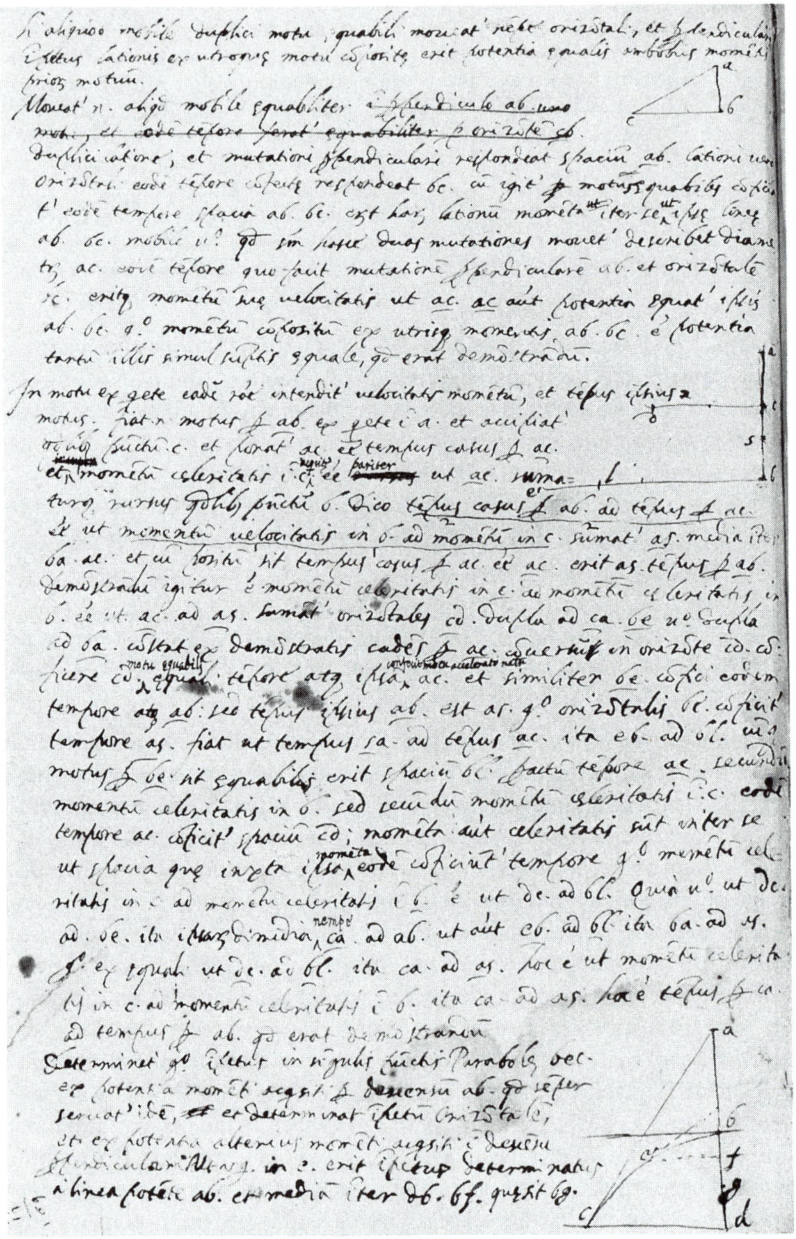

Plate IV. Galileo's argument for time proportionality (fol. 91v). Published with permission of the Biblioteca Nazionale, Florence.

Aristotelian proportions, the moment of speed in c is to the moment of speed in b as dc to bl. In the short third and final step of his proof, Galileo makes use of the law of fall to show that dc and bl, and thus the moments of speed at c and b, are in the same proportion as the times of fall through ac and ab. Here is the text of his proof:

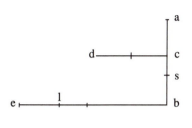

Fig. 3.20 (fol. 91v)

Assume the horizontal [line] cd to be double ca [Fig. 3.20], but be to be double ba: it follows from what has been shown, that the [body] falling through ac, deflected into the horizontal cd, will traverse cd in uniform motion in an equal time as it also traversed ac in naturally accelerated motion; and, similarly, it follows that be is traversed in the same time as ab: but the time of ab itself is as: therefore, the horizontal [line] be is traversed in the time as.

Let eb be to bl as the time sa is to the time ac; and since the motion through be is uniform, the space bl will be traversed in the time ac according to the moment of speed in b: but according to the moment of speed in c, in the same time ac the space cd will be traversed; but the moments of speed are to one another as the spaces, which according to these moments are traversed in the same time: therefore the moment of speed in c is to the moment of speed in b as dc to bl.

But as dc to be, so are their halves, i.e., ca to ab; but as eb to bl, so ba to as; therefore, by the same [ex aequali], as dc to bl, so ca to as: that is, as the moment of speed in c to the moment of speed in b, so ca to as, that is, the time through ca to the time through ab. Quod erat demonstrandum.

Galileo's conclusion, that the moments of velocity are in the same proportion as the times of fall, amounts to a contradiction, since he had previously used the proportionality between the moments of velocity and the distances of fall in order to derive the law of fall and the Double Distance Rule, i.e., two propositions from which he now concluded that the moments of velocity grow in proportion to the times of fall.

Galileo probably first discovered the incompatibility of the proportionality between moments of velocity and distances with the law of fall and the Double Distance Rule by a brief argument traces of which are found in fol. 152r.[117]

---

[117] According to Drake (1990, p. 101) the notes in fol. 152r (reconstructed in the following) refer to a study by Galileo of accretions of impetus to the natural tendency downward, following what Drake describes somewhat vaguely as the "medieval impetus theory as mathematized by Albert of Saxony." In Drake 1978, p. 93, he claims that the result of this study appeared to vindicate the argument from which, in fol. 163v, the Double Distance Rule was derived. In the following, however, we shall argue that the argument documented in fol. 152r presupposes the Double Distance Rule and obtains a conclusion that is incompatible with the Sarpi principle, which in turn was used as a premise in the derivation of the Double Distance Rule in fol. 163v.

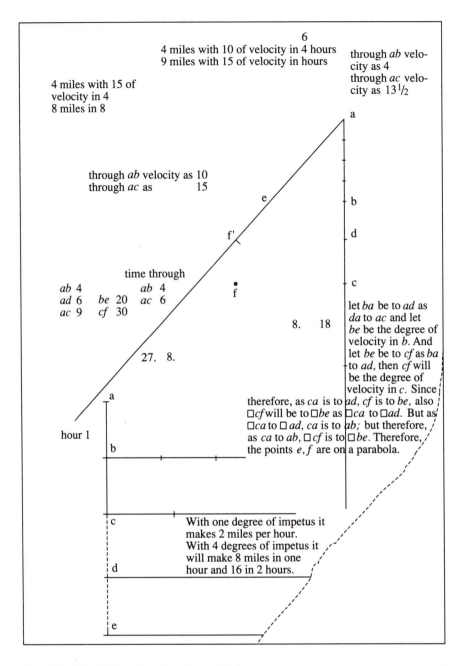

4 miles with 10 of velocity in 4 hours
9 miles with 15 of velocity in hours

through *ab* velocity as 4
through *ac* velocity as 13½

4 miles with 15 of velocity in 4
8 miles in 8

a

through *ab* velocity as 10
through *ac* as   15

e

b

d

f'

c

time through

ab 4		*ab* 4
ad 6	*be* 20	*ac* 6
ac 9	*cf* 30	

f

let *ba* be to *ad* as *da* to *ac* and let *be* be the degree of velocity in *b*. And let *be* be to *cf* as *ba* to *ad*, then *cf* will be the degree of velocity in *c*. Since therefore, as *ca* is to *ad*, *cf* is to *be*, also □*cf* will be to □*be* as □*ca* to □*ad*. But as □*ca* to □ *ad*, *ca* is to *ab*; but therefore, as *ca* to *ab*, □ *cf* is to □*be*. Therefore, the points *e, f* are on a parabola.

8.   18

27.   8.

a

hour 1

b

c

With one degree of impetus it makes 2 miles per hour.
With 4 degrees of impetus it will make 8 miles in one hour and 16 in 2 hours.

d

e

Fig. 3.21 (fol. 152r); point *f'* has been added.

These traces consist of a diagram and a brief text in the lower part of the manuscript. The other diagram and the texts in this manuscript document Galileo's attempts to explore the consequences of his discovery.

In a brief text (see Fig. 3.21) in the lower part of fol. 152r, he relates 1 and 4 degrees of impetus to certain distances traversed in certain times:

> With one degree of impetus it makes 2 miles per hour.

> With 4 degrees of impetus it will make 8 miles in one hour and 16 in 2 hours.

The distances and times to which the degrees of impetus are related in this statement are compatible with the Double Distance Rule if it is assumed that a falling body acquires 1 degree of impetus after falling through 1 mile in 1 hour. It then follows from the Double Distance Rule that this degree of impetus is characterized by a uniform motion along the horizontal in which a body traverses "2 miles per hour," as Galileo's text has it. Four degrees of impetus, on the other hand, correspond to a motion which is 4 times quicker than a motion characterized by 1 degree of impetus; from one of the Aristotelian proportions it thus follows that "With 4 degrees of impetus it will make 8 miles in one hour." If this statement is referred to a uniform motion, it follows in fact from the Archimedean proposition on uniform motion also used in fol. 91v that "16 [miles are traversed] in 2 hours," as Galileo states.

The diagram in the lower part of fol. 152r is similar to the diagram in fol. 91v; it suggests that the degrees of impetus mentioned in the brief text refer to a motion of fall for which a deflection into the horizontal at different points is considered. In the upper part of this diagram 2 points along the horizontal roughly mark 2 units of distance as defined by the vertical line ab. Next to this line Galileo wrote "hour 1" which apparently means that a fall from a to b would take 1 hour. The vertical line is divided in 4 roughly equal parts, marked by the points a, b, c, d, and e. From each of the latter 4 points a horizontal line is drawn, indicating that at these points Galileo considered, as he did in his argument in fol. 91v, a deflection of the motion of fall into the horizontal.

If Galileo in fact referred his brief text to the diagram just described, he must immediately have concluded that the degree of impetus cannot be proportional to the distance of fall. For, if the degree of impetus were to grow in proportion to distance, a falling body would acquire the 4 degrees of impetus mentioned in the text after falling through the 4 units of distance that are marked in Galileo's diagram. According to the law of fall, the body would take 2 hours in order to traverse this distance, since 1 hour was needed to fall through a distance of 1 mile. The Double Distance Rule then implies that, if the body is deflected into the horizontal after these 2 hours, it will traverse 8 miles in 2 hours in a uniform motion. Therefore, 4 degrees of impetus are characterized by a uniform motion that traverses 8 miles in 2 hours. But, according to the kinematic proportion that led to the second proposition of this text, 4 degrees of impetus correspond to 8 miles traversed in 1 hour or in other words (for uniform motion) 16 miles traversed in 2 hours. Hence, the assumption of a proportionality between the degree of impetus and distance leads to a contradiction.[118]

In a similar way, Galileo could conclude that the assumption of a proportionality between degree of impetus and time is compatible with the different possibilities to pass from 1 degree of impetus to 4 degrees of impetus, given by applying the rules of kinematics, the law of fall and the Double Distance Rule. From the Double Distance Rule and the rules of kinematics it follows as before that "With 4 degrees of impetus it will make 16 [miles] in 2 hours." But since now the degree of impetus is assumed to grow in proportion to time, it follows from the law of fall that 4 degrees of impetus are acquired after falling through 16 miles in 4 hours. According to the Double Distance Rule, with these 4 degrees, a body traverses in a uniform horizontal motion 32 miles in 4 hours, or 16 in 2 hours, which is compatible with the previous calculation.

Galileo's derivation of the proportionality between the moment of velocity and time in fol. 91v is a more explicit and more general version of the same argument, using a somewhat different terminology for what are essentially – at least in this context – the same concepts. The argument documented in fol. 91v truly represents a new achievement when compared to Galileo's attempted derivations of the law of fall reconstructed in the previous section. Whereas in these earlier derivations, it was possible to apply kinematical relationships to the accelerated motion only after the transition from the single degrees of velocity to the overall velocity of the motion had been performed by means of the medieval representational technique, the argument in fol. 91v allows for an application of the Aristotelian proportions directly to a single degree of the accelerated motion. This is particularly evident from Galileo's short-hand characterization in fol. 152r of a degree of impetus in terms of a distance and a time: "With one degree of impetus it makes 2 miles per hour." This short-hand formulation makes it also clear that this advance is due to the Double Distance Rule by which a single degree is identified with a uniform motion.

On the other hand, as illustrated by Galileo's argument in fol. 91v, comparisons of uniform motions can easily be based on the traditional kinematic proportions, even when the motions compared do not satisfy the restrictions imposed by the Aristotelian proportions, i.e., either traverse equal distances or employ equal times. In fact, Galileo compared the uniform motions along cd and be (see Fig. 3.20), which neither take the same time, nor traverse the same distance, by simply introducing a third motion along bl which has the same speed as the motion along be, being part of the same uniform motion, but which takes the same time as the motion along cd. Now the Aristotelian proportions suffice to compare all motions involved, and, since the uniform motions are associated to degrees of velocity, these degrees can also be compared to one another even though they are the maximal degrees of two accelerated motions – one along ac,

---

[118] A similar refutation of the proportionality between the degrees of velocity and distances has been identified by Drake (1970a, pp. 39-41) in a later published work Giovanni Andres,(1779) *Raccolta di opuscoli scientifici, e letterarj,* Ferrara, p. 64.

the other along *ab* – which neither take the same time nor traverse the same distance. Hence, the Double Distance Rule is the tool by which a whole new set of conclusions can be reached about degrees of velocity in accelerated motion, although this rule itself was obtained by strict application of the medieval representation of velocity in connection with the Aristotelian proportions.

3.4.1.2 A Reexamination of the Law of Fall

Although Galileo could, as we have seen, derive the proportionality between moments of velocity and times from the law of fall, the Double Distance Rule and the traditional rules of kinematics, he could not, however, consider this proportionality as a well established result, since he had previously derived the law of fall from the contrary assumption of a proportionality between moments of velocity and distances.

In order to resolve this paradox, Galileo therefore had to reconsider his derivation of the law of fall. In the remaining part of fol. 152r, we can see him exploring the consequences for the concepts and representational techniques applied in his earlier derivation of the assumption that the degrees or moments of velocity grow in proportion to time.[119]

---

[119] Most interpretations of fol. 152r have concentrated on this part of the manuscript. Drake (1973b) was apparently the first systematic interpretation. At that time, he saw fol. 152r as "the starting point of the modern era of physics" (p. 90). According to this interpretation, Galileo discovered the law of fall in this manuscript by accident in a search for consistent ratios between speeds, distances, and times (p. 89). This interpretation was, however, based on an erroneous transcription; it has been severely criticized by Wisan (1974, p. 214, note 9), and Naylor (1977a, pp. 367-371). In later publications Drake (1978, pp. 91-93, 1990, pp. 100-102) has corrected his transcription but not changed his interpretation. Since all relationships between distances and times in fol. 152r indicate, contrary to Drake, that Galileo presupposed the law of fall, all other interpretations of this manuscript relate it to a period in which Galileo was attempting to prove the law. The alternative interpretation suggested by Wisan (1974, pp. 210-214) is, however, based on the assumption that Galileo performed erroneous calculations based on "some vaguely recalled medieval formulas for local motion." Her interpretation has been convincingly refuted by Naylor (1977a, pp. 377-380), who (pp. 381-386) has succeeded in reconstructing the figures 4 and $13^1/_2$ in the upper right-hand corner (see Fig. 3.21) as resulting from a calculation of overall velocities based on the traditional represesentation of velocity by an area. The partial interpretation of fol. 152r by Sylla (1986, p. 76), as well as the complete one given here, are indebted to Naylor for stressing this point. In several interpretations, such as those by Naylor (1977a, p. 386), Drake (1978, pp. 135-136), and Romo Feito (1985, p. 107), the significance of the difference between the proportionality between the degrees of velocity and the times of fall, on the one hand, and the proportionality between the degrees and the square roots of the distances, on the other hand, is emphasized. Romo Feito even claims that in fol. 152r Galileo used the latter and attempted to avoid the former. In view of the fact that these

As we shall see, the various notes and the diagram in the upper two-thirds of fol. 152r can be reconstructed as referring to two possibilities of relating the assumption of time proportionality to the traditional representation of velocity:

(1) One possibility is directly based on the traditional triangular diagram as it is interpreted in Galileo's attempted proofs of the law of fall. In this interpretation, the extension represents the distance traversed, an infinite sequence of parallel lines, filling the triangular area, represents the degrees of velocity increasing in proportion to the distances traversed, and the area of the triangular diagram represents the overall velocity. Now, if degrees which increase in proportion to time are represented in this diagram by a sequence of parallel lines, these parallel lines will no longer correspond to a triangular area but to the area of a parabolic section. In fact, it directly follows from the law of fall that the proportionality of the degrees of velocity and the times is equivalent to a proportionality of these degrees and the square roots of the distances traversed. As a consequence, the endpoints of the parallel lines representing these degrees in the traditional diagram whose extension is given by the distance lie on a parabola. Hence, the velocity which is characterized by the assumption that its degrees increase in proportion to time is represented by the section of a parabola.

(2) The second possibility goes back to the intrinsic ambiguity of the traditional representation of velocity analyzed in section 1.2.3. Since the traditional diagram primarily represents the quality that changes, the identification of the extension of this diagram is dependent on the context of application. Although Galileo was committed, as we have seen, to an interpretation of this extension as the distance traversed, it was indeed possible for him to switch to an interpretation of this extension as the time elapsed without leaving the conceptual foundations on which the representation is based. But if the extension in the diagram of Galileo's attempted proofs of the law of fall is thus reinterpreted as representing time rather than distance, the assumption that the degrees of velocity increase in proportion to time can be represented in a much more simple way than described under (1). In fact, if the extension is taken to represent time, the assumption of time proportionality again results in a triangular diagram. Hence, velocity in accelerated motion can again be represented in the traditional way as a uniformly difform quality.

In the following, we shall see that, in fol. 152r, Galileo explored these two possibilities of representing velocity but did not succeed in basing a proof of the law of fall on either of them.

The diagram in the center of fol. 152r shows a vertical line ac and a line ae inclined to it. The lettering of the diagram reappears in the scattered notes surrounding it; Galileo's interpretation of the diagram is most easily inferred from the text in the left middle of the diagram and from the longer text on the right-hand

---

two proportions are mathematically equivalent if the law of fall is taken for granted, this emphasis seems to be misleading.

side of the manuscript. The first of these texts consists of a table, listing distances in the first row, and times in the last row:

		time through
ab 4		ab 4
ad 6	be 20	ac 6
ac 9	cf 30	

According to this table the lengths *ab* and *ac* correspond to successive square numbers and thus denote distances traversed in a motion of fall from the state of rest, while the values given for *ab* and *ac* in the last row can plausibly be identified as the times of fall through these distances. The significance of the middle row is at first sight unclear. If, in analogy to the diagram used in the attempted proofs of the law of fall, the line *ae* is interpreted as marking the endpoints of infinite parallel lines constructed on *ac*, the present diagram again represents the increase of the degrees of velocity with distance, since the line *ac* represents the distances traversed in a motion of fall.[120]

However, the text on the right-hand side of the manuscript identifies the lines *be* and *cf*, listed in the middle row of the little table on the left-hand side, as the degree of velocity in *b* and in *c*, respectively; the table gives the values *be* = 20 and *cf* = 30 for the lengths of these lines which approximately correspond to the distances between the points *b* and *e* and *c* and *f* in the diagram. The values 20 and 30 for *be* and *cf* are obviously not proportional to the corresponding distances of fall *ab* = 4 and *ac* = 9, which is also clear from the fact that the point *f* does not lie along the line *ae* in Galileo's diagram.

Actually, Galileo determined the degrees of velocity represented by *be* and *cf* as being proportional to the corresponding times of fall so that the same diagram contains a representation of space proportionality (indicated by the line *ae*) and of time proportionality (indicated by the points *e* and *f*). The first part of the longer text on the right-hand side shows how he obtained the representation of time proportionality:

> let *ba* be to *ad* as *da* to *ac* and let *be* be the degree of velocity [*gradus velocitatis*] in *b*. And let *be* be to *cf* as *ba* to *ad*, then *cf* will be the degree of velocity in *c*.

The line *ad* (= 6, according to Galileo's table) is determined as the mean proportional between *ab* = 4 and *ac* = 9. According to the Mean Proportional Form of the law of fall, it represents the time of fall along *ac* if the line *ab* is taken as the representation of the time of fall along *ab*. The line *be* represents the

---

[120] The actual distances corresponding to *ab* and *ac* in the diagram of the upper part of fol. 152r are approximately 4 and 8, and not 4 and 9, as indicated by Galileo's table. This feature of the diagram suggests that it may initially have been drawn with the intention of representing subsequent equal distances traversed in a motion of fall, as in the diagrams accompanying Galileo's attempted proofs discussed in the previous section.

degree of velocity at b, and cf is constructed to be in the same proportion to ad as be is to ab. Hence, the degrees are in the same proportion as the times.

So far, only degrees of velocity at two points have been determined under the hypothesis that the degrees grow in proportion to time, whereas the derivation of the law of fall requires a representation of all degrees of velocity between two given points. In the case of space proportionality, this infinite set of degrees is represented by a triangular diagram; in the present case, however, this representation cannot be used since it is obvious from Galileo's diagram that the line be is not contained in the triangle acf.

In the continuation of the text on the right-hand side of the manuscript, Galileo determined the general character of the curve through the points e and f that marks the endpoints of the lines representing degrees of velocity which increase in proportion to time:

> Since therefore as ca is to ad, cf is to be, also the square of cf will be to the square of be as the square of ca to the square of ad. But as the square of ca is to the square of ad, so is ca to ab; but therefore as ca is to ab, so is the square of cf to the square of be. Therefore, the points e, f are on a parabola.

The parabola through the points a, e, and f will mark the endpoints of the infinite degrees of velocity which the motion of fall passes through between the points a and c.

The relationship between degrees and distances represented by this parabola can also be expressed as Galileo did in a brief note in another manuscript, fol. 164v[121]: "The moments of the velocities of a body falling from the highest point are to one another as the square roots of the distances traversed, that is in half proportion of these [distances]."

It was claimed above that Galileo had two possibilities of representing the assumption of time proportionality in his diagram. So far, we have examined the representation of time proportionality by a parabolic line as an addition to the usual triangular diagram. But he also had the possibility of directly adapting the triangular diagram to the assumption of time proportionality by reinterpreting the extension of this diagram as time rather than as distance. In fact, in order to apply this reinterpretation to the diagram in fol. 152r, the times of fall would first have to be marked along the vertical. According to the little table, the times for ab and ac are represented by the lines ab and ad. It is therefore sufficient to take the lines which already represent degrees of velocity that are in the same proportion as the times, i.e., the lines be and cf, and to attach them to the points b and d, marking the times ab and ad. For the line be this operation does not result in any change, since the line ab was taken to represent both a distance and a time. For the line cf, this simply means that it is shifted upwards so as to connect the point d to the oblique line ae and to restore the triangular diagram which now represents a proportional increase of the degrees of velocity with time.

---

[121] MS, fol. 164v (EN VIII, 380).

The two possibilities of representing the proportional increase of velocity with time so far discussed throw light on the relationship between time proportionality and space proportionality; however, they do not by themselves constitute a new derivation of the law of fall. But scattered notes and numbers in fol. 152r indicate that Galileo tried to use both approaches in order to construct such a derivation.

Once a representation of the increasing degrees of velocity by a geometrical figure is established, Galileo's earlier attempt at a proof of the law of fall suggests the next step of a new derivation: the passage from degrees of velocity to overall velocities between certain points of the motion of fall. According to the line of argument followed in this type of proof, the relationship between such overall velocities is given by the relationship between the corresponding areas in the diagram. For this reason, according to Galileo's first representation of time proportionality, overall velocities are represented by sections of a parabola.

Since the areas of parabolic sections are in the same proportion as the inscribed triangles, the area of the parabolic section denoted by the points *acf* in fol. 152r is to the area of the section denoted by *abe* as the area of the triangle *acf* is to that of the triangle *abe*:

$$\Delta acf : \Delta abe = (ac \times fc) : (ab \times eb) = (9 \times 30) : (4 \times 20) = 270 : 80 = 27 : 8$$

Therefore, by the above argument, the overall velocity through *ac* is to the overall velocity through *ab* as 27 : 8. Folio 152r contains evidence that Galileo indeed determined these overall velocities. He wrote the figures 27 and 8 at the bottom of his diagram, and in the upper right-hand corner he made the note:

through *ab* velocity as 4,
through *ac* velocity as $13^{1}/_{2}$.

Since $27 : 8 = 13^{1}/_{2} : 4$, this statement agrees with the above calculation.

The second possibility of representing the proportionality between the degrees of velocity and the times of fall, i.e., the triangular diagram in which the extension is given by the time, also allows for a determination of the overall velocities. But if the extension of the diagram is interpreted as representing time, these overall velocities are, as in Galileo's first proof of the law of fall, represented by triangular areas covering the degrees of velocity passed through until a certain point of the motion. The relationship between the velocities through *ab* and *ac* is thus given by the proportion between the area of the triangle *abe* and that of the corresponding triangle which has *ad* as one of its sides and which can be denoted by adf' [122]:

$$\Delta abc : \Delta adf' = (4 \times 20) : (6 \times 30) = 80 : 180 = 8 : 18.$$

---

[122] The point f' is not marked or labeled in Galileo's original diagram; it has been added to the transcription in order to make the following calculation more transparent.

The numbers 8 and 18 are in fact found in the proximity of the points f and c in Galileo's manuscript, showing that he also considered this second alternative of accommodating his geometrical representation of velocity to the proportionality between velocity and time.

Remarkably, Galileo's two ways of determining the relationship between the overall velocities through ab and ac yield different results: $4 : 13^1/_2$ in the first case, and $8 : 18$ or $4 : 9$ in the second case. Since both proportions are different from the proportion obtained in Galileo's earlier attempted proof of the law of fall – here the proportion of the overall velocities was that of the squares of the corresponding distances, i.e., in the present case, equal to $16 : 81$ – neither of the two representations of velocity in fol. 152r can be completed to a proof of the law of fall by following the line of argument used in this first proof.

Moreover, the application of kinematical rules to the overall velocities resulting from the representation in which the extension is interpreted as time even leads to an absurd conclusion. In this representation, the overall velocities through ab and ac are in the proportion of $4 : 9$. According to the rule that if the velocities are in the same proportion as the distances then the times must be equal, the time of motion through the distance ab must be the same as that through ac, or, in the other words, the motion of fall must be instantaneous.[123]

Although the argument reconstructed from fol. 152r and fol. 91v shows that the proportionality between the degrees of velocity and times can be derived from the law of fall, we have now seen that the different determinations of overall velocities which Galileo obtained on the basis of his traditional representational technique are clearly not compatible with this law, even when the assumption of time proportionality is used. If the elements entering the argument in fol. 91v, i.e., law of fall, time proportionality, and Double Distance Rule, are not themselves called into question, it becomes a plausible conclusion, in view of the paradoxes resulting from the use of the traditional representation of velocity in fol. 152r, that it is precisely the representation of overall velocity by an area in this diagram which is the key problem. But although Galileo may have indeed drawn this conclusion himself, this would not have helped him much without a viable alternative for analyzing accelerated motion.

However, such an alternative possibility of analysis was offered to him precisely by the Double Distance Rule as it was used in the argument in fol. 91v. As we have seen in the above discussion of this argument, the Double Distance Rule allows one to correlate a degree of velocity with a uniform motion to which the Aristotelian proportions can then be applied. The degree of velocity is here the maximum degree of the accelerated motion, and the Double Distance Rule demonstrates that the uniform motion characterized by this degree has double the overall velocity of the accelerated motion. Hence, a uniform motion

---

[123] The kinematic rule used in this reconstruction had actually also been applied by Galileo to accelerated motion, as for instance in the derivation of the Isochronism of Chords; for references, see note 71.

characterized by half the maximum degree should have exactly the same overall velocity as the accelerated motion. Although based on a different argument, this result is quite similar to the traditional Merton Rule, in which the role of the half maximal degree is played by the mean degree (see section 1.2.3).

Now, if overall velocities over different times of the accelerated motion have to be compared with one another, this comparison can be based directly on a comparison of half the maximal degrees. Although this comparison is based on a result essentially equivalent to the Merton Rule of the traditional configuration of qualities doctrine, the comparison of overall velocities of accelerated motion by this argument has an implication that is actually incompatible with the traditional representation of velocity. In fact, the overall velocities can no longer be represented by the areas in the traditional diagram, since they increase exactly in the same proportion as the degrees of velocity, and hence as the times of the accelerated motion, i.e., the overall velocities are proportional to the extension of the diagram and no longer to its area.

In fact, some notes in fol. 152r indicate that Galileo examined this possibility of determining overall velocities. In the little table on the left-hand side of this manuscript, the degree of velocity attained after a motion of fall through the distance ab is given as $be = 20$, and the degree of velocity attained after a fall through the distance ac as $cf = 30$. Hence, by the above argument, the velocities through ab and ac should be as 10 and 15. On the left-hand side of the page slightly above this table, Galileo in fact states:

> through ab velocity as 10,
> through ac as 15.

These values for the velocities through ab and ac appear again in the uppermost text of 152r (in the middle):

> 6
> 4 miles with 10 of velocity in 4 hours,
> 9 miles with 15 of velocity in hours.

The values for the times and distances considered in this text coincide with those of the motion of fall represented by the main diagram and by the table of times and distances in the left middle. In agreement with the numerical values given in this table, the "6" written on top of the first line of this text represents the time for the motion traversing the distance of 9 miles. The fact that the figure "6" is not written in its proper place in the second line of the text suggests that it was calculated on the basis of the other numerical values given in this text and was later filled in. In the following we shall see that the calculation documented by this text was indeed a control of whether or not the values for the velocities are compatible with the law of fall.

The following reconstruction of this calculation as a check of compatibility between the values 10 and 15 for the overall velocities through ab and ac and the law to fall is based on the identification of the principle difficulty of this check and of the identification of an argument suitable to overcome this difficulty, traces of which are in fact documented in Galileo's manuscript. The principal difficulty of the calculation for Galileo was that none of the traditional kinematical rules applied to these velocities yields the desired result, i.e., the time of fall through 9 miles. In fact, since the two motions compared by Galileo neither traverse the same distances, nor have the same velocity, nor take the same time, the conditions for applying one of these traditional rules are not given.

It seems plausible, however, to attempt a solution of this problem along the line of reasoning employed in the argument quoted from fol. 91v, i.e., in this case, by introducing a third motion having the same velocity as the second motion (15 units), but traversing the same distance as the first motion (4 miles). Hence this third motion can be characterized by:

$$4 \text{ miles with } 15 \text{ of velocity in } x \text{ hours,} \tag{1}$$

where x is a time to be determined as follows. If in the first motion a body travels 4 miles with 10 of velocity in 4 hours (as stated in the first line of fol. 152r), the same *distance* of 4 miles is traversed with 15 of velocity in a time that results from the appropriate medieval rule (for equal distances) as $(10 : 15) \times 4 = \frac{8}{3}$ hours, i.e.,

$$4 \text{ miles with } 15 \text{ of velocity in } \frac{8}{3} \text{ hours.} \tag{2}$$

And if we accept that when the velocities are equal, the times are in the same proportion as the distances traversed, we can now calculate the time for the motion which traverses 9 miles with 15 of velocity. Since the distances are as $9 : 4$, the time can be found as $(9 : 4) \times \frac{8}{3} = 6$ hours, i.e.,

$$9 \text{ miles with } 15 \text{ of velocity in } 6 \text{ hours} \tag{3}$$

which exactly corresponds to the second line in fol. 152r. However, the inference that, if the velocities are equal, the times are in the same proportion as the distances traversed, is not one of the two medieval proportions which are restricted to either equal distances or equal times. It corresponds rather to a property of uniform motions which Archimedes described without using the concept of velocity,[124] and on which Galileo based (as we shall see in the next section) a definition of equal velocities that he also applied to accelerated motion. But before turning to this definition and its problematical character, we shall see that an argument similar to the one just proposed was in fact used by Galileo in order to check the compatibility of the values 10 and 15 for the overall velocities with the law of fall.

Somewhat below the text quoted above, Galileo wrote:

4 miles with 15 of velocity in 4,
8 miles in 8.

---

[124]   See section 1.2.1

If one attempts a reconstruction based on the conjecture that this text refers to a third motion introduced in order to compare the first with the second motion, it is puzzling that Galileo claims the time for a motion through 4 miles with 15 of speed to be 4 hours, and not $8/3$ hours as we calculated in (2). If this text really serves the same function as the auxiliary calculation which resulted in (2), we must assume that Galileo used a false value for the time of the third motion. This perplexing problem for a reconstruction of Galileo's reasoning can, however, be solved by paying attention to mathematical techniques of the time. It was in fact a common mathematical procedure in Galileo's time, to solve what today correspond to systems of linear equations by introducing a "false value." This technique is explained at length in Tartaglia's standard textbook, from which Galileo, too, apparently learned mathematics.[125]

Keeping this in mind, one can reconstruct Galileo's line of thought as follows: He first introduced a third, uniform motion having the *same velocity* as the second and traversing the *same distance* as the first motion, and he simply assumed this motion to take a time of 4 hours, thus giving a value that is too large in the proportion of 3 : 2 with respect to the true value $8/3$:

4 miles with 15 of velocity in 4.

If now this third motion is assumed to be a uniform motion, then the Archimedean rule – that in uniform motion the distances traversed are always in the same proportion as the times – can be applied to it. The fact that Galileo did make this assumption is indicated by the next line of text; for if a body in uniform motion traverses 4 miles in 4 hours, it will traverse

8 miles in 8.

Accordingly, this motion will also traverse 9 miles in 9 hours. But since Galileo had committed an "error" by taking a value for the time of the third motion that is too large in the proportion of 3 : 2, also the time of 9 hours resulting from the assumption of this false value must be too large in the same proportion. Consequently, the correct value is 6 hours so that we finally have:

9 miles with 15 of velocity in 6 hours.   (4)

But, properly speaking, this statement still refers to the third (uniform) motion (after it has been corrected for the false value of the time assumed initially), since it was obtained by the Archimedean rule that, for uniform motion, distances are in the same proportion as the times. Only on the assumption that the Archimedian rule characterizes "equal velocities" no matter whether the motion is uniform or accelerated, does it follow that the values for overall velocities are compatible with the law of fall. In the next section, a document will be analyzed in which Galileo made explicit use of this definition of the equality of velocities.

---

[125] See Tartaglia 1556a, in particular Chapter 16, pp. 240ff. For evidence of the role played by this book in Galileo's mathematics education, see Renn 1988, Appendix B.

The above reconstruction has shown that, if this definition of velocity is assumed, the law of fall can in fact be demonstrated to be compatible with the statement that the overall velocities for particular intervals in a motion of fall are in the same proportion as half the maximal degrees acquired at the last instant of the interval. Since the degrees of velocity are in the same proportion as the times of fall, this implies that the law of fall is compatible with the statement that the overall velocities are in the same proportion as the times. On the other hand, Galileo's exploration of the geometrical representation of velocity, traces of which are found in fol. 152r, makes it clear that these statements are not compatible with the traditional representation, because the representation of overall velocity by an *area* implies that the overall velocities are not in the same proportion as the maximal degrees, their halves, or the times of fall.

Hence, the result of the intricate study documented by fol. 152r was that a derivation of the law of fall could possibly make use of the proportionality between overall velocities, degrees of velocity, and times of fall but not of the relationship between overall velocity and degrees of velocity as it is established by the geometrical representation of velocity.

3.4.2   A New Approach to the Derivation of the Law of Fall and a New Paradox

The Aristotelian kinematic proportions are restricted to a comparison of motions characterized by equal times or by equal distances. In the preceding section, however, we have encountered arguments documented by Galileo's working papers, in which even motions that do not satisfy these restrictions are compared. How did Galileo achieve such comparisons? The reconstruction of these arguments in the preceding section has shown that he did not do this by introducing the concept of velocity of classical mechanics (the quotient of a distance and a time) which is not so restricted in application and allows us to compare velocities whatever the given times and distances might be. In the argument in fol. 91v, Galileo introduced a third motion in order to compare the velocities of two uniform motions characterized by unequal times and unequal distances. With this introduction of a third motion of comparison, the Aristotelian proportions, in connection with the Archimedean proposition on the proportionality of distances and times in uniform motion, were now sufficient to carry out the comparisons. The situation in fol. 152r was different. Here two *accelerated* motions characterized by unequal distances and unequal velocities were given. As we have seen, the Aristotelian proportions were insufficient to determine one of the times of motion if the other is given, unless it is assumed that two velocities are equal if the distances traversed are in the same proportion as the times of motion. For uniform motion this relationship corresponds directly to the Archimedean proportion, while for accelerated motion it corresponds to a gen-

eralized definition of the equality of two velocities that is *not* covered by the
Aristotelian proportions but that is, in the case of equal distances or times, compatible with them.

The argument, in the context of which we have referred to the role of this generalized definition of velocity, had to be reconstructed from some difficult to
interpret scattered notes in fol. 152r. There is, however, direct evidence from
Galileo's manuscripts that he explicitly introduced this definition of velocity. In
this section we shall analyze two brief memoranda found in Galileo's manuscripts: Galileo's "Proportion Proof" of the law of fall and the "Mirandum
Paradox."[126] The first memorandum outlines a derivation of the law of fall
without recurring to the concept of degree of velocity. We shall see that a derivation following this outline presupposes the definition of equal velocities discussed above. On the other hand, the second memorandum, explicitly states this
generalized definition but reveals a paradox that results precisely from the
application of this definition to accelerated motion. As a consequence of this
paradox the role of this relationship in the Proportion Proof of the law of fall is
also problematical.[127]

In fol. 152r, Galileo had shown that the law of fall is in agreement with the
statement that overall velocities increase in proportion to the times of fall. If
this argument can be turned around, the law of fall can be derived directly from
the time proportionality of velocities in the Aristotelian sense, avoiding the
representation of these velocities by areas. In one of Galileo's manuscripts, the
outline of a derivation of the law of fall is found in which the assumption of a
proportionality between velocity in its Aristotelian sense and time is indeed the
key premise and in which no diagram is used to represent the changing velocity
in terms of its degrees[128]:

> When the velocity is the same and uniform, the spaces traversed have to one
> another the same proportion as the times; and when the time is the same and
> the velocities are different, the spaces traversed are to one another as these
> velocities. If therefore the velocity should grow according to the proportion
> of the increase of the time, the spaces traversed would grow with the double
> proportion of that by which the time grows.

Here, Galileo apparently claims that the law of fall follows from the
Archimedean proposition and one of the two Aristotelian proportions in connec-

---

[126] Mss. Gal., part VI, vol. III, fol. 61r-64r (EN VIII, 614; copy of Galileo's original
in Viviani's hand); and MS, fol. 164v (EN VIII, 375). The first has not been discussed
before in recent literature, while the second was first extensively discussed (under the
designation given here) in Wisan 1974, pp. 201-204.

[127] With respect to these two memoranda we do not wish to establish any particular
chronology. The reconstruction of the arguments documented in these memoranda
aims exclusively at revealing another structural difficulty in Galileo's reasoning
about accelerated motion, a difficulty that he indeed became aware of as his subsequent
research makes evident.

[128] Mss. Gal., part VI, vol. III, fol. 61r-64r (EN VIII, 614); this note contains no diagram.

tion with the assumption of a proportionality between time and velocity. In the context of these two kinematic rules, the concept of velocity must have been used in its Aristotelian sense so that a proportionality between time and velocity in its Aristotelian sense must be meant. The Archimedean proposition referred to by Galileo is restricted to uniform motion, but, as indicated above, we shall see that he actually based a generalized definition of the equality of velocities on this proposition, a definition that is applicable to accelerated motion, too. The second kinematic rule mentioned above refers to the Aristotelian proportionality between distances and velocities for a given time and is not restricted to uniform motion.[129]

In this text, Galileo does not state the argument by which the law of fall can be obtained from the two kinematic rules and the assumption of time proportionality of acceleration. But, assuming that his reference to the Archimedean proposition on uniform motion really includes the generalized definition of equal velocities, such an argument can easily be reconstructed by strictly following the line of reasoning documented in fol. 152r, where Galileo checked the compatibility between his assumption for the overall velocities and the law of fall by introducing a *third motion* that allowed him to compare two motions which are characterized by different velocities, times, and distances.

Consider a falling body which at the time t_1 has covered the distance s_1. In order to derive the law of fall one then has to determine the distance s_2 traversed after a given time t_2. In agreement with Galileo's text, we assume that the velocities at the times t_1 and t_2 are in the same proportion as these times, i.e.,

$$v_2 : v_1 = t_2 : t_1.$$

Consider now a third motion that traverses a distance s' with the speed v_2 in the time t_1. The Aristotelian proportion for motions equal in time then implies that

$$s' : s_1 = v_2 : v_1.$$

The third motion has the same velocity as the second motion but traverses a different distance in a different time. What does it therefore mean to say that it has the same velocity? According to the generalized definition of equality of speeds, which mimicks the Archimedean proposition on the proportionality between distances and times in uniform motion, it follows that:

$$s_2 : s' = t_2 : t_1.$$

If these proportions are assumed, the proportion between the two distances s_2 and s_1 is equal to the compounded proportion between the velocities and the times, i.e.,

$$s_2 : s_1 = (v_2 : v_1) \times (t_2 : t_1).$$

---

[129] See, for instance, the derivation of the Isochronism of Chords; for references, see note 71.

If we now take into account that the velocities were assumed to be in the same proportion as the times, it follows that the distances are in double proportion of the times, or, in other words, in the same proportion as the times squared, which is the law of fall that was to be demonstrated.

The preceding argument apparently represents a consistent derivation of the law of fall from the assumption of a proportionality between velocity (in the Aristotelian sense) and time, if the generalized definition of the equality of velocities is presupposed. This argument does not involve the notion of a degree of velocity nor the traditional graphical representation of velocity. The existence of a sketch of this argument in one of Galileo's manuscripts does not imply that he had resolved the problems related to this traditional represesentation; it just seems to suggest that it was possible to avoid them, perhaps not in all contexts but at least in a derivation of the law of fall.

There is, however, evidence that the generalized definition of the equality of velocities on which this argument was based actually does lead to a contradiction if applied to accelerated motion. In fact, one memorandum found in Galileo's manuscripts makes explicit use of the generalized definition and shows that Galileo encountered a paradox when he applied it to the comparison of two accelerated motions, one along the perpendicular, the other along an inclined plane.[130]

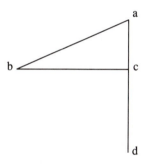

Fig. 3.22 (fol. 164v)

It has to be seen [*Mirandum*], whether the motion along the perpendicular *ad* [Fig. 3.22] is not perhaps faster that along the inclined plane *ab*? It seems so; in fact, equal spaces are traversed more quickly along *ad* than along *ab*; still it seems not so; in fact, drawing the horizontal *bc*, the time along *ab* is to the time along *ac* as *ab* is to *ac*; then, the moments of velocity along *ab* and along *ac* are the same; in fact, that velocity is one and the same which in unequal times traverses unequal spaces which are in the same proportion as the times.

---

[130] MS, fol. 164v (EN VIII, 375). The interpretation given in the following (as well as the translation) essentially follows Wisan (1974, pp. 201-204). For Drake (1978, pp. 124-125) on the other hand, the Mirandum Paradox is not a paradox at all but it is precisely this argument, in connection with the correct proportionality between moments of velocity and square roots of distances, which shows that Galileo "had reached complete clarity on puzzles of accelerated motion that had long plagued him." Accordingly, Drake translates "Mirandum" ("it has to be seen") as "Remarkable!" (1978, p. 125) and even as "Marvelous!" (1989b, p. 55). However, he leaves it unclear how Galileo's supposed resolution of this "seeming paradox" could have made use of the correct proportionality between moments of velocity and square roots of distances; see Drake 1990, p. 105. Galileo's reiteration of this paradox in his published works (see section 3.7.1) does not support this interpretation.

In the first part of this text, Galileo uses the Aristotelian concept of velocity to conclude that the motion along the vertical must be faster than the motion along the inclined plane, since "equal spaces are traversed more quickly" along the vertical than along the inclined plane. In the second part of his text, he first uses the Length-Time-Proportionality to infer that the distances traversed along the perpendicular and along the inclined plane are in the same proportion as the corresponding times. He then expresses the generalized definition of equal velocities in the form of the statement that "that velocity is one and the same which in unequal times traverses unequal spaces which are in the same proportion as the times," and finally he infers from this definition that the moments of velocity are equal along the inclined plane and along the perpendicular, obviously relating the moments to the velocities in the Aristotelian sense.[131]

The proposition that motion along the vertical is quicker than motion along an inclined plane, including its formulation in terms of the medieval kinematic proportions, was central to Galileo's studies of motion along inclined planes, from the time he gave a first account of these studies in *De Motu*. His key theorem on motion along inclined planes, formulated in *De Motu* and extensively used in his subsequent research – for instance, in the derivation of the Isochronism of Chords – states that the velocity of motion along an inclined plane is to the velocity of fall along the height of the plane as the height is to the length of the plane.[132] Galileo's conclusion that these velocities are actually equal, obtained in the text quoted above by applying his definition of equal velocities to both these motions, flatly contradicts this theorem.

Hence, Galileo's Mirandum Paradox consists in the fact that, on the background of his studies of motion along inclined planes, the two incompatible statements, *motion along the perpendicular has the same velocity as motion along an inclined plane* and *motion along the perpendicular is faster than motion along an inclined plane* are equally plausible and well founded. But since one of the two conflicting statements depends on Galileo's generalized definition of equal velocities, while the other depends on the more traditional Aristotelian concept of velocity, the paradox casts doubts on the applicability to accelerated motion of the statement that equal velocities are characterized by equal proportions between distances and times. As a consequence the Proportion Proof, Galileo's argument for the law of fall based on the generalized definition, also becomes doubtful.

Had it not been for the problems revealed by the Mirandum Paradox, the generalized definition of the equality of velocities together with the two Aristotelian proportions would have provided Galileo with more than just an argument by which the law of fall can be derived. They would have also provided him with a set of kinematic rules, applicable to uniform as well as to accelerated

---

[131] For an analysis of the conceptual background of Galileo's argument, and, in particular, of the role of the concept of moment in this argument, see Galluzzi 1979, pp. 292-293; see also Souffrin 1988.

[132] See DM, Chapter 14, pp. 63-69 (EN I, 296-302).

motion, independent of whether or not the motions compared by these rules share the same times, velocities, or distances. As the reconstruction of the Proportion Proof illustrates, such kinematic rules can in fact be derived from the generalized definition together with the Aristotelian proportions and take the form of compounded proportions such as

$$s_2 : s_1 = (v_2 : v_1) \times (t_2 : t_1).$$

Compounded proportions between times, velocities, and distances such as this one, can be considered as the expression in terms of proportions of the definition of velocity as the quotient of distance and time in classical mechanics. However, the Mirandum Paradox makes it clear that these compounded proportions do not represent a consistent set of rules in Galileo's physics.

3.4.3 Summary

In his letter to Sarpi in 1604 Galileo had suggested a principle which he used in his attempts to derive the law of fall and the Double Distance Rule within the conceptual framework of the medieval configuration of qualities. In the course of these studies, he discovered that these two propositions imply that in a motion of fall, the degrees of velocity grow in proportion to the times, which, Galileo realized, is incompatible with the principle he suggested to Sarpi.

He subsequently explored the possibilities of adapting the traditional arguments used in his attempted derivations of the law of fall to this new insight and discovered new paradoxes: in particular, it turned out that the medieval geometrical representation of velocity implies values for the velocities (in the Aristotelian sense) of the motion of fall taken over different times that are incompatible with the law of fall.

Galileo examined yet another relationship between the degrees of velocity and velocity in the Aristotelian sense, which was suggested by the Double Distance Rule but was incompatible with the medieval configuration of qualities. It followed from this relationship that velocities in the Aristotelian sense grow in the same proportion as the times of fall. If this proportionality is assumed as a premise, the law of fall can be derived from it (without the configurations doctrine) by an argument sketched in one of Galileo's manuscripts.

This argument presupposes, however, a definition of the equality of velocities that results from a generalization of an Archimedean proposition for uniform motion. Galileo recognized that the application of this definition to a comparison of accelerated motions is in conflict with an application of one of the traditional kinematic proportions to the same comparison but apparently did not find a resolution of this paradox. This paradox, based as it is on the Aristotelian concept of velocity, undermines Galileo's argument for the law of fall and prevents the development of a consistent set of kinematical rules applicable to uniform as well as to accelerated motion.

The conceptualization of velocity according to the medieval configuration of qualities and the Aristotelian understanding of velocity are thus seen to be at the origin also of the new paradoxes that emerged in the course of Galileo's studies of motion.

3.5 Projectile Motion in Galileo's Manuscripts after ca. 1604

Although Galileo could determine the shape of the trajectory of a projectile, the real problem for a cannoneer was to determine the range of a shot, given its angle and power. As we shall see in the following, Galileo's Double Distance Rule allowed him for the first time to calculate this range, but only for the practically unimportant case of horizontal projection. During the years in which he struggled with the derivation of the law of fall, he also worked on techniques to solve the problem of determining the range of an oblique shot. In an analysis of two manuscripts from this period, we shall see two attempts by Galileo to construct the trajectory for oblique projection which lead to mutually incon-sistent results, both of which are also incompatible with classical mechanics. Galileo's alternative attempt (also documented in his manuscripts) to solve the gunner's problem by simply assuming oblique projection upward to be the inversion of horizontal projection downward required him, however, to be abel to determine not the impetus *causing* the shot but rather the impetus resulting from its impact. We shall see that his first attempt at a solution to this problem leads to contradiction and that his second attempt rests on an additional unproven assumption. It will also be clear that Galileo was aware of these problems.

3.5.1   The Size of the Trajectory for Horizontal Projection

Galileo's attempts to develop a theory of projectile motion, documented by his manuscripts as well as by his correspondence, centered around his key insight that the form of the trajectory is parabolic. In section 3.3.1 we saw that he probably arrived at this insight before 1604 in connection with his discovery of the times squared relationship for the motion of free fall. The first dated docu-ment testifying to his systematic efforts to study projectile motion is a letter to Antonio de' Medici of February 11, 1609[133]:

---

[133]   EN X, 229-230; translation adapted from Drake 1973a, pp. 303-304.

I am now about some questions that remain to me concerning the motion of projectiles, among which are many that bear on artillery shots. And even recently I have found this: that putting the cannon on some elevated place above the plane of the field, and aiming it exactly level, the ball leaves the cannon, driven by much or very little gunpowder, or even just enough to make it leave the cannon; and yet it always goes declining and descending to the ground with the same speed, so that the ball will arrive on the ground at the same time for all level shots, whether the shots are very long or very short, or even if the ball merely emerges from the cannon and falls plumb to the plane of the field. And the same happens with shots at an elevation; these are all completed in the same time provided that they are lifted to the same vertical height. Thus, for example [Fig. 3.23], the shots *aef*, *agh*, *aik,* and *alb*, contained between the same parallels *cd* and *ab*, are all completed in the same time; and the ball consumes the same time in traversing line *aef* as *aik*, or the others – and in consequence, in their halves; i.e., the parts *ef*, *gh*, *ik,* and *lb* are all made in equal times, which correspond to level shots [from *e*, *g*, *i*, and *l*, respectively].

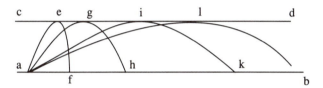

Fig. 3.23 (EN X, 229)

In this letter, Galileo presents his new finding – which we may call "the Isochronism of Projectile Motion" - that all projectile motions that attain the same vertical height take the same time. He introduces the Isochronism first for level shots, i.e., for horizontal projections, and then generalizes it to shots made at arbitrary elevations, emphasizing that their downward halves are covered in the same time as the corresponding horizontal projections from the vertex.

In this letter, Galileo neither provides a justification for his claim of the Isochronism of Projectile Motion nor does he specify the form of the trajectory. For the special case of horizontal projection, however, he could derive the isochronism as well as the form of the trajectory from the superposition of the two motions generating the projectile motion. The superposition of two mutually orthogonal motions follows in fact from Galileo's early analysis of projectile motion in *De Motu*.[134] As we saw in section 3.2.3, horizontal projection was an exceptional case for which the impressed force communicated to the projectile

---

[134] The possibility of this inference proves Hall (1965, p. 187) incorrect when he claims that it is inconceivable that anyone could have formulated a proposition such as the Isochronism of Projectile Motion who had not transcended the limits of medieval and 16th century mechanics.

and the natural tendency downward do not disturb each other at all. If Galileo could take the superposition of vertical and horizontal motion for granted, then his law of fall determined the pattern of motion along the vertical, i.e., the double proportionality between distances and times, and his studies of motion along inclined planes provided him with a justification of the assumption that the motion along the horizontal is uniform. The parabolic shape of the trajectory for horizontal projection is then an immediate consequence of these presuppositions.

The analysis of projectile motion in *De Motu* did not provide Galileo with an analogous argument justifying the superposition of the two motions generating oblique projection, but in the letter to Medici he nevertheless claims that the isochronism also applies to oblique projection. From other documents presented in this section, it will become clear that Galileo indeed assumed a complete analogy between horizontal and oblique projection, also with respect to the form of the trajectory. This conviction could in fact have a number of bases: heuristic arguments in connection with simple experiments on projectile motion, such as the one described in the Guidobaldo manuscript, suggesting the symmetry of the trajectory, an experimental test of the isochronism for oblique projection, or other, possibly more sophisticated experiments on projectile motion. Galileo expressed the intuitive plausibility of the assumption that oblique projection is the inverse of horizontal projection in the following brief dialogue found in fol. 106v[135]:

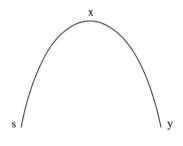

Fig. 3.24 (fol. 106v)

Simpl.: That the ball hurled up again describes the same line *sx* appears difficult to me.

Sagr.: But if it does not appear difficult to you that [Fig. 3.24], if it describes the whole parabola *yxs*, it can redescribe the line *sxy*, don't you see that it describes the line *sx* with necessity?

In this dialogue, Galileo justifies the symmetry between horizontal projection from the vertex and oblique projection up to the vertex by first referring to the motion along the entire parabola, and then to the possibility of inverting this motion. But this does not of course represent a derivation of the trajectory for oblique projection.

---

[135] MS, fol. 106v (EN VIII, 433). This manuscript also contains a table of contents for a book on projectile motion similar to the one later published in the *Discorsi*. Two other manuscripts (MS, fol. 81r and MS fol. 114v) which deal with oblique projection have been discussed extensively in the literature. See Hill 1988 for a recent review.

Galileo's agenda for the further elaboration of a theory of projectile motion after establishing the form of the trajectory can be inferred from one of his manuscripts.[136] In this document he lists the topics to be dealt with in a planned treatise on artillery, covering external as well as internal ballistics. This list covers much the same ground as traditional treatises on artillery such as Tartaglia's *Nova scientia*.[137] The list shows that, even after Galileo had derived the parabolic trajectory for horizontal projection, a number of questions essential to the field of external ballistics were still left open. In order to be able to answer a question such as, "At what elevation do you shoot farther and why?" he had to establish a relationship between the force of a shot and the size of the trajectory, because this question obviously presupposes that the angle of the shot be varied while the force generating the shot is held constant. The simple derivation of the parabolic trajectory in horizontal projection, sketched above, determines only the shape but neither the size of the trajectory nor its dependence on the forces and the quantities of motion that are involved in its generation.

However, Galileo's Double Distance Rule, discussed in section 3.3.3.2, offered him the possibility of establishing a quantitative relationship between the two components of horizontal projection, the accelerated motion along the vertical, and the uniform motion along the horizontal. The uniform horizontal motion appearing in the Double Distance Rule is generated by a preceding fall along the vertical; if the horizontal component of projectile motion is generated in an analogous way by a motion of fall along the vertical which is deflected into the horizontal, the magnitude of this component can be characterized by the height of the preceding vertical fall. The trajectory of a horizontal projection is therefore fully determined if the height of the fall generating its horizontal component is given as well as the height of the subsequent motion of fall along the vertical; and vice versa, it should be possible to determine the height of the fall generating the horizontal component from the range of the trajectory.

Galileo demonstrates these relationships in a manuscript that can be dated approximately to the period of his letter to Medici. In this manuscript, he constructs the "elevation" needed for a body to describe a given semiparabola as the third proportional to half the base and the height of the semiparabola. He then uses the Mean Proportional Form of the Law of Fall, the Double Distance Rule, and the proportionality between distances and times in uniform motion in order to show that the parabola is actually generated by a body falling from this elevation. It then follows as an obvious corollary of Galileo's argument that half the base of the semiparabola (or a quarter of the amplitude of the full parabola) is the mean proportional between this elevation and the height of the parabola. In other manuscripts the elevation is called the"sublimity" and thus we may refer to this relationship as the "Sublimity Theorem."[138]

---

[136] MS, fol. 193r (EN VIII, 424).

[137] Tartaglia 1537.

[138] MS, fol. 87v (EN VIII, 428-429). This manuscript has not received much attention from historians of science, who have focused their attention on Galileo's

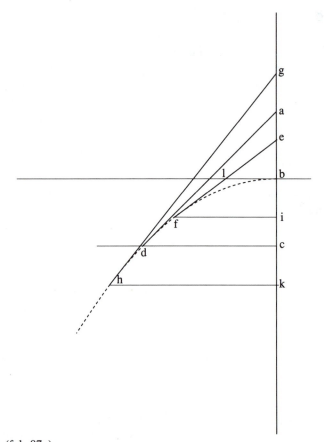

Fig. 3.25 (fol. 87v)

To find the elevation of a parabola descending from which a body describes the given parabola. Let the given parabola be *bf* [Fig. 3.25], whose height is *bi* and whose amplitude is *if*; if the horizontal *bl* is drawn, assume in the vertical *be* to be equal to *bi*, and connect *elf*, which touches the parabola in *f* and intersects with the horizontal at *l*; let *lb* be to *ba* as *be* to *bl*: I say that *ab* is the elevation falling from which the body, deflected in *b*, will describe the parabola *bf*. For if it is understood that the same *eb* is the time of fall along *eb*, and the same *eb* is the moment of velocity in *b*, then *bl* will be the time and the moment in *b* of the body falling from *a*: hence, falling from *a* to *b*, deflected along the horizon, it will in the time *bl* traverse double *ba*; therefore, in the same motion, in the time *eb*, it will traverse double *bl*: for as the time *eb* is to the time *bl*, double *bl* is to double *ba*. But double *bl* is precisely *fi*; therefore the horizontal *if* will be traversed by the body falling from *a* in the time *be*: but in the same time *eb* the vertical *bi* is traversed from the state of

---

discovery of the *shape* of the trajectory and neglected the question of the conceptual tools required to determine its *size*. Galileo's two results are later published in the *Discorsi* as Prop. V and its Corollary in the book on projectile motion, see *Discorsi*, p. 243; EN VIII, 294-295.

rest in *b*: therefore the body falling from *a* and deflected in *b* traverses in the same time the horizontal *if* and the vertical *bi* from the state of rest in *b*: it will therefore describe the parabola *bf*. It is therefore true, that half the base is the mean proportional between the height of the parabola and the elevation above the parabola, falling from which the projectile describes it.

Another of Galileo's manuscripts (fol. 116v) shows that he actually performed an experiment on horizontal projection in which he compared theoretical values following from the above argument with numbers obtained by his experiment, finding a relatively close agreement.[139]

3.5.2 Contradictions in the Analysis of Oblique Projection

As we have seen in the previous section, the Sublimity Theorem provides a derivation of the trajectory for horizontal projection in which its size, and hence the range of the shot, is determined by the vertical height – the sublimity – from which the motion is generated. It thus solves, in a sense, the gunner's problem of determining the relation of range and power (represented by the sublimity) of a shot but only for the case of horizontal projection. In order to solve this problem for the case of oblique projection as well, a construction analogous to that in the proof of the Sublimity Theorem would have to be considered.

In the collection of Galileo's manuscripts, there are, in fact, two which document his attempts to derive the trajectory for oblique projection in a way analogous to the argument for the Sublimity Theorem, i.e., from the generation of the projectile motion by a motion of fall and a subsequent deflection of this motion into the direction of the shot: fol. 175v and fol. 171v.[140]

---

[139] MS, fol. 116v. This manuscript was first published by Drake (1973a) and has since been the object of numerous controversies; for a recent critical review, see Hill 1988. Most interpretations relate the experiment documented by this manuscript not to a technical result of Galileo's theory of projectile motion, as is suggested here, but to more fundamental insights he supposedly achieved by this experiment. Drake (1973a, p. 303; 1978, pp. 127-132; 1985, p. 10; 1989b, p. 58) relates the experiment to a test of what he claims is a principle of horizontal inertia in Galileo's physics or, alternatively, to what he sees as a principle of composition in Galileo's mechanics, a test in the course of which Galileo accidentally hit on the parabolic trajectory. Naylor (1974a, p. 116) relates fol. 116v to a confirmation of the law of fall (see also Naylor 1977a, pp. 389-391), while Wisan (1984, pp. 276ff) relates it to a discovery of the proportionality between the degrees of velocity and the times in free fall. Hill (1986 and 1988) criticizes Wisan's interpretation but also relates the experiment to a test of a fundamental principle related to the law of fall. In spite of obvious indications that the drawing in the manuscript is in fact related to an experimental situation, even this general character of the manuscript has been doubted; see Costabel 1975, where, however, no convincing alternative is offered.

[140] MS, fol. 175v, MS, fol. 171v. Folio 175v was first published in Drake 1973a and has been discussed in the literature; fol. 171v has not been dealt with. Both

3.5.2.1 Oblique Projection and Inclined Plane: Folio 175v

Folio 175v presents a diagram that is neither accompanied by a text nor by calculations. At first glance, the diagram seems to represent a sketch for an experimental device that is conceived for the study of oblique projection.[141]

The diagram shows a plane that is steeply inclined (almost vertical) with
respect to a horizontal line at the bottom of the manuscript. To the steeply
inclined plane, a deflector is connected that transforms a falling motion along the
inclined plane into a motion along the trajectory indicated by a dashed line in the
sketch. For the second part of the trajectory, two dashed lines indicate two
alternative continuations. Two small circles in Galileo's diagram, marked by the
numbers "1" and "3" respectively, indicate two positions of a falling body.

A closer look at Galileo's drawing shows features which indicate that its
meaning cannot be limited simply to that of a realistic portrait of an experimental apparatus and its use.[142] Above the dashed line, a straight line tangential
to the deflector at the bottom of the steeply inclined plane is drawn (the line OV_1
in our transcription).[143] Apparently, this line does not represent a part of an
apparatus but indicates the original direction of the shot. A series of marks along
the steeply inclined plane and beyond, placed at quadratically increasing distances
from the uppermost point, marked A in the transcription, provides further
evidence for the interpretation of the diagram as being related to a theoretical
study. In addition to this sequence of quadratically increasing distances, two other

---

manuscripts contain diagrams but no text. Based on the watermarks, Drake (1979,
pp. li, lxi, and lxvi) dates fol. 175v to about 1609 and fol. 171v to a much later
period (after ca. 1626).

[141] This interpretation was suggested by Drake (1973a, p. 296), and by Drake and
McLachlan (1975a, p. 104). According to a conjecture by Drake (1985, p. 12) this
device was used by Galileo when he also tested the Isochronism of Projectile Motion
for oblique projection.

[142] These features were first identified by Naylor (1980a, pp. 557-561). According to
Naylor's interpretation of this manuscript, it represents a theoretical analysis by
Galileo of the trajectory for oblique projection. The interpretation given here follows
this general idea but reconstructs the reasoning that guided Galileo in his analysis in
a different way. Naylor interprets the construction in fol. 175v as an examination of
the conservation of horizontal inertia and of the principle of superposition, an examination leading to a confirmation of these principles because they alone were capable
of accounting for the experimentally determined form of the trajectory. As we have
seen, however, the physical properties Naylor refers to by these principles were, for
horizontal projection, unproblematic *consequences* of Galileo's physics, while Naylor does not provide any evidence that Galileo ever used both principles to derive the
trajectory for oblique projection. Drake (1990, p. 121) claims that the marks identified by Naylor actually document traces of an experiment and, contrary to Naylor, that
the trajectory shown in this manuscript is parabolic, which is certainly not the case.

[143] The transcription is based on the one given in Naylor 1980a, but has been confirmed and improved by a comparison with the original in Florence.

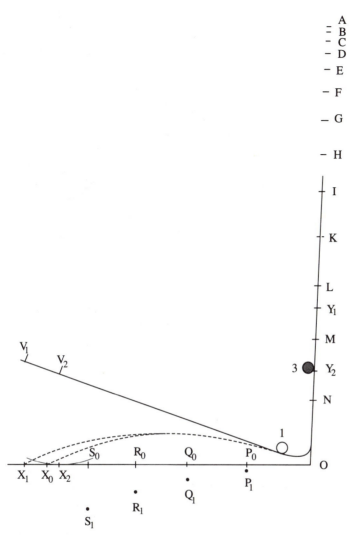

Fig. 3.26 (fol. 175v; letters added)

marks (indicated by Y_1 and Y_2 in our transcription), which were cancelled by Galileo, are found on the same line.

The theoretical purpose of Galileo's diagram on fol. 175v can be reconstructed from the numerous construction marks in the diagram that are not drawn with ink and hence cannot be seen in a facsimile reproduction. In the following, we shall discuss three aspects of Galileo's construction:

(1) The basic mathematical pattern used in Galileo's construction is the quadratically increasing sequence of distances drawn along the steeply inclined plane. He marked this same sequence along the horizontal line (OX_1) by compass marks, the distances decreasing from right to left. The *first* interval along the

horizontal corresponds to the *last* interval along the steeply inclined plane. Galileo also used his compass to plot points under each of the thus obtained four marks along the horizontal, points whose vertical distances from the horizontal follow, increasing from right to left, again the original quadratic sequence along the steeply inclined plane AO. Hence, the vertical distance from the horizontal of the first point (P_1) marked below OX_1 is equal to the first interval (AB) along the steeply inclined plane, the second equal to the second, and so on.

(2) The trajectory in Galileo's diagram is constructed in analogy to the series of points below the horizontal. Although no analogous construction marks could be identified in his diagram, the first part of the trajectory can in fact be obtained by first transferring the quadratic sequence to the line OV_1 tangential to the deflector (again starting with the larger intervals) and by then taking this line as the line of reference for the series of quadratically increasing *vertical* distances.

(3) A compass mark is found at the point V_2 in Galileo's diagram, forming the center of a circle whose circumference touches the horizontal at the point X_2, close to the point X_0 marking the bottom of the lower part of the trajectory. The distance of the point V_2 from the origin O is exactly equal to the space traversed in the *last five* units of time (IO) of the quadratic sequence along the steeply inclined plane AO. The radius of the circle with its center at V_2 which touches OX_1 at X_2 is somewhat less than the space traversed in the *first six* units of the quadratic sequence, measured from the starting point of the sequence.[144]

If the sequences of distances in Galileo's diagram are conceived as sequences of distances traversed in equal times, the construction corresponds to a procedure for deriving the trajectory in the case of oblique projection. The quadratically increasing sequence of distances along the steeply inclined line represents the accelerated motion along an inclined plane that serves a function analogous to that of the sublimity in the Sublimity Theorem. The quadratically decreasing sequence of distances along the line OV_1 represents a uniformly decelerated motion along an inclined plane which plays exactly the same role as the horizontal plane in the construction of the trajectory for horizontal projection. Precisely following the construction in the case of horizontal projection, Galileo first marked, in an auxiliary construction, the distances traversed by the motion of fall in four subsequent equal intervals of time under the horizontal line. In the case of oblique projection, however, these distances follow a quadratically decreasing sequence. Galileo then drew the trajectory such that its vertical distances from the plane OV_1 correspond to the distances traversed in the motion of fall that occurs simultaneously to the motion along OV_1.

The two lines Galileo drew in order to sketch the final part of the trajectory are indicative of his attempts at estimating when and where the trajectory meets the horizontal. In these attempts, the same principle that led to the construction of the first part of the trajectory is employed, i.e., the assumption that a point of the trajectory must be reached by two motions, each of which takes the same

---

[144] These marks were not identified in Naylor 1980a.

time: a decelerated motion along OV_1 and an accelerated motion downward. On the basis of an estimate about the approximate course of the trajectory based on his construction of four points underneath the horizontal, Galileo first assumed that the trajectory would meet the horizontal after approximately five units of time and thus drew the lower part of the trajectory ending in X_0. But then he measured the distance corresponding to five units of time in a decelerated motion along the line OV_1, thus reaching the point V_2. Measuring the height of the point V_2 above the horizontal by posing a compass at V_2, Galileo recognized that this vertical distance does not correspond to five units of time, traversed in a uniformly accelerated motion, but rather to a little less than six units of time. He then measured the distance corresponding to six units of time, traversed in a decelerated motion, along the line OV_1, thus reaching the point V_1. He finally drew the second continuation of the trajectory in such a way that its endpoint lay exactly underneath this point. The height of V_1 above the horizontal indeed corresponds to about 6 units of time (as measured from the beginning of the quadratic sequence).

Although it is not immediately obvious from the diagram on fol. 175v, the trajectory resulting from Galileo's construction is not a symmetric curve. This construction therefore violates the heuristic arguments in favor of a complete analogy between oblique and horizontal motion. In fact, according to the construction on fol. 175v, oblique projection cannot be conceived as "inverted" horizontal projection. The lack of symmetry is a strong objection to Galileo's construction, although this construction otherwise closely follows the rationale of his construction of the trajectory of horizontal projection.

In addition to its incompatibility with Galileo's arguments in favor of a symmetrical shape of the trajectory, a derivation along the lines of the construction in fol. 175v also has to confront a problem which does not present itself for a derivation of the trajectory for horizontal projection: the composition of two nonorthogonal motions. In *De Motu*, the composition of a horizontal with a vertical motion was a special case, in which no mutual disturbance between the two causes of motion occurs, whereas in the general case of oblique projection the two causes were considered to be in conflict. The two motions compounded in the construction in fol. 175v, however, are assumed not to disturb each other and to proceed as if they were mechanically superimposed on each other. Therefore, accepting this construction as a derivation of the trajectory for oblique projection requires a nontrivial revision of the way in which the composition of motions was originally conceived by Galileo.

3.5.2.2 Oblique Projection and Uniform Horizontal Motion: Folio 171v

Another manuscript dealing with projectile motion, fol. 171v, shows that Galileo later explored an alternative to the composition of motions underlying the construction in fol. 175v. In our interpretation of this manuscript Galileo studied the question whether or not the composition of a uniform horizontal motion with a decelerated motion along an inclined plane would lead to the desired symmetry of the trajectory for oblique projection, if one assumes the horizontal motion to be unaffected by the motion along the inclined plane.

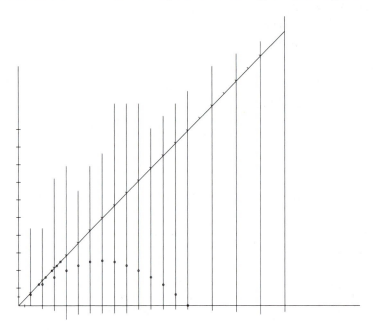

Fig. 3.27 (fol. 171v)

In fol. 171v (see Fig. 3.27), a horizontal line at the bottom and a vertical line on the left-hand side define the extension of a construction that includes a line at an angle of 45° with respect to the horizontal and a sequence of parallel vertical lines. The first 14 vertical lines are spaced at equal distances from one another; the last 4 have twice this distance from one another; half way between the "vertical axis" and the first line to the right of it a small mark is found on the horizontal. Various sequences of points and marks are found on the lines described. Along the leftmost line, the vertical axis, a sequence of 10 equidistant marks can be found; another mark is placed half way between the horizontal axis and the first of these marks. Another series of equidistant marks indicates the crossing points of the first 14 vertical lines with the oblique line. The sequence of these marks is continued with constant spacing beyond these 14 lines. The distance between these marks is the same as that between the marks on the vertical axis; they seem to be compass marks. Along the oblique line a sequence of 6 points is plotted. The distances between these points become shorter as they ascend in a roughly quadratic sequence. Between the horizontal and the oblique line, a sequence of 14 points is marked on the 14 equally spaced vertical lines. The height of these points above the horizontal corresponds ascending and descending to that of the points on the oblique line. The shape of the curve that can be drawn through these points is approximately parabolic. The distance between the origin and the last of these points is exactly equal to the distance between the origin and the uppermost point on the vertical line. Numerous additional marks and a small and very rough sketch (not transcribed in Fig. 3.27) are scattered on the manuscript.

We can reconstruct the process by which Galileo obtained the curve in this diagram as follows: He first marked the equidistant points along the vertical axis and then along the oblique line. The first of the sequence of vertical lines (after the axis) was drawn in such a way as to intersect the first mark along this oblique line. The operation was then repeated generating the other vertical lines. As the next step, Galileo marked the sequence of points in quadratically decreasing distances along the oblique line such that the first interval of this sequence is a little smaller than the equidistant lengths marked on this line. The first point almost lies on the first vertical line, and next to it Galileo marked another point that lies exactly on this vertical line. He then transferred the second point of the quadratic sequence along the oblique line to the second vertical line. The third point of the quadratic sequence along the oblique line is treated in the same way, i.e., on the third of the series of vertical lines a point is marked that has roughly the same height above the horizontal as the third point of the quadratic sequence. This procedure is repeated until the last point of this sequence is transferred. The position of the vertex of the parabolic curve is estimated and marked along the next vertical line. The rest of the points along the descending part of this curve are marked by the same procedure, starting now with the uppermost point of the quadratic sequence along the oblique line.

The diagram reveals itself to be a construction of the trajectory in the case of oblique projection if sequences of distances are interpreted as sequences of distances traversed in equal times. The points along the oblique line then depict a decelerated upward motion (or an accelerated downward motion) along an inclined plane, and the sequence of equidistant points along the horizontal represents a uniform motion along this line. The motion along the parabolic trajectory results from the composition of these two motions. The body describing the parabolic trajectory starts moving along the inclined plane represented by the oblique line with a certain velocity. If the body were to maintain this velocity, it would ascend the oblique line in such a way that the sequence of equidistant marks along this line would be passed in equal times. However, the upward motion along an inclined plane is uniformly decelerated, and, therefore, the body traverses a shorter distance along the inclined plane in the first interval of time, reaching the first mark slightly before the intersection of inclined plane and vertical line.

During the same first interval of time, the body is also assumed to have moved along the horizontal. The magnitude of the velocity along the horizontal follows from the magnitude of the aggregate initial velocity of which the horizontal velocity is a component. If the aggegate initial velocity was such that, in a unit of time, one of the regularly spaced distances along the oblique line is traversed, the horizontal velocity is such that, in the same time, one of the regularly spaced distances along the horizontal is traversed. Thus, by a uniform velocity along the horizontal and a decelerated motion along the inclined plane, the first point of the series of points that forms the trajectory is reached. The deceleration along the inclined plane continues up to the turning point, and so does the uniform motion along the horizontal. At the turning point of the motion along the inclined plane the trajectory reaches its vertex, and the body starts an accelerated motion downward. After the vertex the two motions continue, generating a symmetric trajectory.

In his construction of the trajectory of oblique projection in fol. 171v, Galileo does not factor in a decelerated horizontal component of the motion along the inclined plane, as he had done in fol. 175v, but rather compounds (as in the case of horizontal projection) a uniform horizontal with an accelerated motion. In the present case, however, the uniform horizontal motion is compounded not with the motion of free fall but with the vertical component of the motion along an inclined plane. In spite of their differences, the two constructions of the trajectory in folios 175v and 171v are equally plausible as generalizations of the derivation of the trajectory for horizontal projection.

Neither one of the two constructions can, however, be justified on the basis of a traditional understanding of compounded motions. Moreover, unlike the composition of motions assumed in fol. 175v, the composition of motions essential to the construction in fol. 171v cannot be realized by a mechanical device in which the two motions are completely independent (such as, for instance, by an inclined plane moveable along a horizontal plane). But whereas the composition underlying the construction in fol. 175v results in a decelerated horizontal motion that prevents the trajectory from being symmetrical, the composition of motions underlying the construction in fol. 171v preserves the uniform horizontal component unchanged and results in a symmetrical trajectory.

Although the shape of the resulting curve is parabolic in the case of fol. 171v, this construction of the trajectory for oblique projection is neither compatible with the Isochronism of Projectile Motion explained in the letter to Medici nor with the conclusion that oblique projection is just the reverse of horizontal projection. The reason for this lack of compatibility is that the time for the motion from the starting point to the vertex, which is equal to the time the body takes to descend to the horizontal from the vertex, depends on the inclination of the shot. According to the analysis of the trajectory as reconstructed from fol. 171v steep shots take less time in order to reach the same vertical height than flat shots. The time for the motion along the trajectory depends on the inclination of the plane that is used in its construction because this plane determines the acceleration along the vertical. Therefore, the construction in fol. 171v implies, just as does the construction in fol. 175v, a fundamental difference between horizontal and oblique projection and is hence not compatible with Galileo's heuristic arguments in favor of a complete analogy between oblique and horizontal projection such as the one given in fol. 106v (see section 3.5.1) or the experimental evidence supporting this analogy.

Since both constructions of oblique projection take a decelerated motion along the direction of the shot as their starting point, they both seem to be based on the assumption of a gradually diminishing impetus as the cause of motion. These constructions are hence in agreement with the dynamical explanation of vertical projection as it is described in *De Motu*, in the letter to Sarpi, and in numerous other documents. On the other hand, in Galileo's analysis of horizontal projection, the original cause of motion was taken to remain unchanged because of the special character of horizontal motion as a motion which is neither natural nor forced.

3.5.3 The Search for a Composition Rule for Impetuses

In the previous section, we reconstructed Galileo's unsuccessful attempts to construct the trajectory for oblique projection in analogy to the construction of that for horizontal projection. The failure of these attempts made it impossible for him to solve the gunner's problem for oblique projection, i.e., to determine the range of a shot given its power and its angle, in the same way as he had solved it for the case of horizontal motion. But if it is assumed that oblique projection can be considered as an "inverted" horizontal projection, even though Galileo had only heuristic arguments but no explicit derivation to support this conclusion, it was plausible to attempt a solution of the gunner's problem on this basis. In this perspective the determination of the initial impetus by which an oblique shot is generated becomes the problem of calculating the impetus resulting from the impact of the corresponding inverted horizontal projection.

3.5.3.1 Scalar Addition of Impetuses: Folio 90^{a}r

In fol. 90^{a}r,[145] Galileo tried to answer the traditional question, "At what elevation do you shoot farther and why?" by testing the conjecture that, for the same initial force of the shot, the maximum range will be attained by a projectile fired at an angle of 45°. The conjecture that the maximum range is attained for 45° had been made already by Tartaglia in his *Nova scientia*;[146] it is correct according to classical mechanics and will be referred to as the "Maximum Range Theorem."

The following two passages of Galileo's text in fol. 90^{a}r will allow us to reconstruct his argument:

---

[145] MS, fol. 90^{a}r (EN VIII, 429-430); see document 5.3.16.
[146] Tartaglia 1537.

A body falling [Fig. 3.28] from *a* to *c* will – having been deflected – describe the parabola *cd*; but if the moment of velocity in *c* were double, it would describe the parabola *ce*, for which *eg* would be double *gd*: for double the impetus in *c* traverses double the space along the horizontal in the same time. But in order to acquire double the moment in *c*, it is necessary that the fall takes place from four times the height, that is, from *cb* [...].

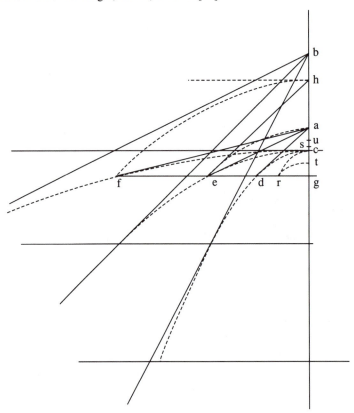

Fig. 3.28 (fol. 90ᵃr)

Let the impetus in *c* of a body falling from *a* be 100; that of a body falling from *b* will be 200: the impetus in *d* will be 200; the impetus in *e* will be 300.

The impetus of a body falling from *h* to *a* will be 141 in *a*; deflected along the parabola *ae*, however, the impetus will be doubled in *e*, that is, 282. It is therefore true that the impetus of the body coming along the parabola *ce* to *e* will be larger than that of the body coming along the parabola *ae*. And if the projectile coming from *e*, according to the elevation *eh*, has an impetus as 282, it will traverse the parabola *ea*; but according to the elevation *ea* the projectile traverses the parabola *ec*, if it has an impetus as 300. Hence it will be thrown farther by the same force along the elevation *eh* of half a right angle than along the elevation *ea*, which is smaller than half a right angle.

Galileo considers three different projectile motions, all of which are horizontal projections that are supposed to be parabolic. The first is generated by a vertical motion of fall along *ac*, the deflection of which into the horizontal yields the parabola *cd,* which has the line *da* as its tangent in *d*; the second is generated by a fall from four times the height *ac*, i.e., from the point *b*, yielding the parabola *ce* which has the line *ea* as its tangent in *e*; the third is generated by a fall along *ha* and results in a parabolic trajectory along *ae* which has the line *eh*, making an angle of 45° with the horizon, as its tangent in *e*. The third and the second trajectories have the same amplitude but different heights; the height *ha* is equal to 2*ac*.

The basic features of the trajectories shown in the diagram are obviously determined by a line of reasoning similar to that of the Sublimity Theorem. Although these trajectories are constructed for horizontal projections, the final conclusion is about oblique projection and corresponds to a special case of the Maximum Range Theorem mentioned above:

> Hence it will be thrown farther by the same force along the elevation *eh* of half a right angle than along the elevation *ea*, which is smaller than half a right angle.

But actually, in the argument we have quoted, this conclusion is only based on a comparison of two specific horizontal projections (along the lines *ae* and *ce*, respectively) which reach the same endpoint (namely the point *e*), and which, at this endpoint make different angles with the horizon, one of them (along *ae*) an angle of 45°. Galileo claims that the motion which makes an angle smaller than 45° with the horizon (the one along *ce*) will have acquired a larger impetus at this point than the one making an angle of 45° (along *ae*):

> It is therefore true that the impetus of the body coming along the parabola *ce* to *e* will be larger than that of the body coming along the parabola *ae*.

Hence, this manuscript is seen as an attempt to solve the gunner's problem by drawing conclusions about oblique projection from a consideration of horizontal projection. Galileo's calculations in this manuscript presuppose the inference that, if among all horizontal projections having the same range, the one making an angle of 45° with the horizon has the least impetus at its endpoint, then among all oblique projections generated by the same initial impetus, the one at 45° will have the longest range.

But how does Galileo determine the impetus at the endpoint of a horizontal projection? This impetus cannot be identical to the initial impetus of the projection, which is determined by its sublimity, because the fall along the sublimity only generates the horizontal contribution to the projectile motion. In order to take both components of the projectile motion into account in his test calculations, Galileo adds the impetus corresponding to the horizontal component of a horizontal projection to the impetus corresponding to the vertical motion of fall. He adds these impetuses as numbers and not as quantities which have a direction (such as the vectors representing instantaneous velocities in classical mechanics); this procedure will be called Galileo's "Scalar Addition Rule."[147] Galileo determines the impetuses or moments of velocity of the component motions by making use of the proportionality between moment of velocity and time, the discovery of which we have reconstructed in the previous section. If the impetus generated by the fall along ac is set to be 100, then the impetus generated by the fall along bc, i.e., along four times this distance, will be 200, since the fall along bc takes double the time of the fall along ac. Similarly, the impetus generated by a fall along ha will be given by the mean proportional between 100 and 200, which is approximately 141.

Using the Scalar Addition Rule, Galileo determines the impetus in d after the motion along the parabola cd to be $200 = 100 + 100$, the impetus in e after the motion along ce to be $300 = 200 + 100$, and the impetus in e after the motion along ae to be $282 = 141 + 141$, and concludes:

> It is therefore true that the impetus of the body coming along the parabola ce to e will be larger than that of the body coming along the parabola ae.

The Scalar Addition Rule is obviously incompatible with the vectorial addition of instantaneous velocities in classical mechanics. But, as we shall see in the following, it is also in conflict with other arguments employed in Galileo's analysis of oblique projection and was, in fact, eventually abandoned by Galileo and replaced by a different composition rule.

3.5.3.2 Vertical Height and Compounded Impetus: Folio 80v

In his *Nova scientia*, Tartaglia had not only suggested that the maximum range is attained at an angle of 45° but also that there are two different angles under which shots made by the same force will reach the same distance. Among Galileo's manuscripts several drafts are to be found in which he attempted to prove this conjecture, which we may call the "Equal Amplitude Theorem." More precisely, he tried to show[148]

> that the amplitudes of parabolas are equal which are made by projectiles along elevations differing by equal amounts from half a right angle.

In the following we shall analyze Galileo's attempt to prove this conjecture as it is documented by fol. 80r. We shall see that his argument refers, as do his calculations in fol. 90[a]r, to horizontal projection, while the Equal Amplitude Theorem, a statement about oblique projection, only follows if it is assumed that a shot along the horizontal can be inverted so as to yield the same trajectory

---

[147] Fol. 90[a]r was discussed by Caverni (1895, pp. 534-537) and Wisan (1974, p. 269) without mention of Galileo's Scalar Addition Rule. This was first discussed in Renn 1990a. Other manuscripts containing applications of the Scalar Addition Rule are MS, fol. 110v and MS, fol. 115v.

[148] MS, fol. 80r (EN VIII, 433, table III).

for oblique projection. The text of fol. 80r exclusively refers to the geometrical properties of the triangles in the accompanying diagram, it mentions neither the parabolas nor the impetuses of the corresponding projectile motions. But the relationship between Galileo's geometrical argument and projectile motion can be reconstructed by referring the argument to his Sublimity Theorem and by taking into account the geometrical properties of a parabola[149]:

> In the right triangle bcd [Fig. 3.29] let the angle d be equal to the angle cbe and let eb be drawn. The two triangles dcb, ebc will thus be similar.

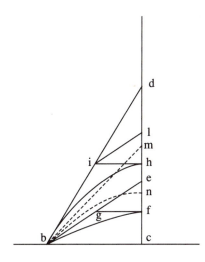

Fig. 3.29 (fol. 80r)

> Let the whole dc be divided in the middle in h; and hi be parallel to the line cb. Also let ec be divided in the middle in f and draw fg parallel to bc; and as dh is to hi, let hi be to hl, and let li be drawn. The triangle lih will be similar to the triangle dhi, and for the same reason also similar to efg. But hi is equal to gf (the double of both is namely bc). Therefore, the remaining sides hl, fe will be equal, so that the third proportional of these lines lh, hi, namely hd, will be equal to the third proportional of the lines ef, fg. But the third proportional of the lines lh, hi is hd; which is half the total dc. Therefore, the third proportional of the lines ef, fg will be equal to the half of cd, that is, to ch.

> But ch is equal to fl, because cf is equal to hl and fh is common. Therefore, the third proportional of ef, fg will be fl, which finishes in the point l, where the third proportional of dh, hi finishes.

> From this it will be shown that the amplitudes of parabolas are equal which are made by projectiles along elevations differing by equal amounts from half a right angle.

The argument in fol. 80r consists of three parts: First the two similar right triangles dcb and ebc are constructed with side bc in common. The lines be and bd make angles with the horizon that differ by equal amounts from 45°. These lines represent the tangents to the two parabolas to be constructed. These parabolas have the heights hc and fc, which are – by a geometrical property of the parabola – the halves of the lines dc and hc, respectively. In the second part, the sublimities of these parabolas are found as they follow from the Sublimity Theorem for horizontal projection. These sublimities have the lengths of the lines hl and hc, respectively. The third part shows that $hc = fl$ and that,

---

149 MS, fol. 80r (EN VIII, 433, table III); the division into paragraphs has been added.

consequently, the sum of sublimity and height is the same for both parabolas, that is, $hc + hl = fc + fl$.

Although taking into account the Sublimity Theorem allows us to reconstruct Galileo's text as an argument about horizontal projection, the role of impetus in this argument remains at first sight unclear. It is obvious however that the force of a shot uniquely depends on its direction and its amplitude. But whereas the construction in fol. 80r concerns horizontal projection so that the impetus of a shot is indeed implicitly determined by its direction and its amplitude (which are both given), the final conclusion concerns oblique projection. Hence it should explicitly involve a statement about the relationship between the impetuses of the two projectile motions compared in this theorem because it is now the impetus which is the given variable of the problem. From later and more elaborated formulations of the same theorem,[150] it is in fact clear that Galileo supposed the two impetuses to be equal. If we now turn back to Galileo's construction in fol. 80r, it follows that this equality of impetus must be represented by the equality of the sum of the sublimity and the height of the two parabolas, which is the geometrical property of his construction revealed by his last conclusion:

> Therefore, the third proportional of ef, fg will be fl, which finishes in the point l, where the third proportional of dh, hi finishes.

We may therefore conclude that the equality of the sum of sublimity and height implied, for Galileo, the equality of the impetuses generated by the two horizontal projections. In other words, these two motions yield the same impetus at their terminal point d because the total vertical heights from which these motions are produced are equal. Now, if the additional assumption is made, that oblique projections generated by the same impetus along two angles differing by equal amounts from 45° describe the same parabolas in the reverse direction of the corresponding horizontal projections, then the Equal Amplitude Theorem indeed follows from what has been shown.

The argument in fol. 80r presupposes no specific rule for the composition of impetuses as do the calculations in fol. 90ªr. But it does presuppose a general property of the compounded impetus of a projectile which any such rule should satisfy, i.e., the property that the compounded impetus of a projectile stays the same if the total vertical height from which different projectile motions are generated (i.e., the sum of sublimity and height) remains the same.[151]

Galileo could easily have checked whether or not the rule for the composition of impetus implied by his calculations in fol. 90ªr is compatible with the general property of the compounded impetus presupposed in fol. 80r by considering some numerical examples. He would have found that his rule for the

---

[150] This theorem corresponds to Proposition VIII of the book on projectile motion in the *Discorsi* (EN VIII, 297/246).

[151] For a more detailed analysis of the role played by this purported general property of projectile motion in Galileo's physics, see Renn 1990a.

composition of impetus does not, in general, have this property and that, consequently, the arguments on fol. 90ᵃr and fol. 80r are not compatible with one another.

3.5.3.3 Vectorial Composition of Scalar Impetus: Folio 91v

Indirect evidence that Galileo did indeed discover that his reasoning about compounding impetuses involved such an inconsistency is provided by numerous manuscripts documenting his use of a composition rule other than the one implied by the calculations in fol. 90ᵃr in order to derive the Maximum Range Theorem. An explicit formulation of this rule, including an argument for its justification, is found in one of the manuscripts that we have already encountered, fol. 91v.[152] The text presented earlier is the second one of three texts found on the same page, which, taken together, provide a heuristic argument for a composition rule of impetuses. The first of the three texts in fol. 91v describes the composition of uniform motions along the vertical and along the horizontal, the second text derives the proportionality between moment of velocity and time in free fall, and the third text shows how to determine the impetus in each point of a parabolic trajectory employing the results of the first two arguments. Here is the first text of fol. 91v:

> If some moveable is equably moved in double motion, i.e., horizontal and vertical, then the impetus of the movement compounded from both will be equal in the square to both moments of the original motions.

> Let some moveable be equably moved in double motion, and let the space ab [Fig. 3.30] correspond to the vertical displacement, and let bc correspond to the horizontal movement carried out in the same time. Since spaces ab and bc are thus traversed in the same time, in equable motions, the moments of those motions will be to one another as ab is to bc. But a moveable that is moved according to these displacements will describe the diagonal ac in the same time in which it makes the displacements along the vertical ab and along the horizontal bc, and its moment of velocity will be as ac. But ac is equal in the square to ab and bc; therefore the moment compounded from both moments of ab and bc is equal in the square to both of them taken together; which was to be shown.

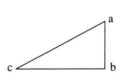

Fig. 3.30 (fol. 91v)

The third text of fol. 91 reads:

---

[152] MS, fol. 91v; see document 5.3.13 and plate IV. The first text corresponds closely to Prop. II of the book on projectile motion in the *Discorsi* (EN VIII, 280/229); the translation is adapted from Drake's translation of the *Discorsi*. The third text is transcribed in EN VIII, 427.

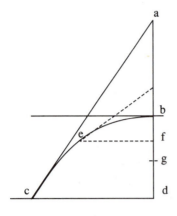

Fig. 3.31 (fol. 91v)

The impetus in the single points of the parabola *bec* [Fig. 3.31] is therefore determined by the square [*potentia*] of the moment acquired in the descent along *ab*, which will always be the same and which determines the horizontal impetus, and by the square of the other moment acquired in the descent along the vertical. Thus, for instance, in *e* the impetus will be determined by the square root of the sum of the squares of [*linea potente*] *ab* and the mean proportional between *db*, *bf*, which is *bg*.

In the first passage Galileo derives the composition of the impetuses corresponding to uniform vertical and uniform horizontal motion, respectively, from the composition of the spaces traversed in the same time by the vertical and the horizontal motion, respectively. These spaces not only represent the velocities of the two motions in the Aristotelian sense but also their degrees or moments of velocity since, for uniform motions lasting the same amount of time, the velocities are in the same ratio as the intensities of these motions.[153] Since the compounded motion is represented by the diagonal, the moment of velocity of this compounded motion is obtained as the square root of the sum of the squares of the component moments.

In the second passsage of fol. 91v, quoted in section 3.4.1.1, Galileo showed that, in the motion of free fall, the moments of velocity are proportional to the times of fall. He had presupposed this property already in his earlier calculations of the compounded impetus in fol. 90^{a}r. But now, in the third passage of the present manuscript, he uses the proportionality between the moment of velocity and time in connection with the rule for the composition of impetuses for uniform motions demonstrated in the first text. Galileo's new rule for the composition of impetuses is formulated in close analogy to the composition rule for uniform motions: the square of the compounded impetus is obtained by

---

153 For a systematic analysis of Galileo's use of the concepts *moment, degree of velocity*, and *impetus* in his theory of projectile motion, see Galluzzi 1979, pp. 372-383. On p. 383, note 36, Galluzzi remarks that in Galileo's work impetus almost always refers to the physical effect of velocity. This acute observation supports the interpretation that Galileo's central propositions on projectile motion are pronounced for oblique projection but proven for horizontal projection because of the problems he encountered in deriving the trajectory for oblique projection. In fact, this transition from oblique to horizontal projection transforms impetus conceived as a cause of horizontal projection into impetus conceived as the effect of oblique projection. Of course, this interpretation does not preclude that the understanding of impetus as the effect of projection was also part of its traditional meaning.

adding the square of the moment of velocity acquired in the fall along the sublimity and the square of the moment of velocity acquired during the fall from the height of the parabola.

The analogy is, however, incomplete. Contrary to what is the case for uniform motions, the moments of velocity to be compounded in projectile motion are not given by distances simultaneously traversed along the vertical and along the horizontal. Hence, their composition cannot be justified as for uniform motion by a geometrical argument based on distances traversed. Consequently, Galileo's third text only formulates the rule of composition but provides neither a derivation for this rule nor an interpretation of the compounded impetus as a distance traversed by the compounded motion. The resultant impetus is not even represented in the accompanying diagram. In later applications of this rule Galileo does provide such a representation, but only as an illustration of the algebraic rule yielding the compounded impetus as an intensive quantity of the projectile and not as a distance related to the projectile motion. For future reference this composition rule may be called Galileo's "Vectorial Composition of Scalar Impetus."[154]

Galileo could easily have convinced himself by numerical calculations that his new composition rule is compatible with the assumption underlying the derivation of the Equal Amplitude Theorem, because this rule does have the property that projectile motions generated by the same vertical height (= sum of sublimity and height of the parabola) have the same compounded impetus. But although the problem, which according to our interpretation, led Galileo to develop his new composition rule is thus resolved, he did not achieve a satisfactory derivation of this rule, as we shall see in the following.

3.5.3.4 Compounded Impetus and Distance Traversed

Galileo's manuscripts document a number of attempts to clarify the crucial problem in the derivation of his composition rule, i.e., relationship between the impetus resulting from his rule and the distance traversed by the projectile, but these efforts, too, remained without success. In a short note on fol. 110v, Galileo formulates the assumption that the compounded impetus grows in proportion to the amplitude of the parabola during the projectile motion and tests

---

[154] Most interpretations of Galileo's science of motion – with the remarkable exception of Dijksterhuis (1924, p. 275) – have confounded this composition rule with the vector addition of velocities in classical mechanics, ignoring the fact that the geometrical representative of Galileo's compounded impetus does not actually lie along the tangent to the parabola as does the vector sum of two instantaneous velocities in a given point of the trajectory; see, e.g., Drake 1978, p. 135; see also Hill 1979, p. 270, and Naylor 1980a, p. 565.

this assumption by some trial calculations which demonstrate that this assumption is incorrect[155]:

> Let the impetus in *b* from *a* be 100 [Fig. 3.32], and let *bc* be equal to the same *ba*; the impetus in *d* through *abd* will be approximately 142, and the distance *cd* will be 200. While the impetus in *f* will be 125, the distance *fi* will be 150; but it ought to be about 176, so that the ratio of the impetus in *d* to its distance *dc* be preserved. The impetus in *h* is about 160, its distance *hk* is 250.

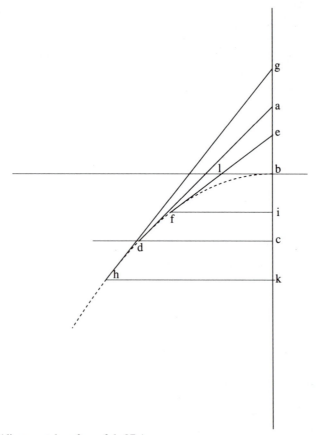

Fig. 3.32 (diagram taken from fol. 87v)

Here Galileo considers points on a parabola whose horizontal distances differ by equal amounts of 50 from the value *cd* = 200 for the amplitude assumed in the beginning, i.e., *fi* = 150 and *hk* = 250. According to his composition rule,

---

[155] MS, fol. 110v (EN VIII, 428). The corresponding diagram, as identified by Drake (1979, p. XXXVI), is found in fol. 87v. Wisan (1974, pp. 271-272) claims that in fol. 110v Galileo is searching for a proposition relating the impetus in projectile motion to the impetus acquired in free fall, but in fact there is no indication of such a search in this manuscript.

the projectile acquires an impetus of approximately 142 in the point d, and impetuses of approximately 125 and 160 in the points f and h. If the distance fi were 176 instead of 150 (as Galileo states it ought to be) we would then have

$$200 : 142 = 176 : 125,$$

or, in other words, the ratio of the horizontal distance traversed by the projectile and the impetus would be the same at different points of the projectile motion, but this is actually not the case. In some of the calculations on fol. 110v, Galileo examined the question of whether or not this relationship holds true for the scalar composition rule employed in fol. 90^{a}r, also finding a negative result. He was thus unable to obtain a simple relationship between the compounded impetus and the range of the projectile.

Another manuscript that only contains calculations and diagrams but no text, fol. 117r [see Fig. 3.33],[156] can be seen to show traces of another attempt by Galileo to explore the relationship between the compounded impetus and distances traversed by the projectile. One of the diagrams found on this page shows a parabola that is obviously generated by a uniform motion along the horizontal and a uniformly accelerated motion along the vertical.

The horizontal line representing the uniform motion is subdivided into four equidistant intervals. Along the verticals, Galileo marked the distances which are traversed by the accelerated motion in the corresponding equal intervals of time. A careful reconstruction of the figures and calculations found in this manuscript shows that he determined the lengths of the four diagonals which correspond to distances between points passed through by the projectile after equal intervals of time, obtaining, e.g., $\sqrt{10^2 + 40^2} = 41^{19}/_{83}$ for the first of the four inter-

---

[156]   MS, fol. 117r. This manuscript was first published by Drake (1973a, pp. 293-294) and interpreted as documenting the results of a measurement of the deceleration of horizontal motion by friction; see also Drake and McLachlan 1975a, p. 104. In spite of Naylor's identification of the "measurements" as the results of a geometrical construction, Drake (1985, p. 11) proposed this interpretation once again. The construction in this manucript that is the object of our subsequent discussion was first identified in Naylor 1975; see also Naylor 1980a, pp. 562-566. Several other geometrical features in this manuscript were identified in Hill 1979. Naylor (1980a, p. 562) interprets fol. 117r as an attempt by Galileo to establish "the way in which the average velocity of the projectile was changing in relation to time." According to Naylor, Galileo's study of the increase of this "average velocity" (a notion that Galileo in fact did not possess) led him to the insight that something was wrong with the principle of acceleration mentioned in the letter to Sarpi, i.e., the proportionality of the degrees of velocity and the distances traversed. In fact, however, Naylor's interpretation is made very implausible by the existence of Galileo's Scalar Addition Rule, documented in fol. 90^{a}r, since this rule shows that Galileo already adhered to the correct proportionality of degrees of velocity and times in free fall, at a time when he did not yet compound velocities or impetuses in projectile motion by the Pythagorean addition (extracting the square root of a sum of squares) applied in the Vectorial Composition of Scalar Impetus as well as in fol. 117r.

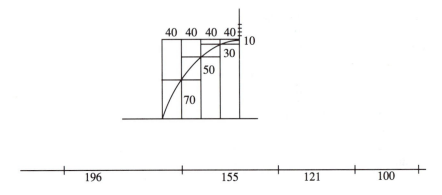

Fig. 3.33 (fol. 117r)

vals.[157] Since the intervals of time between these points are equal, their distances can be interpreted as a sequence of velocities in the Aristotelian sense, a sequence which Galileo represented underneath the main diagram along a horizontal line (changing all intervals by a constant factor so that $41^{19}/_{83}$ becomes 100).

In a certain respect, this sequence is similar to the sequence of compounded impetuses which can be determined for each of the four points according to Galileo's composition rule: Each of these impetuses is also given by the square root of a sum of squares, one of which stays the same (the moment of velocity of the horizontal motion) while the other grows (the moment of velocity of the vertical motion). Nevertheless, the sequence of velocities in the Aristotelian sense determined in fol. 117r and the sequence of compounded impetuses are essentially different because, in the first sequence, the growing components are to each other as the distances traversed in equal times in a motion of fall, while in the second sequence, the growing components are to each other as the times of the motion of fall. Hence, the analysis in fol. 117r, too, could not provide Galileo with a clue for the interpretation of the impetus resulting from his composition rule in terms of a distance traversed by the projectile.

In a number of manuscripts dating from a later period, Galileo again tackled this problem and formulated a "Theorem of Equivalence" between the time of motion along a semiparabola and the time along the inclined plane connecting the highest point of the sublimity with the endpoint of the parabola (e.g., points a and f in Fig. 3.32) beginning and end points. If this theorem had been correct, it would have allowed him to identify the compounded impetus of a projectile with a distance traversed by an equivalent motion along an inclined plane, but since it is actually incorrect, the derivation and interpretation of the composition

---

[157] See Naylor 1975 and 1980a, pp. 562-566. Our interpretation that these calculations are related to a study of the composition of impetuses is confirmed by the existence of similar calculations in MS, fol. 115v, which also contains an application of the Scalar Addition Rule.

of impetuses in projectile motion remained a missing central element of Galileo's theory of motion.[158]

3.5.4 Summary

Galileo elaborated the consequences of his insight that the trajectory for projectile motion is a parabola, following a research program that was shaped by traditional treatises on ballistics. In particular, he tried to solve the "gunner's problem," i.e., the problem of determining the range of a shot, given its initial impetus and angle. But he fully succeeded only in the practically unimportant case of horizontal projection.

Galileo's manuscripts document several attempts to cope with the problem of oblique projection. Among these attempts, we have identified two distinct constructions of the trajectory for oblique projection. Both are plausible generalizations of the analysis in the case of horizontal projection, based on determining the gradually decreasing violent impetus causing the oblique projectile motion by the pattern of motion along an inclined plane. Both constructions are, however, incompatible with other aspects of Galileo's knowledge about oblique projection, aspects which indicated to him that oblique projection can be considered as an inverted horizontal projection.

In another approach to the gunner's problem, Galileo merely assumed that oblique projection is the inversion of horizontal projection, and attempted to determine the initial impetus of oblique projection by determining the final impetus of horizontal projection. He first determined the compounded impetus simply as the sum of the two impetuses corresponding to the two components of horizontal projection, but he subsequently discovered that this composition rule is in conflict with other arguments on oblique projection. He eventually formulated a composition rule that does not lead to contradictions and that does determine the impetus of a projectile in such a way that its magnitude actually corresponds to the correct scalar magnitude of the instantaneous velocity in classical mechanics; however, the geometrical representation of the rule does not indicate the instantaneous direction of motion. Galileo could neither provide a derivation of his composition rule nor could he relate the resulting compounded impetus to a distance traversed by the projectile.

The derivation of the shape and size of the trajectory in the case of horizontal projection, the treatment of oblique projection by an inversion of horizontal pro-

---

[158] See fols. 83v and 86v (EN VII, 427-428). Although in the course of his studies on the Theorem of Equivalence, Galileo discovered that it was incorrect, he did succeed in proving that his Vectorial Composition of Scalar Impetus is indeed compatible with the assumption underlying his derivation of the Equal Amplitude Theorem; see Renn 1990a and for the complete manuscript evidence Renn 1984 and Renn 1988.

jection, and the introduction of a composition rule for impetuses as an ad hoc assumption allowed Galileo to contribute to the solution of several traditional problems of ballistics, but the contradiction in his analysis of the trajectory for oblique projection left his theory of projectile motion in a problematical state. Galileo still lacked the general principles of inertia and superposition that would have allowed him to understand oblique projection.

3.6 The Laws of Free Fall and Projectile Motion in Galileo's Published Works

Galileo's published works, the *Dialogo* and the *Discorsi*, on which his reputation as the founder of classical physics is based, appeared about a quarter of a century after his intensive work on the foundations of the theory of motion. There is ample evidence that he studied the manuscripts that he had preserved from his earlier studies when he prepared these works for publication.[159] Focusing on his final work on motion, the *Discorsi*, we shall see that the two central problems we have discussed in his manuscripts, the derivation of the law of fall and of the trajectory of oblique projection, remain unsolved.

In this section, we shall first see that, in the *Discorsi*, the law of fall is rigorously derived from the correct proportionality of velocities and times and from a proposition which is similar to the traditional Merton Rule but which itself rests on a problematic argument. The proof technique used in the derivation of Galileo's Merton Rule is, in fact, different from its traditional proof, as becomes evident in the context of other applications in the *Discorsi*. In order to avoid the paradoxes that arose in the course of his earlier attempted derivations of the law of fall, Galileo had invented a new technique for dealing with problems involving infinitesimals. Instead of representing velocities by the areas in the traditional triangular diagram, he now attempted to use a one-to-one correspondence between lines representing single degrees of velocity and to compare velocities operating with aggregates of lines in the place of areas. But although

---

[159] See, e.g., the letter to Elia Diodati, Dec. 6, 1636 (EN XVI, 524) where Galileo writes: "... I experience how much old age takes away from the mind's vitality and quickness, when I have a hard time understanding not a few of the things discovered and demonstrated by me in a fresher stage of life." There is an early draft of the beginning of the book on accelerated motion in the *Discorsi* entitled "Liber secundus: in quo agitur de motu accelerato" (EN II, 261-266). This draft, which can be dated about 1630, was bound together with Galileo's early treatise *De Motu*. Fredette (1972, pp. 329-330) argues that it was filed by Galileo himself together with this treatise. If Fredette's reconstruction is correct, it suggests that Galileo kept the manuscript of his first treatise in order to exploit its results and to integrate them into his late treatise on motion. For a detailed study of the phase in which Galileo completed the *Discorsi*, see Wisan 1974, section 8.

similar considerations are at the root of Cavalieri's theory of indivisibles,[160] Galileo's technique, as applied in the context of his theory of motion, is an ad hoc device that cannot be justified within the traditional understanding of the triangular diagram representing accelerated motion; it is incompatible with basic features of integration as well and even leads, if explicated systematically, to absurd consequences.

We shall also see that, at the point in the systematic treatment of projectile motion in the *Discorsi* where oblique projection is actually dealt with and correctly stated to yield a parabolic trajectory, there is simply a gap in the argumentation, and no derivation is offered for this claim.

3.6.1 A Failed Revision of the Derivation of the Law of Fall

Galileo's proof of the law of fall is presented in the "Third Day" of the *Discorsi*. Just as the *Dialogo* published five years earlier, the *Discorsi* is written as a dialogue continued over a number of "Days" which correspond to chapters of a book. The *dramatis personae* of Galileo's dialogues, Salviati, Sagredo, and Simplicio, usually represent positions that can be associated, respectively, with the author himself, an open minded supporter of the new science, and a scholar uncritically adhering to the Aristotelian natural philosophy. Sometimes, however, it seems as if all three figures represent Galileo himself at different stages of his intellectual development. Galileo's own systematic treatise on motion is presented as a book within the book: In the *Discorsi*, his spokesman Salviati quotes from a treatise on motion which is presented as a deductive theory and is written in Latin, while the discussions between Salviati, Sagredo, and Simplicio are conducted in Italian. This treatise, entitled *De motu locali*, is comprised of three books, the first two of which are contained in the Third Day of the *Discorsi*, while the Third Book is read in the Fourth Day. The First Book deals with the basic rules of kinematics, the second with the motion of fall and motion along inclined planes, and the third with projectile motion.

3.6.1.1 A New Proof Technique for Deriving the Merton Rule

In the Second Book of *De motu locali*, read on the Third Day of the *Discorsi*, the law of fall is derived in two steps. Proposition I of the Second Book claims that a motion uniformly accelerated from the state of rest and a uniform motion whose degree of speed is one-half the maximal degree of the accelerated motion traverse the same distance in the same time. Proposition I is then used in the proof of Proposition II, which demonstrates the law of fall in the form of the

---

[160] Cavalieri 1635.

assertion that the spaces traversed by the uniformly accelerated motion from rest are in the duplicate ratio of the corresponding times. The key step in the proof of Proposition II is the application to uniformly accelerated motion of a composite kinematical proportion, according to which, in uniform motion, the spaces traversed in two motions with differing speeds are to each in other in the compounded ratio of these speeds and the times of the two motions. The crucial role of Proposition I in the proof of the law of fall is precisely to justify the application to accelerated motion of a kinematical rule for uniform motion.

Galileo's formulation of this proposition reads as follows[161]:

PROPOSITION I. THEOREM I

The time in which a certain space is traversed by a moveable in uniformly accelerated movement from rest is equal to the time in which the same space would be traversed by the same moveable carried in uniform motion whose degree of speed is one-half the maximum and final degree of speed of the previous, uniformly accelerated, motion. (EN VIII, 208/165)

Proposition I is essentially equivalent to the medieval Merton Rule. But in contradistinction to the standard version of the Merton Rule, Galileo denotes the degree characterizing the uniform motion not as the mean degree of the accelerated motion but as half its maximum degree. We have earlier seen that, in his manuscripts, Galileo had proven a similar theorem, the Double Distance Rule, which in fact concerns the maximum degree of the accelerated motion, and which can be interpreted as stating that the overall velocity of uniformly accelerated motion, i.e., the space traversed in a certain time, is half of the overall velocity of a uniform motion equal in degree to the maximum degree of the accelerated motion and lasting as long. As our reconstruction of his studies in fol. 152r has shown (section 3.4.1.2), Galileo could easily have derived the result presented as Proposition I in the Second Book of *De motu locali* from the Double Distance Rule.[162]

However, although Galileo's theorem is distinguished from the traditional Merton Rule only by minor differences in its formulation, in his proof, which is based on a representation of the increase of velocity by the traditional diagram, Galileo strictly avoids the identification of the area with velocity, thus drawing

---

[161] In the following, references to the *Discorsi* will be given in parentheses after the quotation.

[162] The problem of the relationship between Proposition I and the Double Distance Rule has been much discussed by historians of science precisely because it is related to the question of whether or not Galileo had knowledge of the traditional Merton Rule; for a recent account see Sylla 1986, p. 89. Sylla tries to find an answer to the question why Galileo did not formulate Prop. I as referring to the mean degree, i.e., in precise analogy to the Merton Rule. But the close similarity in the formulations of Prop. I and the Double Distance Rule (both refer to the maximal degree), the ample documentation of the crucial role played by the Double Distance Rule in Galileo's studies of motion, and the fact that Galileo actually could derive Prop. I from it, make the answer to the question obvious, independent of the whether or not he knew about the original Merton Rule.

the consequences of the contradictions revealed by the analysis in fol. 152r.[163]
This proof begins as follows:

> Let line AB [Fig. 3.34] represent the time in which the space CD is traversed
> by a moveable in uniformly accelerated movement from rest at C. Let EB,
> drawn in any way upon AB, represent the maximum and final degree of speed
> increased in the instants of time AB. All the lines reaching AE from single
> points of the line AB and drawn parallel to BE will represent the increasing
> degrees of speed after the instant A. Next, I bisect BE at F, and I draw FG and
> AG parallel to BA and BF; the parallelogram AGFB will [thus] be constructed,
> equal to the triangle AEB, its side GF bisecting AE at I. (EN VIII, 208/165)

The parallelogram AGFB represents a uniform motion characterized by half the
maximum degree of the accelerated motion. Galileo states that this parallelogram
is "equal to the triangle AEB" representing the accelerated motion, obviously referring to the equality of the areas of the triangle and the rectangle.

Galileo does not, however, identify these areas directly as distances as would be
appropriate in classical mechanics, nor does he identify them as representing the
overall velocities of the corresponding motions and hence, on the basis of the
Aristotelian concept of velocity, indirectly as distances traversed in equal times.
In fact, in order to represent the distance traversed, a separate diagram is used.
Precisely because Galileo does not consider the equality of these areas as evidence

---

[163] There have been numerous attempts to reconstruct the meaning of the proof of
Prop. I, and there seems to be no general agreement whether or not this proof is still
rooted in the same conceptual framework as Galileo's earlier attempts to prove the
law of fall. Drake (1970a, 1972a) has emphasized the difference between this proof
and earlier arguments including the Merton Rule, claiming that the *Discorsi* proof is
not based on a comparison of areas but on a one-to-one comparison of degrees of
velocity. Although Drake's emphasis on the conceptual aspects of the difference
between this proof and the traditional proofs is, as we shall see, problematical, his
acute identification of the existence of a difference was an important starting point
for the analysis given here. Sylla follows Drake's interpretation of the proof of Prop.
I (Sylla 1986, pp. 85-89) and also reconstructs it as a reaction to difficulties Galileo
had previously encountered (Sylla 1986, p. 77, note 79). Different interpretations of
the proof of Theorem I, which link this proof to Galileo's earlier proofs or to the
Merton Rule have been proposed, among many others, by Clavelin and Ogawa. But in
view of the results by Drake and Sylla, Clavelin's (1974, pp. 298ff) claim that, in
Theorem I the area is treated as a sum of lines, and the degrees of velocities are
summed to an overall velocity as in Galileo's earlier proofs seems to be just as
unacceptable as Ogawa's (1989, p. 48) anachronistic treatment of Theorem I and the
Merton Rule on the same footing. Settle 1966, pp. 172-183, presents an ingenious
reconstruction of Galileo's argument in the proof of Prop. I on the basis of
infinitesimal considerations, but this reconstruction presupposes that Galileo's
problem was, as Settle puts it, to prove that an area can represent a distance (p. 166),
whereas the analysis by Drake indicates that areas do not crucially enter Galileo's
argument. Nardi (1988, pp. 49, and 51-52) even argues that Galileo may have used $s \propto
vt$ in his proof. In support of this anachronistic interpretation he appeals to the role
of the concept of moment in Galileo's proof and to the implausible claim of a
similarity between the diagram to Prop. I and a balance.

for his claim that the two motions traverse the same distance, he enters a detailed analysis of the aggregates of parallel lines representing the degrees of velocity:

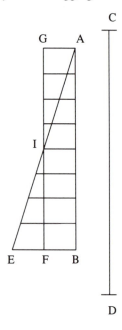

Fig. 3.34 (EN VIII, 208)

Now if the parallels in triangle AEB are extended as far as IG, we shall have the aggregate of all parallels contained in the quadrilateral equal to the aggregate of those included in triangle AEB, for those in triangle IEF are matched by those contained in triangle GIA, while those which are in the trapezium AIFB are common. Since each instant and all instants of time AB correspond to each point and all points of line AB, from which points the parallels drawn and included within triangle AEB represent increasing degrees of the increased speed, while the parallels contained within the parallelogram represent in the same way just as many degrees of speed not increased but equable, it appears that there are just as many moments of speed consumed in the accelerated motion according to the increasing parallels of triangle AEB, as in the equable motion according to the parallels of the parallelogram GB. For the deficit of moments in the first half of the accelerated motion (the moments represented by the parallels in triangle AGI falling short) is made up by the moments represented by the parallels of triangle IEF.

It is therefore evident that equal spaces will be run through in the same time by two moveables, of which one is moved with a motion uniformly accelerated from rest, and the other with equable motion having a moment one-half the moment of the maximum speed of the accelerated motion; which was [the proposition] intended. (EN VIII, 208f/165f)

In the first part of this passage Galileo argues that the aggregate of the parallels contained in the rectangle AGFB is equal to the aggregate of the parallels contained in the triangle AEB because a one-to-one correspondence can be established between these two sets of lines. From the equality of the aggregates of parallel lines representing the degrees of velocity of the two motions, the equality of the moments of velocity is then inferred on the basis of the assimilation of the concept of moment to the concept of degree, familiar from Galileo's earlier attempts to prove the law of fall. Having concluded that the (infinite) numbers of moments of velocity are equal for the two motions, Galileo finally claims to have shown that in these two motions equal spaces are traversed in equal times. This last inference presupposes that the equality of the two infinite sets of moments of velocity establishes the equality of the corresponding overall velocities, to which then kinematic rules can be applied. Once this equality of overall velocities is established, the remainder of the argument is, in fact, straightforward: It immediately follows from one of the Aristotelian proportions that the distances traversed in the same time by two motions having the same velocity are equal (the conclusion drawn in the last paragraph of the proof).[164] In Galileo's systematic theory of motion, this kinematic rule is presented as Prop. II of the First Book of *De Motu locali*:

> If a moveable passes through two spaces in equal times, these spaces will be to one another as the speeds. And if the spaces are as the speeds, the times will be equal. (EN VIII, 193/150)

This rule indeed applies to the present case, since it is not restricted to uniform motion as most of the other kinematic rules contained in the First Book.

The truly problematical key step in Galileo's proof is the inference from the matching of two infinite sets of lines to the matching of the two infinite sets of moments which constitute the overall velocities of the motions compared.[165] Although the transition from an infinite set of moments to the overall velocity of a motion, to which the Aristotelian proportions can then be applied, is plausible on the background of the traditional conception of moment as the indivisible element of motion, the very notion of the equality of two such infinite sets is ambiguous, as Galileo himself recognized. The discussion of indivisibles in the First Day of the *Discorsi* reveals in fact that he was familiar with the problem that, if a finite quantity such as a line is conceived as being constituted by "indivisible" elements such as points, there is more than one way to use these elements in order to establish a one-to-one correspondence between different finite quantities.[166] Hence, the conclusion that the matching of two infinite sets of moments follows from the matching of two infinite sets of lines, is intrinsically ambiguous.[167]

---

[164] The formulation of the proposition and the final conclusion of the proof of Prop. I are actually different; for an interpretation of this fact, see section 3.7.1. This reconstruction disagrees with Wisan (1974, p. 220), who claims that the proof of Prop. I does not presuppose kinematic proportions which (in classical mechanics) are restricted to uniform motion.

[165] This identification of the key problem of Galileo's proof essentially agrees with Galluzzi 1979, p. 354 and Blay 1990, p. 3. Galluzzi supports his interpretation by a careful analysis of Galileo's infinitesimal considerations presented in the First Day of the Discorsi and by an examination of Galileo's correspondence, in particular with Cavalieri. Blay confirms this identification of the problem in Galileo's proof by referring to the contemporary reception of Galileo's proof by Torricelli and Varignon.

[166] The connection between Galileo's proof of Prop. I and the infinitesimal considerations of the First Day of the *Discorsi* was suggested in Settle 1966, Chapter IV; for a comprehensive discussion, see Galluzzi 1979, Chapter V.

[167] There is no general agreement among historians of science on the question of whether or not Galileo's proof is correct, either as a proof in classical mechanics, or at least within Galileo's conceptual system. But contrary to the position defended here, most interpretations tend to represent this proof as an actual solution to Galileo's earlier problems, even if they reconstruct it in radically different terms. While according to Wisan (1974, p. 214) in Theorem I Galileo has finally resolved the problem of foundations, Drake (1978, p. 371) goes as far as to claim that "Proposition One is perhaps the only theorem capable of rigorous proof relating the

3.6.1.2 Derivation of the Law of Fall

Galileo's Merton Rule provides a means by which to derive the law of fall, stated in Proposition II of the Second Book of *De Motu locali* :

PROPOSITION II. THEOREM II

If a moveable descends from rest in uniformly accelerated motion, the spaces run through in any times whatever are to each other as the duplicate ratio of their times; that is, are as the squares of those times.

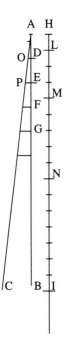

Fig. 3.35
(EN VIII, 209)

Let the flow of time from some first instant A be represented by the line AB [see Fig. 3.35], in which let there be taken any two times, AD and AE. Let HI be the line in which the uniformly accelerated moveable descends from point H as the first beginning of motion; let space HL be run through in the first time AD, and HM be the space through which it descends in time AE. I say that space MH is to space HL in the duplicate ratio of time EA to time AD. Or let us say that spaces MH and HL have the same ratio as do the squares of EA and AD.

Draw line AC at any angle with AB. From points D and E draw the parallels DO and EP, of which DO will represent the maximum degree of speed acquired at instant D of time AD, and PE the maximum degree of speed acquired at instant E of time AE. Since it was demonstrated above that as to spaces run through, those are equal to one another of which one is traversed by a moveable in uniformly accelerated motion from rest, and the other is traversed in the same time by a moveable carried in equable motion whose speed is one-half the maximum acquired in the accelerated motion, it follows that spaces MH and LH are the same that would be traversed in times EA and DA in equable motions whose speeds are as the halves of PE and OD. Therefore, if it is shown that these spaces MH and LH are in the duplicate ratio of the times EA and DA, what is intended will be proved.

Now in Proposition IV of Book I ["On Uniform Motion," above] it was demonstrated that the spaces run through by moveables carried in equable motion have to one another the ratio compounded from the ratio of speeds and from the ratio of times. Here, indeed, the ratio of speeds is the same as the ratio of times, since the ratio of one-half PE to one-half OD, or of PE to OD, is that of AE to AD. Hence the ratio of spaces run through is the duplicate ratio of times; which was to be demonstrated.

---

law of fall to the definition of uniform acceleration." Similarly, according to Giusti (1981, p. 39), in his proof of Prop. I Galileo has definitely overcome his previous difficulties. The extreme opposite view is held by Clavelin (1983, p. 47), who does not even grant the derivations of Galileo's theorems the status of a proof but just that of "simple ordered recapitulations of the main reasons for their acceptance."

It also follows from this that this same ratio of spaces is the duplicate ratio of the maximum degrees of speed; that is, of lines PE and OD, since PE is to OD as EA is to DA. (EN VIII, 209f/166f)

Like the proof of Prop. I, the proof of Prop. II is accompanied by two diagrams, one representing the increasing degrees of the accelerated motion, the other the distances traversed in different times. In the first part of this proof, the meaning of the two diagrams is explained, and the proposition is restated with reference to two arbitrary times of a uniformly accelerated motion starting from rest. The second part of the proof applies Galileo's version of the Merton Rule to these arbitrary times with the result that two uniform motions are found which traverse the same distances as the accelerated motion in the two given times. The speeds of these two uniform motions are characterized by half the maximum degrees of velocity of the accelerated motion, reached at the two instants of time under consideration. The third and final part of the proof makes use of a compounded kinematic proportion which we encountered earlier in the reconstruction of Galileo's Proportion Proof of the law of fall but which is now restricted to uniform motion, as Prop. IV, Theorem IV in the First Book of *De Motu locali*:

If two moveables are carried in equable motion but at unequal speeds, the spaces run through by them in unequal times have the ratio compounded from the ratio of speeds and from the ratio of times. (EN VIII, 194/151)

Since the speeds of the two uniform motions introduced in the proof of Prop. II are characterized by half the maximum degrees of velocity of the accelerated motion and since these degrees are assumed to be in the same proportion as the corresponding times by Galileo's definition of uniform acceleration, it follows from Prop. IV that the space traversed by the two uniform motions, and hence the spaces traversed by the accelerated motion in the two corresponding times, are in the duplicate proportion of these times.

This derivation of the law of fall is indeed free from the paradoxes Galileo had earlier encountered, since now the overall velocity of accelerated motion, to which kinematic proportions can be applied, is no longer represented by the area of the triangular diagram but by half the line representing the maximum degree reached at a given time. As a consequence, the overall velocities of an accelerated motion taken over different times are always in the same proportion as the degrees of velocity reached at these times. This conclusion bluntly contradicts the traditional interpretation of the triangular diagram, according to which these velocities are in the proportion of the squares of the degrees, but it is in agreement with the use of a one-to-one correspondence between degrees of velocity in a comparison of overall velocities such as the one performed in Galileo's version of the Merton Rule, and it is also in agreement with the concept of velocity in classical mechanics.

3.6.1.3 Refutation of Space Proportionality by the New Proof Technique

Nevertheless, Galileo's derivation of the law of fall does not rest on the same conceptual foundations as classical mechanics. Evidence for this is provided by his refutation of the proportionality between the degree of velocity and the distance traversed, i.e., the principle of acceleration he had formulated in his letter to Sarpi, written more than 30 years before the *Discorsi* was published (see section 3.3.3.1).[168]

In the *Discorsi* Galileo refutes the assumption of space proportionality in the course of his definition of uniformly accelerated motion, before the derivation of Proposition I. The assumption of space proportionality is brought into discussion by Sagredo (the open-minded supporter of the new science), who considers it as equivalent to the proportionality between the degree of velocity and time:

> Sagr. By what I now picture to myself in my mind, it appears to me that this could perhaps be defined with greater clarity, without varying the concept [as follows]: Uniformly accelerated motion is that in which the speed goes increasing according to the increase of space traversed. Thus for example, the degree of speed acquired by the moveable in the descent of four braccia would be double that which it had after falling through the space of two, and this would be double of that resulting in the space of the first braccio. For there seems to me to be no doubt that the heavy body coming from a height of six braccia has, and strikes with, double the impetus that it would have from falling three braccia, and triple that which it would have from two, and six times that had in the space of one. (EN VIII, 203/159)

Galileo, in the person of Salviati, admits that he himself had adhered to this assumption:

> Salv. It is very comforting to have had such a companion in error, and I can tell you that your reasoning has in it so much of the plausible and probable, that our Author himself did not deny to me, when I proposed it to him, that he had labored for some time under the same fallacy. (EN VIII, 203/159f)

---

[168] Galileo's refutation of the proportionality between the degrees of velocity and the distances traversed was often criticized as a fallacious argument (see, e.g., Hall 1958), until Drake (1970a, pp. 28-36) proposed a reconstruction of this argument that makes it at least plausible. Drake's reconstruction, on which the interpretation given in the following is based, is similar to the earlier interpretation of Dijksterhuis (1924, pp. 246-250); this interpretation makes use of a one-to-one correspondence between aggregates of infinite velocities. Drake supports this reconstruction citing an interpretation of the argument along the same lines by a contemporary of Galileo's (Tenneur 1649, p. 8). A criticism of Galileo's argument was reproposed by Finocchiaro (1972) and convincingly refuted by Drake (1973c). In the *Discorsi* (EN VIII, 203/160, see below) Galileo characterizes the principle he had earlier suggested to Sarpi by a quotation from Vergil (Aeneid, IV, 175): "vires acquirat eundo." For Descartes' use of the same quotation see document 5.1.3.

Simplicio (usually the orthodox Aristotelian) supports this suggestion by adducing the effect of percussion:

> Simp. ... That the falling heavy body *vires acquirat eundo* [gathers strength as it goes], the speed increasing in the ratio of the space, while the moment of the same percussent is double when it comes from double height, appear to me as propositions to be granted without repugnance or controversy. (EN VIII, 203/160)

In the same way, Galileo had referred to percussion as providing evidence for the proportionality between the degree of velocity and the distance of fall in his proof of the law of fall in fol. 128[169]:

> This principle appears to me very natural, and one that corresponds to all the experiences we see in instruments and machines that work by striking, where the percussent works so much the greater effect, the greater the height from which it falls.

In the *Discorsi*, Galileo refutes the proportionality between the degree of velocity and distance as well as the proportionality between the effect of percussion and distance by showing that the assumption of these proportionalities leads to the conclusion that the motion of fall is instantaneous. He does not, however, reexamine the proof of the law of fall based on the assumption of space proportionality mentioned in his letter to Sarpi, but he uses his new technique of comparing aggregates of degrees of velocity to refute Sagredo's propositions:

> Salv. And yet they are as false and impossible as [it is] that motion should be made instantaneously, and here is a very clear proof of it. When speeds have the same ratio as the spaces passed or to be passed, those spaces come to be passed in equal times; if therefore the speeds with which the falling body passed the space of four braccia were the doubles of the speeds with which it passed the first two braccia, as one space is double the other space, then the times of those passages are equal; but for the same moveable to pass the four braccia and the two in the same time cannot take place except in instantaneous motion. But we see that the falling heavy body makes its motion in time, and passes the two braccia in less [time] than the four; therefore it is false that its speed increases as the space. (EN VIII, 203f/160)

The plural "speeds" in Galileo's formulation "the speeds with which the falling body passed the space of four braccia" shows that speed is here to be understood as the degree of velocity at a certain point of the motion. On the other hand, the argument makes use of the kinematic rule presented as Prop. II in the First Book of *De Motu locali*, formulated here as the inference that "when speeds have the same ratio as the spaces passed or to be passed, those spaces come to be passed in equal times." The speeds to which this kinematical proposition refers are obviously velocities in the Aristotelian sense of the term.

The conclusion that the motion under study here is an instantaneous motion presupposes that the proportionality between distance and velocity in the sense of the Aristotelian concept follows from the proportionality between distance and

---

[169] MS, fol. 128, EN VIII, 323; see also section 3.3.3.1.

velocity in the sense of the degree of velocity. This conclusion becomes plausible if a one-to-one correspondence between the degrees of velocity of the first part of the motion and those of the entire motion is established along the line of reasoning on which the proof of Prop. I is based. Such a one-to-one comparison of the degrees of velocity passed through in the first two braccia with the degrees of velocity passed through in the entire four braccia indeed suggests that the overall velocities corresponding to these two infinite sets of degrees are in the same proportion as the maximum degrees reached after the fall through two braccia and four braccia, respectively, and hence are in the same proportion as the distances traversed.

In Galileo's attempts at a proof of the law of fall associated with the letter to Sarpi, on the other hand, the overall velocities up to given points of the motion of fall were, on the basis of the traditional understanding of the triangular diagram, inferred to be in the same proportion as the *squares* of the maximum degrees of velocity reached at these points. The refutation of space proportionality in the *Discorsi* is thus no refutation of these proofs but of the proportionality between the degree of velocity and distance under the presupposition of a proportionality between overall velocity and degree of velocity. But Galileo's argument is also incompatible with classical mechanics, since, in classical mechanics, it does not follow that a motion whose velocity increases in proportion to the distance traversed must be instantaneous but rather that such a motion cannot begin at all from the state of rest.[170]

3.6.1.4 The Problematical Role of the Area in the New Proof Technique

In the preceding analysis of Galileo's refutation of space proportionality, it has been emphasized that the ratio of the overall velocities to which kinematic proportions are applied is different from the ratio of the areas which would correspond to these overall velocities in the traditional diagram. This feature of the technique of one-to-one comparisons of degrees of velocity was less obvious in the case of the proof of the Merton Rule since, in this case, the ratio of the areas was the same as that of the velocities. The final example of a one-to-one comparison of degrees of velocity, also taken from the *Discorsi*, will make it clear that, from this technique, an equality of two overall velocities may even follow if the corresponding areas of the traditional diagram are unequal. In the course of the analysis of this example, we shall see that it is precisely this feature of Galileo's technique which has absurd implications. The example showing that Galileo's comparison of aggregates of degrees of velocity may lead to the conclusion that the velocities of two motions are the same even if the areas in the traditional diagram are different is the proof of Proposition III of the Third Day, the Length-Time-Proportionality:

---

[170] See the discussion of this implication in section 1.5.3.1.

PROPOSITION III. THEOREM III

If the same moveable is carried from rest on an inclined plane, and also along a vertical of the same height, the times of the movements will be to one another as the lengths of the plane and of the vertical.

Let the inclined plane AC [Fig. 3.36] and the vertical AB each have the same altitude above the horizontal CB, that is, the line BA. I say that the time of descent along plane AC has, to the time of fall of the same moveable along the vertical AB, the same ratio that the length of plane AC has to the length of vertical AB. Assume any lines DG, EI, and FL parallel to the horizontal CB; it follows from our postulate that the degrees of speed acquired by the moveable from the first beginning of motion, A, to the points G and D, are equal, since their approaches to the horizontal are equal; likewise, the speeds at points I and E are the same, as are the speeds at L and F. Now, if not only these, but parallels from all points of the line AB are supposed drawn as far as line AC, the moments or degrees of speed at both ends of each parallel are always matched with each other. Thus the two spaces AC and AB are traversed at the same degrees of speed. But it has been shown that if two spaces are traversed by a moveable which is carried at the same degrees of speed, then whatever ratio those spaces have, the times of motion have the same [ratio]. Therefore, the time of motion through AC is to the time through AB as the length of plane AC is to length of vertical AB; which was to be demonstrated. (EN VIII, 215f/173f)

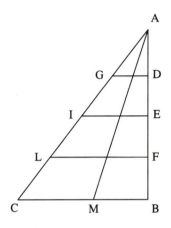

Fig. 3.36 (EN VIII, 211)

In the beginning of the Third Day of the *Discorsi*, Galileo had introduced the postulate on which the present proof is based:

I assume that the degrees of speed acquired by the same moveable over different inclinations of planes are equal whenever the heights of those planes are equal. (EN VIII, 205/162)

This postulate makes it possible to match degrees of velocity along the vertical and along the inclined plane at points that have equal distances from the horizontal. Since the vertical AB is the height of the inclined plane AC, a one-to-one correspondence between two infinite sets of degrees of velocity can be established in this way so that "the two spaces AC and AB are traversed at the same degrees of speed."

If these two infinite sets of degrees of velocity, both of which grow proportionally with time, are graphically represented as in the diagram accompanying Prop. I of the Second Book of *De Motu locali*, two triangles are obtained having an equal base (the maximum degree of velocity attained in both motions is the same) but unequal heights (the times of the two motions are unequal). Hence, these triangles have different areas but nevertheless represent the same degrees of speed because a one-to-one correspondence between the single degrees contained in the two triangles can be established.[171] Before turning to the implications of this peculiar feature of Galileo's argument, his line of argument in the above proof will be reconstructed.

In this proof, he claims that "it has been shown that if two spaces are traversed by a moveable which is carried at the same degrees of speed, then whatever ratio those spaces have, the times of motion have the same [ratio]." This claim is based on two presuppositions: first, it is assumed that the equality of the aggregates of degrees implies the equality of the overall velocities; the second presupposition corresponds to the generalized definition of equal velocities Galileo had earlier used in arguments such as the derivation of the Mirandum Paradox: if the velocities are equal, then the ratio of the distances is equal to the ratio of the times. The first presupposition corresponds to the essential hypothesis underlying the technique applied in Galileo's proof of the Merton Rule and in the refutation of space proportionality. The second presupposition, however, is apparently in contradiction to the restriction to uniform motion of the corresponding kinematic rule in the *Discorsi*. The First Book of *De Motu locali*, presented in the beginning of the Third Day, does not in fact contain Galileo's generalized definition of velocity which was at the origin of the Mirandum Paradox. Rather, this deductive treatise on kinematic rules contains a proposition that is closer to the original unproblematic Archimedean proposition. Proposition I of the First Book states in fact that

> if a moveable equably carried with the same speed passes through two spaces, the times of motion will be to one another as the spaces passed through. (EN VIII, 192/149)

---

[171] In the draft of the *Discorsi* mentioned in note 159 (EN II, 261-266) Galileo illustrates the differences in acceleration along differently inclined planes by a diagram in which equal aggregates of infinite moments of velocity are represented by unequal areas. This point has been stressed in Galluzzi 1979, pp. 338-340. Contrary to Galluzzi (1979, p. 340), our interpretation

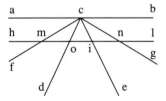

Fig. 3.37 (EN II, 264)

attempts to show that the problems implied by this representation of velocity are not overcome in the *Discorsi*. In Galileo's interpretation of the diagram [Fig. 3.37], the line *hl* is in uniform motion downwards, and its distance from the horizontal *ab* represents the flow of time. Galileo argues that all the parallel lines contained in the triangle *oci* are also contained in the triangle *mcn* during the descent of the line *hl*, though acquired in less time. Hence, if the parallel lines represented degrees of velocity, the same degrees of velocity will be attained along *cm* and *co* but in different times, while the corresponding aggregates of degrees will be represented by unequal areas. (For a translation of Galileo's argument see Drake 1978, pp. 315-318.)

But although this rule, as formulated, applies only to uniform motion, this restriction can be overcome with the help of the Merton Rule. Galileo's Prop. I in fact allows him to extend the above rule to uniformly accelerated motion, too, because it allows him to replace a uniformly accelerated motion with a uniform motion having the same overall velocity. Hence, by means of the Merton Rule, the expression "the same speed" in the above kinematical rule can in fact be related directly to accelerated motion so that Galileo's generalized definition of the equality of velocities makes its reappearance in the *Discorsi*.[172] In section 3.7.1.3, we shall return to this point and see that this also brings the Mirandum Paradox back into the *Discorsi*.

This justification of the extension to accelerated motion of a kinematic rule restricted to uniform motion is indeed explained by Sagredo in a subsequent commentary on the above proof:

> Sagr. It appears to me that the same can be very clearly and briefly concluded, since it has already been shown that the overall [*somma del*] accelerated motion of passage through AC (and AB) is that of the equable motion whose degree of speed is one-half the maximum degree, [at] CB. Therefore, the two spaces AC and AB being [considered as] passed with the same equable motion, it is manifest by Proposition I of Book I that the times of [these] passages will be as the spaces themselves. (EN VIII, 218f/176)

This explication of the preceding proof with the help of the Merton Rule confirms that Galileo's argument indeed depends on the equality of two overall velocities (*somma del moto*) which correspond to unequal areas in the traditional diagram. But the fact that the one-to-one comparison of degrees of velocity can establish an equality of overall velocities, even when the areas in the corresponding traditional diagram are unequal, implies a generalization of the Merton Rule with absurd consequences. Closer inspection of his proof of the Merton Rule in the light of the discussion up to this point indeed makes it clear that this proof also works equally well for nonuniformly accelerated motion, although there is no indication that Galileo ever made use of this generalization.

Let, for a nonuniformly accelerated motion, the degrees of velocity be given by a monotonously increasing curve, extended between the times 0 and t_2:

---

[172] Wisan (1974, p. 220) claims that the proof of Theorem III is based on a rule for uniform motion, but according to the interpretation proposed here this was not an error but justified by the relationship Galileo had previously established between accelerated and uniform motion. Galluzzi (1979, pp. 359-362) reconstructs the proof of Theorem III on the basis of the claim that Galileo did not consider a division of the line into an infinite number of points but just into an arbitrary number; however, in his proof Galileo in fact refers explicitly to the "parallels from all points of the line AB." In another reconstruction of the proof of Theorem III, Souffrin (1986) refers to a manuscript (MS, fol. 138v; EN VIII, 372) documenting an attempt by Galileo to generalize the Archimedean proposition on uniform motion to accelerated motion. It is, however, implausible to assume that, in a published text, Galileo referred to an unpublished manuscript.

Let t_1 be the time in which the corresponding motion reaches half the maximum degree of velocity attained at the time t_2 (Fig. 3.38), and consider a uniform motion which is characterized by half the maximum degree of the accelerated motion, and which is extended over the same time as the accelerated motion. If these two motions are compared in the same way as the uniformly accelerated and the uniform motion in the proof of Galileo's Merton Rule, it follows that they traverse the same distance in the same time. Indeed, since it is possible to establish a one-to-one correspondence between the parallels in the two shaded areas, it follows similarly as in Galileo's proof that the deficit of the degrees of velocity in the first part of the accelerated motion (represented by the upper shaded area) is made up by the surplus of the degrees of velocity in the second part (indicated by the lower shaded area).

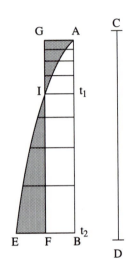

Fig. 3.38

This conclusion is correct in spite of the fact that, in this case, the areas representing the two aggregates of degrees of velocity are unequal, because, as the previous discussion has shown, Galileo's comparison of aggregates of degrees of velocity is independent from the relationship of the areas covering these aggregates. Hence, the above argument implies that the Merton Rule is valid for any monotonically accelerated motion whether uniformly accelerated or not.

As a further consequence it follows that all such accelerated motions attaining the same maximum degree of velocity traverse the same distance in the same time, because they all can be reduced to the same uniform motion characterized by half the maximum degree. Although neither Galileo's manuscripts nor his published texts indicate that he noticed this clearly nonsensical implication of the proof technique by which he had derived the law of fall, there are indications that some of his disciples and colleagues were aware of the problematical nature of his argument.[173]

---

[173] Indeed, in one of Torricelli's manuscripts, an argument concerning a comparison between a nonuniformly accelerated motion and a uniform motion is found, similar to the hypothetical argument reconstructed above; see Blay 1990, pp. 4-5. In Torricelli's argument, too, areas do not represent distances traversed but are related to aggregates of degrees of velocity. In order to avoid paradoxical consequences of this argument such as the ones discussed above, Torricelli introduces an additional ad hoc assumption. Cavalieri's first reaction to the *Discorsi* indicates that he, too, was aware of the problematical character of Galileo's proof technique, because he criticizes Galileo for not having emphasized that the indivisibles have to be taken as equidistant (see Bonaventura Cavalieri to Galileo, June 28, 1639, EN XVIII, 67); for a discussion of Galileo's contemporary correspondence with Cavalieri, as well as for other reactions to Galileo's treatment of indivisibles, see Galluzzi 1979, Chapter V.

3.6.2   Oblique Projection as a Gap in the *Discorsi*

Although various statements are made about oblique projection in Galileo's published works, neither the systematic treatment of projectile motion in the Third Book of *De Motu locali*, presented in the Fourth Day of the *Discorsi*, nor any other of his published writings, contains an explicit derivation of the trajectory for oblique projection.[174]

The first proposition in which Galileo deals with oblique projection is the Corollary to Proposition VII, presented in the Fourth Day of the *Discorsi*. This proposition corresponds to the Maximum Range Theorem discussed in section 3.5.3.1. Proposition VII itself refers only to horizontal projection:

PROPOSITION VII. THEOREM [IV]

In projectiles by which semiparabolas of the same amplitude are described, less impetus is required for the describing of one whose amplitude is double its altitude than for any other. (EN VIII, 294/244)

Although the restriction to horizontal projection of this proposition is not evident from its formulation, it can be inferred from its proof. The proof of Proposition VII in fact explicitly uses the construction of horizontal projection based on the Sublimity Theorem discussed in section 3.5.1.

The Corollary to Proposition VII formulates the analogous result for oblique projection:

COROLLARY

From this it is clear that in reverse [direction] through the semiparabola *db* [Fig. 3.39], the projectile from point *d* requires less impetus than through any other [semiparabola] having greater or smaller elevation than semiparabola *bd*, which [elevation] is according to the tangent *ad* and contains one-half a right angle with the horizontal. Hence it follows that if projections are made with the same impetus from point *d*, but according to different elevations, the maximum projection, or amplitude of semiparabola (or whole parabola) will be that corresponding to the elevation of half a right angle. The others, made according to larger or smaller angles, will be shorter [in range]. (EN VIII, 296/245)

No proof beyond the "from this it is clear" is offered for this corollary, and yet after the statement of the corollary Sagredo remarks:

The force of necessary demonstrations is full of marvel and delight; and such are mathematical [demonstrations] alone. I already knew, by trusting to the accounts of many bombardiers, that the maximum of all ranges of shots, for

---

[174] This lack of a proof of the trajectory for oblique projection was noticed and extensively discussed by Wohlwill (1884, pp. 111-116) and Dijksterhuis (1924, pp. 264-284) but has been ignored by many later interpretations of Galileo's science. Wohlwill, whose interpretation is the starting point for the one given here, explained this gap in the deductive structure of Galileo's theory of motion by the absence of a general principle of inertia and a general principle of superposition from the conceptual foundations of the theory.

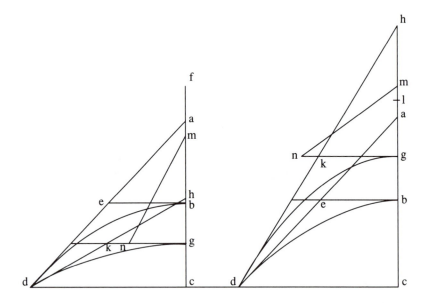

Fig. 3.39 (EN VIII, 295)

artillery pieces or mortars – that is, that shot which takes the ball farthest – is
the one made at elevation of half a right angle which they call "at the sixth
point of the [gunner's] square." But to understand the reason for this phenomenon infinitely surpasses the simple idea obtained form the statements of
others, or even from experience many times repeated. (EN VIII, 296/245)

The subsequent proposition is directly stated for oblique projection (Fig. 3.40)
without a preceding statement of the corresponding result for horizontal projection. Proposition VIII corresponds
to the Equal Amplitude Theorem presented in section 3.5.3.2; it states:

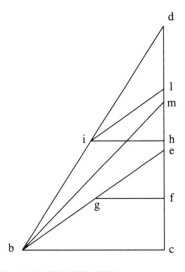

PROPOSITION VIII
THEOREM [V]

The amplitudes of parabolas described by projectiles sent forth
with the same impetus, according
to elevations having angles equidistant above and below half a
right angle, are equal to one another. (EN VIII, 297/ 246)

Galileo's proof of this proposition
closely follows the manuscript version
discussed in section 3.5.3; it again
only employs a construction of the
trajectory for horizontal projection.

Fig. 3.40 (EN VIII, 297)

3.6.3 Summary

The proof of the law of fall in the *Discorsi* depends on a theorem essentially equivalent to the Merton Rule with the help of which Galileo could apply kinematical rules restricted to uniform motion to accelerated motion as well. The derivation of Galileo's version of the Merton Rule is based on a one-to-one comparison between infinite sets of degrees of velocity, a comparison which in general yields results different from those obtained on the basis of the medieval representation of velocities by areas. Galileo's proof technique is also incompatible with classical mechanics and leads, for instance, to the absurd conclusion that all monotonously accelerated motions, which start from rest, last the same time, and attain the same maximum degree of velocity, also traverse the same distance.

Although the *Discorsi* takes it for granted that the trajectory for oblique projection is a parabola, no derivation of this proposition is presented.

We have thus seen that in his published works Galileo did not provide consistent derivations of two of his key contributions to classical mechanics, the law of fall and the parabolic shape of the trajectory of projectile motion.

3.7 Preclassical Limits to Galileo's Mechanics

In this section, we shall try to understand why Galileo, who had arrived at essential results of classical mechanics, was nevertheless unable to construct for them an adequate conceptual framework. We shall therefore reexamine the concepts on which his arguments for the laws of fall and of projectile motion fundamentally depend, in particular the concepts of motion and of velocity, as they are used in his published works. We shall see that his concept of velocity was still based on the Aristotelian concept and that motion is still conceived in the traditional sense as dependent on a cause. We may thus conclude that the reason for Galileo's unsuccessful or nonexistent derivations of the key results of classical mechanics is that his theory of motion remained firmly rooted in preclassical mechanics.

3.7.1 The Aristotelian Concept of Velocity and the Configuration of Qualities in Galileo's Published Works

As we shall see in the following, the conceptualization of accelerated motion in terms of the medieval configuration of qualities and the Aristotelian concept of velocity still provide the background for Galileo's reasoning about accelerated motion in his published works in spite of the considerable improvement in his understanding of motion since the time of his early treatise *De Motu*.

3.7.1.1 Traditional Foundations of Galileo's Derivation of the Law of Fall

In the analysis of the *Discorsi* proof of the law of fall presented in section 3.6.1, it became clear that this proof makes use of the traditional diagram familiar from the medieval configuration of qualities as well as from Galileo's earlier attempts to prove this law. In spite of the modifications with respect to the traditional argument also discussed in the last section, the basic interpretation of this diagram is still rooted in the medieval conceptual framework; in other words, in the *Discorsi* the triangular diagram is still interpreted as representing velocity as a uniform difform quality. This conclusion is confirmed by the terminology used in the discussion leading up to the definition of uniformly accelerated motion in the *Discorsi*:

> And thus it is seen that we shall not depart far from the correct rule if we assume that intensification of speed is made according to the extension of time; from which the definition of the motion of which we are going to treat may be put thus:
>
> [DEFINITION]
>
> I say that that motion is equably or uniformly accelerated which, abandoning rest, adds on to itself equal moments of swiftness in equal times. (EN VIII, 198/154)

As discussed in the Chapter 1, the terms "intensification" and "extension" are in fact part of the technical vocabulary of the medieval configuration of qualities.

The definition of uniformly accelerated motion itself makes use of the concept of moment; immediately before the above quoted passage, Galileo speaks of degree or moment of speed. The formulation of the definition confirms a conclusion already drawn in section 3.6.1, i.e., that the "moments of swiftness" which the accelerating motion adds onto itself have to be conceived as the infinitesimal constituents of an overall velocity which are assimilated to the degrees of an intensive quality in the medieval understanding.[175]

---

[175] In the definition quoted above, the notion of moment may seem to be close to the modern notion of an infinitesimal increment of velocity, while in remarks immediately preceding the definition Galileo provides an explanation of the degrees of velocity in accelerated motion that makes them appear rather close to the modern concept of instantaneous velocity. In fact, in these remarks Galileo draws the consequences of his earlier insight – related to the Double Distance Rule – into the relationship between a degree of velocity in uniformly accelerated motion and uniform motion: " ... if the moveable were to continue its motion at the degree or moment of speed acquired in the first little part of time, and were to extend its motion successively and equably with that degree, this movement would be twice as slow as [that] at the degree of speed obtained in two little parts of time." (*Discorsi*, EN VIII, 198/154)

Galileo's derivation of the law of fall in the *Discorsi* not only employs the concepts of the medieval configuration of qualities doctrine but also the Aristotelian concept of velocity. In the first step of this derivation, the proof of the Merton Rule, the Aristotelian proportion between velocities and distances for equal times is, as shown in section 3.6.1, applied to conclude that, if the overall velocities of the accelerated and the uniform motion are equal, the two motions, which take the same time, also traverse the same distance. This is in fact the final conclusion of the proof of the Merton Rule, according to which the equality of the distances traversed by the two motions follows from the equality of the times:

> It is therefore evident that equal spaces will be run through in the same time by two moveables, of which one is moved with a motion uniformly accelerated from rest, and the other with equable motion having a moment one-half the moment of the maximum speed of the accelerated motion; which was [the proposition] intended. (EN VIII, 209/166f)

But it is odd that this conclusion is just the reverse of the proposition to be proven,[176] since, in Galileo's original formulation of the Merton Rule, it is the equality of the times that follows from the equality of the distances traversed by the two motions:

PROPOSITION I. THEOREM I

> The time in which a certain space is traversed by a moveable in uniformly accelerated movement from rest is equal to the time in which the same space would be traversed by the same moveable carried in uniform motion whose degree of speed is one-half the maximum and final degree of speed of the previous, uniformly accelerated, motion. (EN VIII, 208/165)

From the point of view of the Aristotelian kinematic proportions, however, it is indeed indifferent whether the equality of the times is inferred from that of the distances or vice versa, given the equality of the overall velocities, the establishment of which is the core of Galileo's argument.

3.7.1.2 The Unsolved Mirandum Paradox

In Chapter 1, we saw that the application to accelerated motion of the traditional kinematic proportions may lead to inconsistent conclusions. The Mirandum fragment, discussed in section 3.4.2, documents the fact that Galileo had earlier encountered such inconsistencies. The *Dialogo* and the *Discorsi* provide evidence that, even in his later published works, he was still unable to overcome

---

Galileo's use of "degree" or "moment" of speed as equivalent in certain contexts makes it however clear that neither of these can be identified with the modern notion of an infinitesimal increment of velocity or with that of an instantaneous velocity, but that both are still part of a traditional conceptual system.

[176] This reversal was noticed, for instance, by Wisan (1974, pp. 289-290).

these inconsistencies, precisely because of the essential role played by the Aristotelian concept of velocity in his final theory of motion.

Galileo very explicitly unfolds the difficulties raised by the Mirandum Paradox in a discussion between Salviati, Sagredo, and Simplicio in the First Day of the *Dialogo*. In the course of this discussion, the postulate, mentioned in section 3.6.1.4 in connection with the proof of the Length-Time-Proportionality, is applied to a comparison of motions, one along the perpendicular and the other along an inclined plane having the same vertical height:

> [T]wo equal movable bodies, descending by different lines and without any impediment, will have acquired equal impetus whenever their approaches to the center are equal. (*Dialogo*, EN VII, 47/23)

To Sagredo it seems that this postulate is in conflict with the statement that the motion along the perpendicular is faster than the motion along the inclined plane. In response to Sagredo's objection, Galileo's spokesman Salviati describes the Mirandum Paradox:

> Salv. Then it would seem to you still more false if I should say categorically that the speeds of the bodies falling by the perpendicular and by the incline are equal. Yet this proposition is quite true, just as it is also true that the body moves more swiftly along the perpendicular than along the incline. (*Dialogo*, EN VII, 48/24)

In the ensuing discussion the two incompatible statements are related to two different characterizations of velocity: the first is based on the Aristotelian concept, while the other is Galileo's generalized definition of the equality of velocities. Simplicio, the Aristotelian, introduces the first and more widely known concept:

> Salv. ... Tell me, Simplicio, when you think of one body as being faster than another, what concept do you form in your mind?
>
> Simp. I imagine one to pass over a greater space than the other in the same time, or to travel an equal space in less time.
>
> Salv. Very good. Now as to bodies of equal speed, what is your idea of them?
>
> Simp. I conceive them to pass equal spaces in equal times.
>
> Salv. And nothing more than that?
>
> Simp. This seems to me to be the proper definition of equal motions. (*Dialogo*, EN VII, 48/24)

In the sequel, Sagredo explains the second, more general characterization of velocity, which in conjunction with the Length-Time-Proportionality yields the conclusion that the speeds along the vertical and the incline are equal:

> Sagr. Let us add another, however, and call the velocities equal when the spaces passed over are in the same proportion as the times in which they are passed. That will be a more general definition.

Salv. So it is, because it includes equal spaces passed in equal times, and also those which are unequal but are passed in times proportionate to them. (*Dialogo*, EN VII, 48/24)

In the ensuing lengthy discussion following the introduction of this concept, Galileo illustrates its application to the accelerated motions along the inclined plane and the perpendicular. At the end of this explanation, his spokesman Salviati comes to the conclusion that his concept of velocity does not allow for a consistent comparison of the speeds of these two motions, because such a comparison does not yield unequivocal results:

Salv. ... Now since we can conceive distances and velocities along the incline and along the perpendicular such that the proportion between the distances will be now greater and now less than the proportion of the times, we may very reasonably admit that there are also spaces along which the times of the motions bear the same proportion as the distances.

Sagr. I am already freed from my main doubt, and perceive that something which appeared to me a contradiction is not only possible but necessary. (*Dialogo*, EN VII, 50/26)

In this way, Galileo has identified the application of his kinematic rules to accelerated motion as the source of the contradiction in the Mirandum Paradox. Due to the fact that the two motions compared to each other are accelerated, the result of the comparison depends on the starting and endpoints of the two motions, and is hence ambiguous. This analysis does not, however, offer a solution to the Mirandum Paradox that would allow for an understanding of accelerated motion in terms of a more advanced concept of velocity that does not lead to contradictions or ambiguities. On the contrary, Galileo's analysis of the Mirandum Paradox in the *Dialogo* is evidence to the fact that his reasoning about motion is still shaped by the Aristotelian concept of velocity, since this analysis does not indicate any doubts about the legitimacy of applying the Aristotelian proportions to accelerated motion.[177]

---

[177] This conclusion is in disagreement with Kuhn's interpretation of this passage from the *Dialogo* as well as with the conclusions he draws from it for an understanding of cognitive development. With respect to Galileo's "thought experiment" as well as with respect to a related experiment in child psychology by Piaget, Kuhn (1977, p. 264) writes: "Full confusion, however, came only in the thought-experimental situation, and then it came as a prelude to its cure. By transforming felt anomaly to concrete contradiction, the thought experiment informed our subjects what was wrong. That first clear view of the misfit between experience and implicit expectation provided the clues necessary to set the situation right." For a discussion of the problems of cognitive development in Galileo's science, see also Renn 1989, and 1990b. For references to the historical literature on this problem, see section 3.4.2.

3.7.1.3 The Rules of Kinematics in the *Discorsi*

As will become clear in the following, the *Discorsi* do not show any essential progress in this respect, although, at first sight, it appears that Galileo was able to avoid the problems revealed by the Mirandum Paradox by carefully restricting the application of kinematic rules to uniform motion. However, this restriction mainly concerns Galileo's generalized definition of the equality of velocities, while the traditional Aristotelian proportions remain as they are the basis of his treatment of motion, whether uniform or accelerated.

The First Book of the treatise *De Motu locali*, presented in the Third Day of the *Discorsi*, contains the results of Galileo's efforts to elaborate a deductive theory of kinematic proportions.[178] Although Galileo calls this book a treatise on uniform motion, it in fact forms the basis for all of his subsequent derivations involving kinematical relationships, whether these concern uniform or accelerated motion. In the following we shall see that, in spite of the relatively coherent mathematical structure of this treatise, Galileo's efforts have failed to yield a theory capable of systematically justifying his use of the concept of velocity and of effectively resolving the Mirandum Paradox. In particular, the treatise does not contain a new definition of velocity but is essentially based on the Aristotelian concept of velocity.

Galileo's treatise opens with a definition of uniform motion and then presents four axioms which establish relationships of monotony between velocity, space, and time. The first two axioms are valid only for uniform motion whereas the last two are of a general character. The concept of velocity appears for the first time in this treatise in Axioms III and IV:

AXIOM III

The space traversed with greater speed is greater than the space traversed in the same time with lesser speed.

AXIOM IV

The speed with which more space is traversed in the same time is greater than the speed with which less space is traversed. (EN VIII, 191f/148f)

These two axioms closely parallel the first of the two propositions on the notion of "quicker" in Aristotle's *Physics*. In contrast to Aristotle, but in agreement with medieval rules of kinematics, Galileo makes a point of treating velocity as a quantity that is different from space and time.

---

[178]   This theory has been extensively studied by Wisan (1974, pp. 281-286), Clavelin (1983), Giusti (1986), Souffrin (1988), and Napolitani (1988). The interpretation presented here is indebted, in particular, to Giusti for the clear exposition of the deductive structure of this treatise, and to Souffrin for his insight into the role of the Aristotelian concept of velocity in Galileo's theory. Napolitani's paper contains an important analysis of the parallelity between Galileo's treatment of velocity and his treatment of specific weight.

The four axioms are followed by six theorems, only *two* of which are not explicitly restricted to uniform motion. Galileo's generalized definition of the equality of velocities is not contained in this treatise, while the corresponding Archimedean proposition on uniform motion appears as its first theorem:

PROPOSITION I. THEOREM I

If a moveable equably carried with the same speed passes through two spaces, the times of motion will be to one another as the spaces passed through. (EN VIII, 192/149)

Contrary to this theorem, the formulations of the second and the third theorem do not restrict them to uniform motion; they correspond in fact to the two Aristotelian kinematical proportions:

PROPOSITION II. THEOREM II

If a moveable passes through two spaces in equal times, these spaces will be to one another as the speeds. And if the spaces are as the speeds, the times will be equal. [...]

PROPOSITION III. THEOREM III

Of movements through the same space at unequal speeds, the times and speeds are inversely proportional. (EN VIII, 193/150)

In order to derive the first two theorems, Galileo establishes relationships of proportionality between velocity, distance, and time by following a standard Euclidean method. In anachronistic terms, this method can be described as deriving the linearity of a function $f(x)$ from its monotony and from the property of homogeneity $f(nx) = nf(x)$ for any integer n. In Galileo's treatise, the monotony is guaranteed by the axioms, while the property of homogeneity needed for the derivation of Theorem I follows from the definition of uniform motion.

In the derivation of Theorem II, however, Galileo presupposes a property of homogeneity which is neither guaranteed by the axioms nor by any definition. In modern terms this property can be expressed as $s(nv) = ns(v)$ for integer n, where $s(v)$ is the space traversed in a given time with the speed v, and $s(nv)$ is the space traversed with n times the speed v in the same time. This property, which remains implicit in Galileo's argument, is, however, an obvious consequence of the Aristotelian concept of velocity, which is thus seen to underly Galileo's derivation. Theorem II, corresponding to one of the Aristotelian proportions, is in fact also valid for nonuniform motion, since its derivation does not involve Theorem I or axioms restricted to uniform motion. The derivation of Theorem III, on the other hand, which corresponds to the other of the two Aristotelian proportions, involves Theorem I so that this theorem is actually restricted to uniform motion. Nevertheless, the statement of this theorem suggests, as we have seen, that it is not restricted to uniform motion, just as the corresponding Aristotelian proportion.

Galileo also derives compounded proportions between velocities, spaces, and times. These compounded proportions, taken together, could be read as the expression in the language of proportions of the definition of velocity as the quotient of a distance and a time, as it is used in classical mechanics. An example is provided by the fourth proposition, which plays a role in Galileo's proof of the law of fall:

PROPOSITION IV. THEOREM IV

If two moveables are carried in equable motion but at unequal speeds, the spaces run through by them in unequal times have the ratio compounded from the ratio of speeds and from the ratio of times. (EN VIII, 194/151)

In order to derive these theorems, however, Galileo makes use of the traditional kinematic relationships introduced in the beginning of his treatise, relationships which, as we have seen, are completely rooted in the Aristotelian understanding of velocity. The derivations of the compounded proportions also make use of the introduction of a third motion which Galileo had already employed in his manuscripts. But the derivations of all of these theorems now depend on theorems and axioms which are restricted to uniform motion, and hence the theorems themselves are restricted to uniform motions.

As a consequence, the equality of velocities when distances and times are in the same proportion – an equality that crucially enters the argument of the Mirandum Paradox – now indeed follows from Proposition VI, but only for uniform motions:

PROPOSITION VI. THEOREM VI

If two moveables are carried in equable motion, the ratio of their speeds will be compounded from the ratio of spaces run through and from the inverse ratio of times. (EN VIII, 196/152)

Hence, formally, the Mirandum Paradox can no longer arise for the trivial reason that Galileo's generalized definition of the equality of velocities on which it was based, is – by the construction of Galileo's deductive system – now transformed into a corollary applicable only to uniform motion. However, as we have seen in section 3.6.1.4, Galileo's technique for comparing overall velocities actually undermines this restriction to uniform motion of the kinematical rules stated in the First Book of *De Motu locali*. The conclusion reached in the proof of the Length-Time-Proportionality that the overall velocity of the motion along the inclined plane is equal to the overall velocity of the motion along the vertical is, in fact, one of the two statements constituting the Mirandum Paradox. The other statement follows from the application to these two motions of Proposition II of the First Book of *De Motu locali*, which corresponds to one of the Aristotelian proportions and is in fact not restricted to uniform motion. According to this proposition, the velocities of the two motions compared in the Mirandum Paradox are in the same proportion as the distances traversed in the same time. Hence, the difficulties apparent from Galileo's discussion of the Mirandum Paradox in the *Dialogo* are still present in his treatment of accelerated motion in the *Discorsi*, because this treatment is still based on the Aristotelian concept of velocity.

3.7.2.Impetus as the Cause of Motion in Galileo's Published Works

We have seen that the *Discorsi* does not contain a derivation of the trajectory for the case of oblique projection. In order to reconstruct the reasoning underlying the theory of projectile motion in Galileo's published works, we are therefore limited to an analysis of the derivation of the trajectory for *horizontal* projection as well as of occasional remarks providing clues to Galileo's thinking on oblique projection. We shall first take up some of these clues and then turn to an analysis of the dynamical background of Galileo's treatment of the motion of fall in the *Discorsi*. In this way, we can provide evidence for our claim that the notion of impetus as a cause of motion is at the core of Galileo's dynamics, even in his published works, and that neither a general principle of inertia nor of superposition belong to it.

3.7.2.1 Horizontal Projection and Horizontal Plane

At the beginning of the Fourth Day of the *Discorsi*, Galileo introduces the conservation of uniform motion along a horizontal plane in order to derive the parabolic form of a projectile's trajectory for the case of horizontal projection:

> I mentally conceive of some moveable projected on a horizontal plane, all impediments being put aside. Now it is evident from what has been said elsewhere at greater length that equable motion on this plane would be perpetual if the plane were of infinite extent; but if we assume it to be ended, and [situated] on high, the moveable (which I conceive of as being endowed with heaviness), driven to the end of this plane and going on further, adds on to its previous equable and indelible motion that downward tendency which it has from its own heaviness. Thus there emerges a certain motion, compounded from equable horizontal and from naturally accelerated downward [motion], which I call "projection." (EN VIII, 268/217)

Why does Galileo introduce motion along a plane in order to discuss projectile motion? The explanation is that, even in the *Discorsi*, the conservation of the horizontal component of projectile motion is indeed neither presupposed as an independent assumption nor derived from any principle according to which impressed force *as such* is conceived as being naturally indelible. Galileo's reference to motion along a horizontal plane suggests that he still conceives the conservation of uniform horizontal motion as a consequence of his analysis of motions along inclined planes.

The conservation of uniform motion along a horizontal plane is discussed relatively late in the Third Day, in the context of the treatment of a specific problem, the Scholium to Problem IX:

> It may also be noted that whatever degree of speed is found in the moveable, this is by its nature indelibly impressed on it when external causes of acceleration or retardation are removed, which occurs only on the horizontal plane;

for on declining planes there is a cause of more acceleration, and on rising planes, of retardation. From this it likewise follows that motion in the horizontal is also eternal, since if it is indeed equable it is not [even] weakened or remitted, much less removed. (EN VIII, 243/197)

This passage illustrates that Galileo derived the conservation of uniform motion on a horizontal plane from the properties of motion along inclined planes, essentially following a pattern of argument developed in his early research.[179] If motion along a horizontal plane is conceived as an intermediate case between motions along planes sloping either upwards or downward – or in the terminology of *De Motu*, as a motion which is neither forced nor natural – and if upward motion is decelerated and downward motion accelerated, it follows that horizontal motion must be uniform.

It is obvious from this reconstruction of the reasoning underlying the conservation of horizontal motion in the case of horizontal projection that it must have been difficult for Galileo to derive the trajectory for oblique projection.

3.7.2.2 The Paradoxes of Oblique Projection in the *Discorsi*

The only passage in the *Discorsi* where the relationship between horizontal and oblique projection is touched upon is a discussion between Salviati, Sagredo, and Simplicio about the presuppositions on which the derivation of the trajectory for horizontal projection is based. After Salviati, Galileo's spokesman in these dialogues, has presented the proof that the path is a parabola in the case of horizontal projection, first Sagredo, and then Simplicio express their doubts whether the premises of this derivation can be granted to the author:

> Sagr. It cannot be denied that the reasoning is novel, ingenious, and conclusive, being argued *ex suppositione*; that is, by assuming that the transverse motion is kept always equable, and that the natural downward [motion] likewise maintains its tenor of always accelerating according to the squared ratio of the times; also that such motions, or their speeds, in mixing together do not alter, disturb, or impede one another. In this way, the line of the projectile, continuing its motion, will not finally degenerate into some other kind

---

[179] The present text has often been interpreted as providing evidence for Galileo's recognition of the principle of inertia; for a classic interpretation see Mach 1942, pp. 168-169, and pp. 330-337. In fact, in this passage, Galileo does not use the terminology "natural" and "forced," but characterizes the motion along the horizontal plane by the absence of "external causes of acceleration or retardation" without explaining what these external causes are; see Mittelstrass 1970, pp. 268-282. However, not only is Galileo's formulation in this passage an isolated case in his writings, as has been stressed by Wohlwill (1884, pp. 126-134), but also the crucial point is that the justification of what some historians see as a principle of inertia actually still follows the pattern of an argument derived from the *De Motu* theory; see section 3.3.1. An interpretation similar to that given here, also following Wohlwill, was proposed by Dijksterhuis (1924, pp. 264-271).

[of curve]. But this seems to me impossible; for the axis of our parabola is vertical, just as we assume the natural motion of heavy bodies to be, and it goes to end at the center of the earth. Yet the parabolic line goes ever widening from its axis, so that no projectile would ever end at the center [of the earth], or if it did, as it seems it must, then the path of the projectile would become transformed into some other line, quite different from the parabolic.

Simp. To these difficulties I add some more. One is that we assume the [initial] plane to be horizontal, which would be neither rising nor falling, and to be a straight line – as if every part of such a line could be at the same distance from the center, which is not true. For as we move away from its midpoint towards its extremities, this [line] departs ever farther from the center [of the earth], and hence it is always rising. One consequence of this is that it is impossible that the motion is perpetuated, or even remains equable through any distance; rather, it would be always growing weaker. Besides, in my opinion it is impossible to remove the impediment of the medium so that this will not destroy the equability of the transverse motion and the rule of acceleration for falling heavy things. All these difficulties make it highly improbable that anything demonstrated from such fickle assumptions can ever be verified in actual experiments. (EN VIII, 273f/222f)

Salviati grants all objections that so far have been made to his derivation:

Salv. All the difficulties and objections you advance are so well founded that I deem it impossible to remove them. For my part, I grant them all, as I believe our Author would also concede them. I admit that the conclusions demonstrated in the abstract are altered in the concrete, and are so falsified that horizontal [motion] is not equable; nor does natural acceleration occur [exactly] in the ratio assumed; nor is the line of the projectile parabolic, and so on. (EN VIII, 274/223)

Both Sagredo's and Simplicio's main objections to the assumptions underlying Galileo's treatment of horizontal projection come from viewing this treatment from a cosmological perspective in which the fact that the earth is a sphere is taken into account. According to the traditional understanding of projectile motion on which Sagredo's argument is based, the natural motion would eventually have to dominate over the violent motion and carry the projectile to its natural place, the center of the earth. Simplicio's first argument can be described by saying that, from a cosmological perspective, Galileo's analysis of horizontal projection is defective because one of the elements of its very definition, responsible for the characteristic properties of horizontal motion, ceases to be valid after the first instant of time.

For Galileo, an essential feature of motion along a horizontal plane is the assumption that the distance with respect to the center of the earth remains unchanged. This assumption is, however, only approximately true for a plane, if the fact that the earth is a sphere is taken into account. From a cosmological perspective, the meaning of motion along the horizontal becomes ambiguous. It can either mean "motion along a tangential line" or "motion along the circumference." From Simplicio's discussion of the consequences of this problem, it appears that he excludes the possibility that both characteristics of motion along the horizontal, rectilinearity and uniformity, can be maintained.[180]

In agreement with Galileo's justification of the conservation of uniform motion for the special case of horizontal projection, Simplicio states that if rectilinearity is maintained, the motion ceases to be uniform because it is no longer equidistant from the center of the earth. According to his argument, the motion that was initially considered to be uniform has to diminish continually because it is going uphill. In Simplicio's argument, which is accepted by Salviati, Galileo thus treats this motion as if the line along which it proceeds were a real inclined plane. In his earlier attempts to construct the trajectory for oblique projection, Galileo had similarly treated one of the components of the projectile motion as a motion along an inclined plane (see section 3.5.2).

The analogy between the treatment of the "cosmological picture" of horizontal projection in the *Discorsi* and these earlier attempts (in particular, the one reconstructed from fol. 175v) demonstrates that, in the *Discorsi*, Galileo's earlier arguments are not superseded, although they lead to otherwise unacceptable constructions of the trajectory of oblique projection. Hence, it is still the same contradiction between different, equally plausible statements about the trajectory for oblique projection which accounts for the absence of an explicit construction of the trajectory for oblique projection in the *Discorsi*.

In an insertion intended for publication in a revised edition of the *Discorsi*, Galileo once again grappled with the problem of oblique projection. Whereas his discussion of the cosmological picture of horizontal projection in the *Discorsi* closely follows the line of reasoning underlying the construction of the trajectory for oblique projection as reconstructed from fol. 175v, the argument presented in

---

[180]   The interpretation given here is indebted to the analysis of the discussion between Salviati, Sagredo, and Simplicio given in Dijksterhuis (1924, pp. 271-277), Wisan (1974, pp. 261-263), and Chalmers and Nicholas (1983, pp. 329-333). A passage in the First Day of the *Dialogue* (EN VII, 43/19) shows that Galileo's cosmological views indeed made it impossible for him to accept rectilinear uniform motion as a principle of natural motion, i.e., as a motion which is proper to bodies constituting an ordered universe. According to the argument given in this passage, straight motion is incompatible with the notion of an ordered universe, first, because the *terminus ad quem* is different from the *terminus a quo* so that, if the world was perfect in the beginning, it will no longer be perfect after the motion has taken place, and second, because, properly speaking, there is no *terminus ad quem* for an infinite straight motion so that it cannot be a natural motion aiming for a natural place. This argument illustrates that Galileo's cosmology with its emphasis on circular motion was not only influenced by Platonic philosophy or by his early insight that circular motion is neither natural nor forced, and hence perpetual but also that it was shaped by such aspects of the Aristotelian concept of motion as the determination of a rectilinear motion by its terminal points. There are, however, interpretations that claim that Galileo's cosmological statements should not be taken that seriously; see Coffa 1968, p. 280.

this insertion shares certain features with Galileo's other construction of the trajectory as seen in fol. 171v.[181]

> Simpl.: I ask you, before we continue, to put me in a position to understand how to demonstrate the converse, which the Author presupposes as clear and free from doubt; I mean that the projectile which, when it comes from above to below describes a semiparabola, must return to its starting point along the same line if it is hurled conversely from below to above, by following exactly the same traces without having for this another regulator than the direction of the simple straight line that touches the semiparabola already drawn above; if it describes [this line] from above to below, the tranversal, horizontal impetus makes me quietly admit the great curvature at the vertex, but I can neither understand nor see how the impetus generated from below along a straight tangent can reproduce a transversal impetus that would be suited to regulate the same curvature.

> Salv.: Signor Simplicio, by mentioning the straight tangent, you admit a condition, namely that the line is tangential and inclined; and just this inclination is sufficient to have the effect that the projectile approaches the axis of the parabola in equal times along equal distances along the horizontal, as we will perhaps understand further down.

In this dialogue no construction of the trajectory for oblique projection is given; the only element of an explicit construction that is mentioned in this dialogue is the uniform motion along the horizontal. The existence of this motion also in oblique projection is justified by a reference to the existence of a horizontal component of the original direction of motion. If, as is indeed suggested by this dialogue, it was not obvious to Galileo how the impetus generated from below can reproduce such a transversal impetus, this argument made it nevertheless plausible to construct the trajectory of oblique projection starting from the assumption of a uniform motion along the horizontal. On the basis of this assumption, the trajectory of oblique projection could be compounded of a uniform motion along the horizontal and a motion along the inclined plane that determines the direction of the shot without interfering with the uniform horizontal motion – just as it is documented by the construction in fol. 171v.

3.7.2.3 The Role of Causes and Contrariety in Galileo's Understanding of Projectile Motion

The fact that the earlier attempts at a construction of the trajectory for the case of oblique projection are not superseded in the *Discorsi* also suggests that the dynamical reasoning underlying these earlier attempts still forms the background of Galileo's theory of motion in his published works. As will become clear in

---

[181] Cod. B, fol. 14v (EN VIII, 446-447). The manuscript is in the handwriting of Viviani; the text was to be inserted before Prop. VIII of the 4th Day of the *Discorsi*. See section 3.6.2 and document 5.3.11.

the following, in these works motion is in fact still conceived as being due to a moving cause, and the composition of different moving causes is still treated by a logic of contraries.

In the Second Day of the *Discorsi*, in a discussion of the resistance of materials, Galileo describes an experiment along an inclined plane that essentially corresponds to the Guidobaldo experiment discussed in section 3.3.1, as well as an arrangement with a hanging chain as two convenient means to draw a parabola quickly:

> Salv. There are many ways of drawing such lines, of which two are speedier than the rest; I shall tell these to you. One is really marvelous, for by this method, in less time than someone else can draw finely with a compass on paper four or six circles of different sizes, I can draw thirty or forty parabolic lines no less fine, exact, and neat than the circumferences of those circles. I use an exquisitely round bronze ball, no larger than a nut; this is rolled [*tirata*] on a metal mirror held not vertically but somewhat tilted, so that the ball in motion runs over it and presses it lightly. In moving, it leaves a parabolic line, very thin, and smoothly traced. This [parabola] will be wider or narrower, according as the ball is rolled higher or lower. From this, we have a clear and sensible experience that the motion of projectiles is made along parabolic lines, an effect first observed by our friend, who also gives a demonstration of it. We shall all see this in his book on motion at the next [*primo*] meeting. To describe parabolas in this way, the ball must be somewhat warmed and moistened by manipulating it in the hand, so that the traces it will leave shall be more apparent on the mirror.
>
> The other way to draw on the prism the line we seek is to fix two nails in a wall in a horizontal line, separated by double the width of the rectangle in which we wish to draw the semiparabola. From these two nails hang a fine chain, of such length that its curve [*sacca*] will extend over the length of the prism. This chain curves in a parabolic shape, so that if we mark points on the wall along the path of the chain, we shall have drawn a full parabola ... (EN VIII, 185f/142f)

Just as Guidobaldo had compared the trajectory of projectile motion with a hanging chain in the account of his experiment, and just as Galileo himself had earlier performed a closer comparison of their shapes in his manuscripts, both physical phenomena are also related to each other in the *Discorsi*. In fact, Galileo mentions them not only as different methods for drawing parabolas (although the curve of a hanging chain is actually not a parabola but a catenary) but also in a discussion of the composition of impetuses and of the curvature of the trajectory at the end of the Fourth Day:

> Sagr. I observe that with regard to the two impetuses, horizontal and vertical, as the projection is made higher, less is required of the horizontal, but much of the vertical. On the other hand, in shots of low elevation there is need of great force in the horizontal impetus, since the projectile is shot to so small a height. But if I understand correctly, at full elevation of 90°, all the force in the world would not suffice to shoot the projectile one single inch out of the vertical, and it must necessarily fall back at the same place from which it was shot. Yet I dare not affirm with equal certainty that a projectile, even at zero elevation, which is to say in the horizontal line, could not be shot to some

[little] distance by some [great] force, or that infinite force would be required – as if, for example, not even a culverin had the power to shoot an iron ball horizontally, or "point blank" as they say (that is, at no point [on the gunner's square]), where there is zero elevation. I say that in this case there remains some ambiguity, and that I am unable to deny resolutely either fact, for the reason that another event seems no less strange, though I have a logically conclusive demonstration of it. This is the impossibility of stretching a rope so [tightly] that it shall be pulled straight, and [held] parallel to the horizontal; for it always sags and bends, nor is there any force that will suffice to hold it straight. (EN VIII, 309/255f)

The question Galileo addresses in this passage corresponds to a question which had preoccupied students of projectile motion since Tartaglia: Can there be a straight part of the trajectory or is the trajectory always curved, even for a (horizontal) projection made with great force? Sagredo relates this question to that of the curvature of the hanging rope for which he claims to have a rigorous answer. In response to this comparison, Salviati points to a deeper physical relationship between the two phenomena:

Salv. Well, Sagredo, in this matter of the rope, you may cease to marvel at the strangeness of the effect, since you have a proof of it; and if we consider well, perhaps we shall find some relation between this event of the rope and that of the projectile [fired horizontally].

The curvature of the line of the horizontal projectile seems to derive from two forces, of which one (that of the projector) drives it horizontally, while the other (that of its own heaviness) draws it straight down. In drawing the rope, there is [likewise] the force of that which pulls it horizontally, and also that of the weight of the rope itself, which naturally inclines it downward. So these two kinds of events are very similar. Now, if you give to the weight of the rope such power and energy as to be able to oppose and overcome any immense force that wants to stretch it straight, why do you want to deny this [power] to the weight of the ball? (EN VIII, 309f/256)

In establishing a close similarity not only between the curvature of the trajectory and that of the hanging rope but also between the physical generation of these two phenomena, Galileo follows – even in the *Discorsi* – the tradition of applying concepts of statics to projectile motion. This tradition was illustrated in section 3.2.3 by examples taken from Tartaglia and Benedetti as well as by Galileo's discussion of projectile motion in *De Motu*. The forces generating the projectile motion are conceived as being the same forces that play a role in traditional statics. In the *Discorsi*, however, the forces generating the projectile motion are compared not to forces acting on a balance – as was the case in Tartaglia – nor to forces acting along an inclined plane – as was the case in Benedetti and in *De Motu* – but to the forces causing the curvature of the hanging rope. Nevertheless, the shape of the trajectory results in the *Discorsi* in a similar way from the physical generation of the motion by a horizontal and a vertical force, just as in *De Motu* the properties of the trajectory result from its generation by a force conceived by comparing a cannon to an inclined plane.

Numerous passages in Galileo's published works confirm that the composition of the two moving forces generating projectile motion is in these works as much a delicate problem as it was in the traditional analyses of projectile motion. More precisely, Galileo used arguments from the logic of contraries already displayed in *De Motu* in order to justify superposition without mutual interference in the special case of two orthogonal motions. A detailed discussion of the composition of two orthogonal motions is found, e.g., in the following passage of the *Dialogo*. The example is a stone falling from a ship's mast and thus displaying two motions, that of vertical fall and a circular motion around the center of the earth communicated by the uniform motion of the ship:

> As for the other, the supervening motion downward, in the first place it is obvious that these two motions (I mean the circular around the center and the straight motion toward the center) are not contraries, nor are they destructive of one another, nor incompatible. As to the moving body, it has no resistance whatever to such a motion, for you yourself have already granted the resistance to be against motion which increases the distance from the center, and the tendency to be toward motion which approaches the center. From this it follows necessarily that the moving body has neither a resistance nor a propensity to motion which does not approach toward or depart from the center, and in consequence no cause for diminution in the property impressed upon it. Hence the cause of motion is not a single one which must be weakened by the new action, but there exist two distinct causes. Of these, heaviness attends only to the drawing of the movable body toward the center, and impressed force only to its being led around the center, so no occasion remains for any impediment. (*Dialogo*, EN VII, 175 /149)

In the first part of this explanation, Galileo states that the two motions to be compounded are not contrary to each other. As the last two sentences show, this conclusion is attained by conceiving the two motions as produced by two distinct causes whose "aims" are compatible because they are orthogonal to and therefore do not interfere with one another. This argument is the precise analog to the analysis of (almost) horizontal projection in the second treatment of projectile motion in *De Motu*.

3.7.2.4 The *De Motu* Theory of Motion in Media in the *Discorsi*

The conceptualization of motion as being caused by a moving force is not only the basis of Galileo's analysis of projectile motion in his last published work, it also shapes his understanding of the motion of fall. In the First Day of the *Discorsi*, Galileo analyzes the motion of bodies of different weights falling in different media. This analysis is not part of the systematic treatise *De Motu locali*, but it nevertheless contains a conclusion that enters as a crucial presupposition into this systematic treatise, presented in the Third and Fourth Days: the conclusion that in a vacuum all bodies fall with the same speed.

Galileo's analysis of the motion of falling bodies in media can be divided into three parts: In the first part he refutes Aristotle's theory in exactly the same way as in *De Motu*; in the second part the role of the specific weight for the speed of fall is analyzed with the result that in a vacuum all bodies fall with the same speed whatever their specific weight; in the third part Galileo proposes a rule for the determination of the speed of fall of a given body in a given medium, a rule which for bodies of the same material is identical with the corresponding rule in *De Motu*. The following passage shows that the difference between the weight of the falling body and the weight of the same volume of the medium plays a key role in Galileo's explanation of the differing speeds of bodies falling in different media, just as it did in the Archimedean theory of fall in *De Motu*:

> If we then assume the principle that in a medium no resistance exists at all to speed of motion, whether because it is a void or for any other reason, so that the speeds of all moveables would be equal, we can very consistently assign the ratios of speeds of like and of unlike moveables, in the same and in different filled (and therefore resistant) mediums. This we shall do by considering the extent to which the heaviness of the medium detracts from the heaviness of the moveable, which heaviness is the instrument by which the moveable makes its way, driving aside the parts of the medium. No such action occurs in the void *[nel mezzo vacuo]*, and therefore no difference [in speed] is derived from different heaviness. And since it is evident that the medium detracts from the heaviness of the body contained in it to the extent of the weight of an equal quantity of its own material, diminishing in that ratio the speeds of the moveables which in a non-resistant medium would remain equal (as assumed), we shall have our goal. (EN VIII, 119/78f)

The resemblance of the *Discorsi* theory with that developed in *De Motu* can be seen in the following passage from *De Motu*[182]:

> For, clearly, in the case of the same body falling in different media, the ratio of the speeds of the motions is the same as the ratio of the amounts by which the weight of the body exceeds the weights [of equal volumes] of the respective media.

In both theories, buoyancy accounts for the effect of the medium on the motion of the falling body. Furthermore, in the *Discorsi* just as in *De Motu*, heaviness is conceived as the cause of speed, and speed is understood in the Aristotelian sense, neglecting internal characteristics of the motion such as its acceleration.

The precise mathematical meaning of Galileo's formulation "diminishing in that ratio the speeds of the moveables" in the *Discorsi* theory is at first glance unclear, so that a quantitative comparison with the *De Motu* theory appears difficult. But Galileo's general statement in the *Discorsi* is followed by numerous examples which make it clear that the speed losses of two bodies falling in different media are in the same proportion to the vacuum speed (which is

---

[182] DM, 36; EN I, 272; see section 3.2.1.

supposed to be the same for all bodies) as the specific weights of the media are to the specific weights of the falling bodies.[183]

This rule is incompatible with classical mechanics, but in the case of the same body falling in different media it coincides exactly with that given in *De Motu*, because in this special case the speed losses are proportional to the specific weights of the respective media. The main difference between the analysis of the motion of fall in a medium presented in the *Discorsi* and that developed in the early treatise is the addition in the *Discorsi* of the proposition that in a vacuum all bodies fall with the same speed. As a consequence, Galileo claims in the *Discorsi* that his theory of motion in media now corresponds at least approximately to observation without having to be corrected for the effect of acceleration:

> And reasoning with this rule, I believe, we shall find that experience fits the computation much better than it fits Aristotle's rule. (EN VIII, 120/79)

3.7.2.5 Did Galileo Renounce Causal Explanation of Acceleration in the *Discorsi*?

While in *De Motu* acceleration played the role of an effect that superimposed itself on the laws governing the motion of bodies falling in media and accounted for discrepancies between these laws and observation, it is precisely the phenomenon of acceleration which is now difficult to integrate into an analysis of fall which is based on Archimedean hydrostatics. In fact, in the *Discorsi,* Galileo attempted to adapt the causal explanation of acceleration originally given in *De Motu* to the insights into the nature of acceleration, gained in the meantime, but found the result of this attempt unsatisfactory.[184]

The causal explanation of acceleration in the *Discorsi* is given in the same context in which it originally appeared in *De Motu*, that of a causal explanation of projectile motion by an impressed force. Salviati, Galileo's spokesman, asks his

---

[183] For reconstructions of Galileo's theory of fall in media presented in the *Discorsi*, see Dijksterhuis 1924, pp. 227-234, 1961, p. 336, and Clavelin 1974, p. 333. Dijksterhuis' early interpretation also stresses the similarity between the *De Motu* and the *Discorsi* theories .

[184] Galileo's skeptical discussion of the causal explanations of acceleration has led many historians to assume that Galileo's physics in the *Discorsi* is pure kinematics, abjuring all attempts of a causal explanation (see, e.g., Settle 1966, p. 152, Barbin and Cholière 1987, p. 94, Drake 1990, p. 68). First of all, they have overlooked, the fact that Galileo did highlight one particular explanation as a plausible one and, second, that this explanation is not just a reiteration of the initial explanation of acceleration given in *De Motu*. The *Discorsi* explanation can in fact be considered as a correction of the *De Motu* explanation, and, according to the interpretation presented here, it is precisely this correction that makes it problematical for Galileo.

interlocutors to grant him the assumption that acceleration in vertical fall proceeds according to the same pattern as deceleration in vertical projection:

> Salv. ... I believe you would not hesitate to grant me that the acquisition of degrees of speed by the stone falling from the state of rest may occur in the same order as the diminution and loss of those same degrees when, driven by impelling force, the stone is hurled upward to the same height. (EN VIII, 200/157)

This statement reiterates Galileo's observation in the letter to Sarpi from 1604[185]:

> that the naturally falling body and the violent projectile pass through the same proportions of velocity.

But because the proportionality between degree of velocity and distance proposed to Sarpi was no longer acceptable to Galileo, he could not demonstrate the symmetry between upward projection and free fall in the same way as he had done in the letter to Sarpi.

Nevertheless, since Galileo presupposes in the *Discorsi,* just as he had done in *De Motu,* that the vertically projected body is "driven by impelling force," the symmetry between projectile motion and motion of fall which he claims in the above passage suggests that the same impelling force also accounts for the acceleration in the motion of fall. In the subsequent discussion, this consequence is indeed drawn by Sagredo:

> Sagr. From this reasoning, it seems to me that a very appropriate answer can be deduced for the question agitated among philosophers as to the possible cause of acceleration of the natural motion of heavy bodies. For let us consider that in the heavy body hurled upwards, the force [*virtù*] impressed upon it by the thrower is continually diminishing, and that this is the force that drives it upward as long as this remains greater than the contrary force of its heaviness; then when these two [forces] reach equilibrium, the moveable stops rising and passes through a state of rest. Here the impressed impetus is [still] not annihilated, but merely that excess has been consumed that it previously had over the heaviness of the moveable, by which [excess] it prevailed over this [heaviness] and drove [the body] upward. The diminutions of this alien impetus then continuing, and in consequence the advantage passing over to the side of the heaviness, descent commences, though slowly because of the opposition of the impressed force, a good part of which still remains in the moveable. And since this continues to diminish, and comes to be overpowered in ever-greater ratio by the heaviness, the continual acceleration of the motion arises therefrom. (EN VIII, 201/157f)

Sagredo's explanation of acceleration is essentially identical to the one Galileo had earlier proposed in *De Motu*[186]:

---

[185] EN X, 116; see section 3.3.3.
[186] DM, 100; EN I, 328-329

the [motion of the] body is accelerated because the contrary [i.e., upward] force is continuously diminishing while [in consequence] the natural weight is being attained.

There is, however, a subtle difference between the *Discorsi* and the *De Motu* explanations of acceleration. In the *Discorsi* Galileo presupposes the symmetry between projectile motion and motion of fall as well as the unboundedness of acceleration as consequences of his earlier research. He can therefore no longer derive acceleration from the difference between weight and impressed force, as he had done in *De Motu*. In fact, as we have seen in section 3.3.1, neither the unbounded continuation of acceleration in absence of a medium nor the symmetry between vertical projection and free fall are compatible with the explanation of acceleration by the difference of two opposed moving forces, one constant and the other gradually diminishing. In response to this difficulty, Galileo relates acceleration – in a rather vague formulation – to the ratio rather than to the difference of the two tendencies of motion:

> And since this [the impressed force] continues to diminish, and comes to be overpowered in ever-greater ratio by the heaviness, the continual acceleration of the motion arises therefrom. (EN VIII 201/158)

If the ratio between impressed force and heaviness is assumed to account for acceleration in free fall, then the unboundedness of natural acceleration no longer presents any difficulty to an explanation in terms of impressed force. But unlike the role played by the difference between heaviness and impressed force in *De Motu*, the role played by the ratio between these two tendencies of motion in the *Discorsi* cannot be justified on the basis of a systematic theory of motion such as Galileo's Archimedean theory of fall in *De Motu*. Consequently, Galileo does not elaborate this explanation of acceleration but expresses himself rather skeptically about the possibilities of finding the cause of acceleration:

> Salv. The present does not seem to me to be an opportune time to enter into the investigation of the cause of the acceleration of natural motion, concerning which various philosophers have produced various opinions, some of them reducing this to approach to the center; others to the presence of successively less parts of the medium [remaining] to be divided; and others to a certain extrusion by the surrounding medium which, rejoining itself behind the moveable, goes pressing and continually pushing it out. Such fantasies, and others like them, would have to be examined and resolved, with little gain. (EN VIII, 202/158f)

3.7.3 Summary

The derivations of the law of fall and the trajectory of projectile motion in Galileo's *Discorsi* depend largely on traditional concepts, arguments, and techniques which had already played a role in his earliest writings including *De Motu*.

The derivation of the law of fall makes essential use of concepts pertaining to the medieval configuration of qualities as well as of the Aristotelian concept of velocity. Galileo's *Dialogo* contains a discussion of the Mirandum Paradox that resulted from the application of an Aristotelian kinematic proportion and a more generalized definition of equal velocities to accelerated motion but offers no resolution of this paradox. A systematic treatise on kinematics is presented in the beginning of the Third Day of the *Discorsi* which contains proportions that can be read as expressions in traditional terminology of the modern definition of velocity as quotient of distance and time. But the derivations of these kinematic proportions also depend on the Aristotelian concept of velocity. Although most of them are restricted to uniform motion, so that a reappearance of the Mirandum Paradox seems to be formally excluded, the restriction to uniform motion is overthrown in the subsequent treatment of accelerated motion in the *Discorsi*, with the consequence that this paradox implied by the Aristotelian concept of velocity is actually not resolved.

Galileo's justification of the statement that the horizontal component of horizontal projection is uniform is based on his theory of motions along inclined planes. In his treatment of motions along inclined planes in the Third Day of the *Discorsi*, he follows an argument developed in the course of his elaboration of *De Motu*, according to which motion along a horizontal plane is an intermediate case between motions on planes sloping either upward or downward.

Even in the context of his final theory of motion, Galileo still conceived oblique projectile motion in essentially the same way as he had in his manuscripts, i.e., as being caused by a gradually diminishing impetus. His comparison of the trajectory with a hanging chain in the *Discorsi*, as well as a passage from the *Dialogo*, confirm that in his understanding motion is caused by forces, the composition of which is governed by a logic of contraries and hence is, in general, problematical.

The *Discorsi* analysis of bodies falling in different media is an adaptation of the Archimedean theory of *De Motu* to the insight that in a vacuum all bodies fall with the same speed. It is nevertheless, like the *De Motu* theory, based on the principle that speed is caused by heaviness.

The *Discorsi* proposes a tentative explanation of acceleration which is based on the understanding of projectile motion in terms of an impressed force elaborated in *De Motu*. In the *Discorsi*, this explanation is, however, put forward only as a possible hypothesis about the nature of acceleration because Galileo's insight that acceleration is a continuous process prevented him from accounting for it in terms of a difference of two contrary motive forces and hence from justifying this explanation within the Archimedean theory.

Galileo's published writings are in fact not, as they are often seen, a final synthesis and resolution of the problems raised in his working papers; they are rather a collection of proofs and paradoxes, precisely because both proofs and paradoxes were still conceived in the conceptual framework of preclassical mechanics.

3.8 Descartes' Critique Revisited

The preceding analysis of the development of Galileo's thinking on motion should make it possible to answer the questions raised in the Introduction, 3.1: Can Descartes' critique of Galileo's last work on motion really be discarded as the envious comment of a stubborn philosopher, unable or unwilling to acknowledge that this work inaugurated a new physics? Does the theory of motion presented in the Third and Fourth Days of the *Discorsi* contain – at least implicitly – a derivation of the laws of fall and of projectile motion from principles valid in classical mechanics?

Our study has provided ample evidence that Galileo's work on motion was shaped by the same traditions that the previous two chapters have shown to be characteristic of Descartes' thinking. The foundations of the arguments of both of them were provided, at least in their early work, by the same conceptual framework of preclassical mechanics. Galileo and Descartes used the same concept of motion. They conceived of velocity as a global characteristic of motion which can adequately be conceptualized by means of the Aristotelian definition as space traversed in a given time. Both of them used the Aristotelian proportions to compare motions by their velocities. Both of them tried to solve the problems that occurred when these concepts of motion and velocity were applied to the study of accelerated motion by recurring to the medieval doctrine of the configuration of qualities. Both of them conceived of motion as consisting of indivisible moments with changing degrees of velocity. Both of them considered motions as caused by forces imparted to the moving body and both of them used the medieval term impetus to denote these forces. Both of them related these forces to the changing degrees of velocity which served as the conceptual means to account precisely for the phenomenon of acceleration. Finally, both of them used the same geometrical figures provided by the medieval tradition to represent changing qualities and to infer relations between the changing degrees of velocity and the overall velocity so as to determine the relation between times elapsed and spaces traversed in accelerated motion. Hence both of them associated the area of such geometrical figures with the concept of velocity instead of with the concept of space and failed to find the simple solution of classical physics, i.e., to identify the area of a diagram representing the relation between time and velocity of an accelerated moving body with the space it has traversed.

Thus, the conceptual framework which at the beginning of our study could have seemed to be an idiosyncracy of Descartes' early thinking turns out to be a essential element of preclassical mechanics. There was no direct contact between

Descartes and Galileo, and yet they conceptualized the problems of mechanics in the same categories. Although these categories look strange from our modern perspective and do not fit into the framework of classical mechanics, they must have seemed quite natural to the mathematicians and engineers of Descartes' and Galileo's time.

On the other hand, the situation of Descartes' comments on Galileo's final publication was quite different. Originally, Descartes had faced the same problems as Galileo when he tried to derive the relation between times elapsed and spaces traversed in free fall using certain assumptions about the cause of gravitation. Thus Descartes was not entirely wrong when in his letters to Mersenne he equated Galileo's derivation of the law of fall with his own earlier attempts. However, Descartes did not notice that Galileo was fully aware of the alternative between the assumptions that the velocity of a falling body increases proportionally to the spaces traversed or that it increases proportionally to the times elapsed. Neither Descartes nor Beeckman had seen this alternative as such.

However, by the time Descartes wrote his comments on Galileo's *Discorsi*, the agreement between the conceptual foundations of their efforts was already history. In the meantime Descartes had developed the theoretical framework of *Le Monde* and of his later *Principia*, and this development was connected with a complete reconceptualization of the foundations of mechanics.

According to the medieval tradition, motion is caused by an impetus imparted to the moving body. This tradition was already anti-Aristotelian insofar as the concept of impetus was a unifying concept for natural and forced motion. In particular, the motion of a falling body no longer had to be conceived simply as natural but could be reconstructed as the effect of the impetus which was continuously imparted; and projectile motion could be seen as the effect of the simultaneous action of the different impetuses of gravitation and the motive force of an agent. However, the break with the Aristotelian tradition carried out by Descartes, which culminated a few years later in the *Principia philosophiae,* was much more radical. In this work Descartes introduced a completely different theoretical model of local motion. As a result of this fundamental change in his thinking Descartes' comments on Galileo's *Discorsi* were no longer written on the background of a shared conceptual system.

Galileo, too, did not simply perservere in the framework of preclassical concepts without any alteration of his thinking. Our study of his early treatise *De Motu* has shown that he started out by criticizing Aristotle and formulating a new understanding of gravitation. The motion of a falling body was conceptualized as a phenomenon analogous to the hydrostatic phenomenon of a floating body with a moving force determined according to the theory of Archimedes by the specific weight, its velocity being proportional to the difference of the weight of the falling body and the weight of the medium.

It goes without saying that this is not the solution of classical physics; on the contrary, this model is still completely shaped by the Aristotelian conceptual framework. Every motion needed a force as its cause and the velocity of a moving body was conceived as proportional to this force. For Galileo, as for Aristotle, the motion of fall was the natural motion of heavy bodies to their natural place, the center of the earth. It was conceived essentially as a motion with constant velocity which was accelerated only by accidental circumstances. In spite of the generally anti-Aristotelian tone as well as the numerous anti-Aristotelian arguments of *De Motu*, some of the most important foundational concepts used in this treatise were still Aristotelian and the mechanical explanation of the motion of falling bodies and of projectile motion developed in this treatise were by no means related to the concepts of classical mechanics.

No doubt his later *Discorsi*, on which Descartes was commenting, represented an impressive advance over this early theory of motion. As we have seen in the preceding analysis, Galileo encountered contradictions when elaborating the *De Motu* theory and found its agreement with experience wanting in such cases as projectile motion. Some of the results of his subsequent work, first to correct the theory and later to replace it with a better one, became important elements of classical mechanics, e.g., the statement that in a vacuum all bodies fall with the same speed. Another achievement of Galileo's mature theory of motion was the conception of acceleration as an essential property of the motion of fall characterized by the proportional increase of velocity with time. In the *Discorsi*, Galileo finally states the law of fall and claims that the trajectory of projectile motion is parabolic in shape. The book on uniform motion, also contained in the *Discorsi*, presents a list of kinematic rules which express (in terms of proportions) the modern conception of velocity as the quotient of distance and time. All of these results are among Galileo's lasting contributions to classical mechanics.

The work which yielded these results aimed to find general principles from which the fundamental propositions about the motion of bodies could be derived. In this respect, as our study has revealed, Galileo failed. The reason for this is that basic concepts employed in the Third and Fourth Day of the *Discorsi* come from the same antique and medieval traditions that had also shaped Descartes' early thinking on motion. That is the deeper reason why Descartes could legitimately claim to have the more advanced understanding of motion. The limits of Descartes' own reconceptualization, as we saw in Chapter 2, lay not in the requirements he formulated for physical theory but in the idiosyncratic form in which he attempted to fulfill them.

In his final work, Galileo treated the motion of fall in media on the basis of Archimedian hydrostatics, as he had done more than 40 years earlier. Even in the *Discorsi*, he still considered an impressed force as the cause of projectile motion and its gradual disappearance as a possible explanation of acceleration. He still referred to the action of an impressed force in order to explain the continuation of uniform motion, i.e., in situations where Descartes would already invoke the principle of inertia like a classical physicist. In Galileo's mature theory of motion, the composition of motions in projectile motion was not yet treated on the basis of a general principle of superposition but was still conceived in terms of the logic of contraries which, however, as we have seen in the second chapter, also formed the background for Descartes' thinking on the problem of composition. Finally, it has become clear from our study that the basis for Galileo's understanding of velocity in the *Discorsi* was still provided by the Aristotelian concept characteristic of his earlier work.

Some of the original aspects of Galileo's studies of motion published in the *Discorsi*, such as the hydrostatic analysis of motion in media or the tentative explanation of acceleration by an impressed force, were in fact no longer part of the systematic theory developed in the Third and Fourth Days. In his mature theory, Galileo did indeed adapt some of the consequences of the *De Motu* theory to new insights gained in the course of his research, and as a consequence some assertions have changed their status within the edifice of the theory of motion presented in the *Discorsi*. However, this does not imply that the conceptual foundations of this theory were different from those of the early treatise *De Motu*.

Therefore, the key problems of the deductive structure of the *Discorsi* theory lie not in some traditional statements which have changed their status in Galileo's presentation, but in the derivations of what are usually considered to be his most significant contributions to classical mechanics. These contributions are isolated results which are badly or not at all justified within his theory of motion. Galileo has in fact not succeeded in deriving the parabolic shape of the trajectory for oblique projection because he found no way to avoid the ambiguities he had earlier encountered; his published proof of the law of fall depends on a problematical ad hoc assumption, hardly suited to escape the paradoxes of his earlier attempted derivations; and his application of the concept of velocity to motion along inclined planes leads to a contradiction from which he was unable to free his theory of motion. The ambiguities, paradoxes, and contradictions which characterize Galileo's failed attempts to derive key results of the *Discorsi* indicate that his arguments and deductions are generally not conceived within the conceptual framework of classical mechanics – even where they happen to be unproblematical such as in the case of the derivation of the trajectory for horizontal projection.

How must this theory have appeared from the perspective of a contemporary scientist who – like Galileo – did not yet have classical mechanics at his disposal? Our analysis has shown that Descartes was not only right in asserting that the theory of motion presented in the *Discorsi* did not cohere but also in claiming that some of its foundational concepts were questionable.

Summing up: It is in fact neither fundamentally different attitudes towards experiment nor radically different views on the role of mathematical deduction in theory that distinguish Descartes' research on motion from that of Galileo. When Descartes criticized Galileo for failing to provide convincing arguments for some of his theorems, he was in fact measuring him by shared standards. It was Galileo's primary ambition to prove statements like the law of fall and the parabolic trajectory of projectile motion within a deductive theory. Thus,

Descartes' observation that a proof of the parabolic trajectory of oblique projection was in fact missing from the *Discorsi* reveals his critique to be at the same level as the scientific work at which it was directed. The point, however, is not to debunk myths about one of the heroes of classical mechanics but rather to analyze precisely what his achievement actually was – and thus to illustrate how conceptual development in science actually occurs. Galileo's achievement lay in exploring the limits of preclassical mechanics not in transcending them. He established theorems which implicitly define new concepts and whose appropriation by his successors, as we shall see in the Epilogue, did indeed transcend preclassical mechanics.

4
Epilogue

Classical mechanics indeed has its origin in the 17th century. The period we have studied here is not merely a prelude to the Scientific Revolution but an essential part of it. In the three studies presented here, we have tried to illuminate some aspects of these origins, in particular, the question of what it means for classical mechanics to originate. However, contrary to the widespread view that at least Galileo crossed the borderline between medieval and modern scientific thinking, we have tried to show in the preceding chapters that this period of science was in fact still deeply rooted in medieval traditions. Although in the problems and results we can clearly identify constitutive elements of classical mechanics, the arguments and derivations are still based on the conceptual tools of medieval natural philosophy. Even Galileo's most celebrated achievements turn out to have been obtained by twisting and squeezing a patchwork of inadequate traditional concepts. However, the specific uses of these tools and the knowledge produced cannot be imagined as outcomes of medieval scholarship.

To designate this intermediate stage of development, we have used the term *preclassical mechanics*. Preclassical mechanics in this sense represents a period, in which deductive thinking still depended on the semantics of concepts adapted from the medieval traditions but systematically applied to the solution of a new set of problems. These solutions became the core of classical mechanics.

Does our claim, that the episodes studied in this book still belong to preclassical mechanics, imply that we still have not found the real beginning of classical mechanics and must search for it in a later period? What kind of problems had still to be solved before the law of falling bodies was correctly demonstrated, before the principle of inertia and the principal of the composition of motions and forces changed the structure of the causal explanation, before the theory of mechanics was based on adequate conservation laws? Were the true heroes of the new science yet to appear? Are we to look for the proper revolution in a later era? Do dramatic changes in scientific thinking take place only in the age of Newton or even later?

Such conclusions are by no means the necessary consequences of our analysis. In fact they are quite wrong, as can easily be detected by a follow-up study of the problems faced by Descartes, Galileo, and their younger contemporaries. Even a mere glance at later treatments of these problems reveals the striking fact that none of the mistakes of the preclassical solutions needed to be corrected. As an epilogue to the studies presented in this book, we shall take a short look at three episodes immediately following those we have analyzed.

4.1   First Example: Conceptualization of Free Fall

In February 1643, Descartes described in a letter to Huygens an experiment that he had performed to study the dependency of the velocity of water flowing out of a pipe on the height of the water level above the outlet. His theoretical explanation of the results makes use of the law of fall. In this context Descartes again refers to the proof of the law by means of a geometrical representation. This time however his argument – compared with its earlier presentations – shows some dramatic changes.[1]

> As for gravity, I also consider that it increases the speed of bodies that it makes fall, almost in the same proportion as the times in which they fall; so that if a drop of water falls for two minutes of an hour, it goes almost twice as fast at the end of the second [minute] as it did at the end of the first; from which it follows that the path it covers is almost in the double proportion [i.e., as the square] of the time; that is to say, if during the first minute it falls from the height of one foot, during the first and the second together it ought to fall from the height of four feet.

The relation of the time elapsed to the space traversed which was to be proved is now stated correctly and in general terms: the space traversed is in double proportion to the time elapsed. As we know, Descartes was convinced at that time that gravity was produced by the impact of particles and not by attraction. Nevertheless, he obviously now accepts the Galilean law, which he had rejected so vigorously before, as an approximation nearly conforming to the actual motion of the falling body. He correctly states the presupposition of the law that the velocity of the falling body – in this case of a drop of water – is proportional to the time elapsed and not to the space traversed.

Much more striking, however, is how he relates the concepts used in stating the law to the geometrical representation:

> This is easily explained by the triangle ABC [Fig. 4.1], of which the side AD represents the first minute, the side DE the speed that the water has at the end of this first minute, and the space ADE represents the path it meanwhile covers, which is the length of one foot. Then DB represents the second minute, BC the speed of the water in [i.e., at the end of] this second minute, which is double that of the preceding, and the space DECB the path, which is three times the preceding. And one can also note that if this drop of water continued to move in some other direction with the speed that it acquired by its fall from a height of one foot during the first minute, without gravity's helping it after that, it would cover during one minute the path represented by the rectangle DEFB, which is two feet. But if it should continue to move for two minutes with the speed which it has acquired by descending four feet it would cover the path represented by the rectangle ABCG, which is eight feet. (AT III, 619-620)

---

[1]   Descartes to Huygens, Feb. 18 or 19, 1643 (AT III, 619-620); see document 5.4.1.

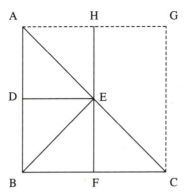

Fig. 4.1 (AT III, 620)

Descartes' conceptualization of the problem, as well as his interpretation of the figure which represents the solution, has changed completely. He no longer conceives the horizontal lines as representing minima of force or degrees of velocity but simply as representing the velocity of the body. The vertical lines are interpreted as representing time intervals. However, the most important new element in Descartes' reinterpretation of his earlier proof is that it no longer shows any uncertainty about the most critical element, i.e., the interpretation of the areas of the figure. Descartes has shifted completely from the traditional interpretation of the area as representing the "quantity of motion" to its direct identification with the space traversed.

Thus, it is not Galileo but Descartes who was the first to conceptualize the relations of time, space, and velocity in a manner consistent with the conceptual framework of classical mechanics. Furthermore, it is noteworthy that Descartes at least does not present this proof as being in any way different from his previous proofs.

4.2 Second Example: Composition of Motions

We have seen that it was essential to Descartes' argument, both about the laws of impact and the law of refraction, that component determinations cannot be taken independently, and that the change of determination of a body's motion in interaction could not be ascertained directly but rather only derived by way of the conservation of motion. The two sides of the parallelogram of determinations were never used to derive the diagonal; on the contrary, one side and the diagonal were used to derive the other side. Descartes' disciples, as we have indicated briefly above, had particular difficulties with this aspect of his system. Even if we examine only the most strictly orthodox tradition: Clerselier – Rohault –

Regis, we find that determination is treated as an independent magnitude and that the change of determination in interaction is not derived, but that the interacting determination is *presupposed* to be *reversed*. As soon as Descartes' most faithful disciples ceased to paraphrase his derivations and began to develop their own explanations they fell into error.

When Clerselier reopened the debate with Fermat on the *Dioptrique* in 1662, he was compelled to answer some new objections by Fermat which Descartes had never dealt with. Taking up a purported counterexample presented by Fermat, Clerselier considers the oblique collision of billiard balls (one at rest, one in motion). Since the opposition can only be along the normal between their centers of gravity, the collision can only result in the resting ball's being pushed in the direction of the opposition.[2]

> It is confirmed by experience that, in whatever manner the ball A [Fig. 4.2] is pushed at point B by the balls C, D, E, F, G, and whatever are the determinations of which one may suppose their route to be composed, they will always push it towards H.
>
> First of all, for the ball E, it is clear that it would push towards H since ball A is totally opposed to its determination; but what is clear for ball E should be similarly extended to the others which, although they come at a slant towards ball A, only touch it at point B and only push it insofar as they descend towards H, and not at all insofar as they move towards I (or towards K). This is why they cannot impress a different motion on this ball except that of making it go towards H. But since the determinations of the balls D and F are opposed insofar as the one goes to the right and the other to the left, they are not at all opposed insofar as they descend, and thus they ought to produce the same effect on ball A, which is to push it towards H.

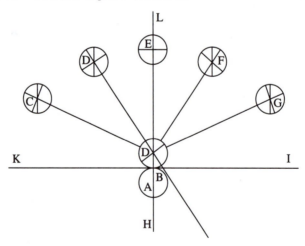

Fig. 4.2 (Fermat 1894, vol. 2, p. 478; figure rotated 180° to agree with text)

---

[2] Clerselier, letter to Fermat, May 13, 1662; Fermat 1894, vol. 2, p. 477-479 (emphasis added).

Clerselier then extends this argument to a surface of reflection by assuming that ball A is firmly fixed in its place so that the colliding ball is reflected conserving its motion.

> But if we suppose that ball A is hard and immobile, all the balls after colliding with it will be constrained *to change the determination they have to go towards H into that to go or reflect towards L* and to keep the other [determinations] if they have them, which it [ball A] cannot cause to change since it is not at all opposed to them in that sense: and this explains the reflection at equal angles.

Clerselier has simply applied the parallelogram rule to resolve the original determination into two components and to recompound the unchanged and the reversed components after impact, thus obtaining the new determination. The conservation of motion plays no role in the argument and Descartes' circles representing scalar motion are not used.

Pierre Silvain Regis, the successor to Clerselier and Rohault as spokesman of orthodox Cartesianism, used this same kind of argument when presenting the laws of reflection and refraction in his Cartesian textbook, *Cours de la philosophie*. The perpendicular determination is reversed in impact[3]:

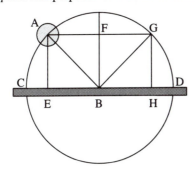

... Body A [Fig. 4.3] when it encounters the ground *should take a determination completely contrary to that which it had,* by which in equal times it should advance equal quantities, that is to say that, if in the time of one minute the body A descended from the line A[G] to the line CD, it should in a second minute climb back from the line CD to the line A[G].

Fig. 4.3 (Regis 1691, vol. 1, p. 352)

Regis does in fact still pay lip service to Descartes by referring to the conservation of motion and constructing the circle that represents it in his derivation; but the operation is redundant. He resolves the determination of ball A into the parallel and perpendicular components AF and AE. The parallel component remains unchanged (FG) while the pependicular component is *reversed* (HG). Although the intersection of FG and HG unequivocally fixes the point G, Regis in a third and superfluous step draws the circle that represents the distance traveled by ball A after reflection: "Accordingly, if we join together these three truths we will be obliged to recognize that the body A has arrived at point G where the circle and the lines AG and GH intersect ..."

We can thus see that even in situations where Descartes' followers are simply presenting material already drafted by Descartes, the conceptualization of the

---

[3] Regis, 1691, vol. 1, pp. 352-353 (emphasis added); see document 5.4.3.

problem has changed. Determination, or whatever it is that the sides of the parallelogram are supposed to represent, is treated as an independent magnitude even when the presentation closely follows Descartes' original.

4.3 Third Example: Conceptualization of Projectile Motion

Even before Galileo's *Discorsi* appeared in print, Bonaventura Cavalieri published a derivation of the parabolic trajectory without restriction to horizontal projection that is consistent with classical mechanics. From the book in which the derivation is given as well as from his subsequent correspondence with the enraged Galileo, who felt deprived of his priority, it is clear that Cavalieri was convinced he had merely repeated a result achieved by Galileo and known among Galileo's disciples to have been achieved by him.[4] Although, as we have seen, Galileo did not, in fact, include such a derivation in his own works, the results published by Cavalieri continued to be ascribed to him – among others, for instance, by his disciple Torricelli. After dealing with the motion of falling bodies in the first part of his book *De Motu Gravium* (1644), Torricelli takes up in the second part the motion of projectiles. In the third theorem of this second part, he shows that the "curved line that is described by a moving body which has been projected according to an arbitrary elevation is a parabola." In his proof, which closely follows Galileo's derivation of the trajectory for horizontal projection, Torricelli introduces a statement that is closer to the general principle of inertia than anything that can be found in either Galileo's published works, or in his letters and manuscripts. He also compounds the vertical motion of fall with a motion along an arbitrary direction, presupposing a general principle of composition of motions.[5]

> Let the horizontal FH [Fig. 4.4] and the vertical AH be drawn, then also FI, IH will be equal as well as AI and IB. Let AB be divided in arbitrary equal parts AL, LI, IM, MB, and erect vertical lines through the points L, I, M. It is manifest that equal spaces AL, LI, IM, MB will be traversed by the moving body in equal times, if it should move with uniform motion and without the addition of a new downward motion from the internal gravity of the proceeding body.

The last statement of this passage shows that the line of direction AB represents the motion that would occur without the effect of gravity. In this statement, Torricelli assumes as evident that the projectile would proceed in uniform motion along a straight line if it were not diverted from that motion by gravity.

---

[4] Cavalieri 1632, Chap. 29 and 30, pp. 153ff (see document 5.4.4). See also Cavalieri's letter to Galileo of Sept. 21, 1632; EN XIV, 394-395. For discussion see Koyré 1978, pp. 237-251.

[5] Torricelli 1644 (1919, p. 156).

Here, gravity plays the same role as it does in classical mechanics, it is a cause of deviations from uniform motion along a straight line in an arbitrary direction. After Torricelli has given a representation of the motion without gravity, he can now refer the action of gravity to this representation[6]:

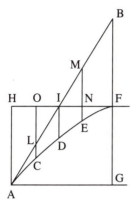

But since [the mobile] is immediately released from the projector in A and the attraction of gravity is added to it, it continually starts to deviate downwards from the line of direction, and the deviations will be such that the line LC of descent in 1 unit of time will be as 1, but the line ID of descent in 2 units of time will be as 4, and ME in 3 units of time as 9, BF in 4 units of time as 16, and so forth according to that law that the spaces of descent are always as the times squared.

Fig. 4.4 (Torricelli 1919, p. 156)

The effect of gravity is determined by the law of fall. The point that the body would have reached along AB in the uniform motion that results if gravity is absent is taken as the starting point for the motion of fall that represents the effect of gravity. Torricelli compounds the uniform motion and the motion of fall, assuming that no mutual interference takes place. He thus analyzes oblique projection along the lines of classical mechanics in the course of what he himself perceived as just a minor reworking of Galileo's theory of motion, which in fact excluded these principles.

Torricelli not only bases his derivation on some of Galileo's results, such as the law of fall, he also stresses the essential identity of his own result, the parabolic shape of the trajectory for oblique projection, with Galileo's result for horizontal projection. Galileo himself had already treated the trajectory for oblique projection as being identical with that for horizontal projection, but this identity was, as Torricelli puts it, "more desired than proven" in Galileo's *Discorsi*. Torricelli's theorem thus appears only to fill a small gap in Galileo's analysis of projectile motion. In the introduction to his book, Torricelli emphasized this understanding of his own contribution[7] :

Here the science of the motion of heavy bodies and of projectiles will be addressed, which has been treated by many but which has, as far as I know, been derived geometrically only by Galileo. I confess that, since he has reaped these crops as with a scythe, nothing remains for us but to collect, as we

---

6 Torricelli 1644 (1919, p. 156).
7 Torricelli 1644 (1919, p. 103).

follow the traces of this diligent reaper, the spicules that remain, whether these have been left by him or simply thrown away; if, however, not even these, then we shall at least pluck the privets and the violets that grow in the ground; but perhaps we shall be able to weave a not contemptible garland from the flowers.

* * * * *

These three examples show that the results achieved by preclassical mechanics could only a short time later be perceived and reproduced in the conceptual framework of *classical* mechanics. This framework is in fact constituted by these results if they are interpreted as determining the meanings of the terms they contain. For Galileo and the young Descartes, physical concepts such as, for instance, velocity were defined independently of and prior to their studies on free fall. These presuppositions of preclassical mechanics also placed limits on the scope of their achievements. After their work, velocity could be taken as that magnitude which increases in proportion to time in free fall. Their writings and arguments could be read from the point of view of their results instead of from the point of view of their presuppositions, and the limits set by these presuppositions could thus be ignored or not even be recognized as such.

5
Documents

A Note on Sources, Translations, and Figures

All sources quoted in the text and documented in this chapter are cited in English translation. Wherever a good translation was already available we have adopted its rendering as far as possible. Almost every adopted translation has undergone some significant changes, which are not indicated individually. The extensive quotations from Galileo's *De motu* (Drabkin) and *Discorsi* (Drake) have, however, been taken over largely unchanged.

The figures in the text have all been redrawn using computer graphics based on scanner copies of the standard published versions or in many cases the original manuscripts. They have been rectified in the sense that straight lines are straight, circles are round, and angles are bisected in the middle: The actual source of the figure is given in parentheses in each case. To avoid additional confusion, all figures are labeled as in the sources; we have neither modernized the labeling of points and lines with upper case and lower case letters nor have we reversed modernizations introduced into the standard editions; our only standardization has been to set all letters in the graphics in 9 point Times Roman type. Elipses in a text that include a paragraph break are indicated by square brackets "[...]"; otherwise by "..."

Close to half the material presented as documents in this chapter 5 has been published in translation in whole or part before. In these cases we have either taken the translation and revised it based on our reading of the original or else made an independent translation and compared versions adopting the other renderings when they seemed better than our first efforts. We have attempted in general to translate as literally as good English usage allows but also to avoid introducing merely stylistic differences to an already available translation. We alone are responsible for the final versions. Wherever our translation is based on or significantly collated with another version, its translator is mentioned at the end of the text.

5.1 Documents to Chapter 1: Concept and Inference

5.1.1 Descartes' Initial Document (1618)

In the proposed question, in which it is imagined that at each single time a new force is added to that with which the heavy body tends downward, I say that this force increases in the same manner [Fig. 5.1] as do the transverse lines *de*, *fg*, *hi*, and the infinite other transverse lines that can be imagined between them. To demonstrate this I take as the first minimum or point of motion, caused by the first attractive force of the Earth that can be imagined, the square *alde*. For the second minimum of motion we have the double, namely *dmgf*; the force, which was in the first minimum persists and a new, equal force is added to it. Thus in the third minimum of motion there will be three forces, namely those of the first, second, and third time minima, and so on. This number is triangular, as I will perhaps explain more fully elsewhere, and it appears that the figure of the triangle *abc* represents it. But, you will say, there are parts *ale*, *emg*, *goi*, etc., which protrude outside of the figure of the triangle. Therefore, the figure of the triangle cannot explain this progression. But I reply that these protuberant parts originate because we have ascribed latitude to the minima which must be imagined as indivisible and as consisting of no parts. This is demonstrated as follows. I divide the minimum *ad* into two equal ones at *q*; then *arsq* will be the

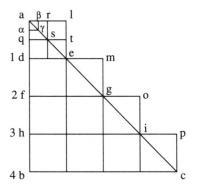

Fig. 5.1 (AT X, 76; J IV, 50)

[first] minimum of motion, and *qted* the second minimum of motion, in which there will be two minima of force. In the same manner we divide *df*, *fh*, etc. Then we will have the protuberant parts *ars*, *ste*, etc. Clearly they are smaller than the protuberant part *ale*. Furthermore, if I take a smaller minimum such as *paα*, then the protuberant parts will be yet smaller, such as *aβγ*, etc. If, finally, I take as this minimum the true minimum, i.e., the point, then there will be no protuberant parts, for they clearly could not be the whole point, but only a half

of the minimum *alde*, and there is no half of a point. From which it clearly follows that if we imagine, for example, a stone which is attracted in a vacuum by the earth from *a* to *b* by a force which flows equally from the Earth while the first persists, then the first motion at *a* will be to the last at *b* as the point *a* is to the line *bc*. The part *gb*, which is half, will be covered by the stone three times more quickly than the other half *ag*, because it will be drawn by the Earth with three times the force. The space *fgbc* is three times the space *afg*, as is easily proved. And one can say this of the other parts proportionately.

This question can be posed in another, more difficult, way in the following manner. Imagine [Fig. 5.2] that the stone is at point *a* and that the space between *a* and *b* is a vacuum; and that all at once, say today at nine o'clock God, should create in *b* a force attracting the stone; and that afterwards he should create

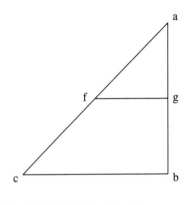

Fig. 5.2 (AT X, 77; J IV, 51)

in every single moment a new force equal to the one he created in the first moment; and that together with the force created before it would attract the stone ever more strongly, because in a vacuum that which is once in motion moves always; and finally the stone which was at *a* will reach *b* today at ten o'clock. And if it is asked, in how much time it will traverse the first half of the space, i.e., *ag* and in how much [time] the remaining [space]; I answer that the stone fell through the line *ag* in 7/8 of an hour; but through the space *gb* in 1/8 of an hour.[1] Now, a pyramid should be constructed on a triangular basis, so that its altitude is *ab*, which is in whatever manner agreed upon divided along with the whole pyramid by transverse lines [planes] parallel to the horizon. The stone will traverse the lower parts of the line *ab* as much more quickly as the sections of the whole pyramid are larger.

Finally the question concerning compound interest may be asked. And if it is imagined that it is augmented in the single moments and if it is asked what is owed in this or that time: this question will be solved by proportions taken from the triangle; but the line *ab* should not be divided into arithmetic, i.e., equal parts, but into geometrical, i.e., proportional parts. I could prove all that most evidently by my geometrical algebra, but it would take too much time.

(AT X, 75-78; J IV, 49-52; translation adapted from Mepham, Koyré 1978, pp. 82-85. For interpretation see 1.3.1 and 1.4.2)

---

[1] Descartes mistakenly writes 1/8 for *ag* and 7/8 for *gb*.

5.1.2 Beeckman's Note in his *Journal* (1618)

Marginal Note: Why does a stone falling in the vacuum fall ever more quickly?

Objects are moved downward toward the center of the Earth, the intermediate space being a vacuum, in the following manner:

In the first moment so much space is traversed as can be by the attraction of the Earth [*per Terrae tractionem*]. [The stone] continues in this motion in the second moment and a new motion of attraction is added, so that in the second moment double the space is traversed. In the third moment, the doubled space is maintained, to which is added a third space resulting from the attraction of the Earth, so that in the one [i.e. the third] moment a space triple the first space is traversed [...].

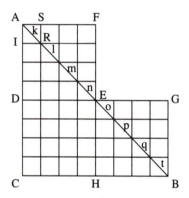

Fig. 5.3 (J I, 262)

Since these moments are indivisibles [*individua*], you will have a space such as *ade* [Fig. 5.3] through which an object falls in one hour. The space through which it falls in two hours doubles [*duplicat*] the proportion of time, i.e., *ade* to *acb*, which is the double proportion of *ad* to *ac*.

For let the moment [*momentum*] of space through which something falls in one hour be of some magnitude, say *adef*. In two hours it will go through three such moments, namely *afegbhcd*. But *afed* consists of *ade* with *afe*, and *afegbhcd* consists of *acb* with *afe* and *egb*, i.e., the double of *afe*. Thus, if the moment be *airs*, the proportion of space to space is as *ade* with *klmn* to *acb* with *klmnopqt*, i.e., double *klmn*. But *klmn* is much smaller than *afe*. Since, therefore, the proportion of space traversed to space traversed will consist of the proportion of triangle to triangle with some equal additions to each term, and since these equal additions continually become smaller as the moments of space become smaller, it follows that these additions would be of no quantity [*nullius quantitatis*] when a moment of no quantity is taken. But such is the moment of space through which an object falls. It remains, therefore, that the space through which an object falls in one hour is related to the space through which it falls in two hours as the triangle *ade* is to the triangle *acb*.

This was demonstrated thus by Mr. Peron, when I posed the problem to him, asking whether one could know how much space an object traverses in a single hour if it is known how much it traverses in two hours according to my principles, viz. that in a vacuum, what is once in motion will always be in motion, and supposing that there is a vacuum between the Earth and the falling stone.

Hence, if it were known by experience that a stone has fallen in 2 hours through one thousand feet, the triangle *abc* would contain 1000 feet. The root of the same is 100 [*sic*] for the line *ac*, which corresponds to 2 hours. If this is bisected in *d*, then *ad* corresponds to 1 hour. Therefore as the double proportion of *ac* to *ad* obtains, i.e., 4 to 1, so is 1000 to 250, i.e., *acb* to *ade*.

But if the minimum moment [*momentum minimum*] of space be of some quantity, the progression will be arithmetic. Nor could it be known from one case alone how much space [the object] would traverse in the single hours; but two cases will be necessary for us to know the quantity of the first moment. I had supposed it this way, but since the assumption of indivisible moments is more appealing, I shall not examine this further.

But in a different way also we see that the space of fall in 1 hour is to the space of fall in 2 hours as *ade* to *acb*, namely if we consider that in arithmetic progression, all the numbers contained under half of the terms are to the numbers of all terms not at all as 1 to 4, even though the proportion grows perpetually. This way the progression of 2 terms, which is 1, 2 is as 1 to 3. Thus 1, 2, 3, 4, [5], 6, 7, 8, is as 10 and 36. Thus these 8 terms are to 16 [terms] are as 36 to 136 [the sum of the first 8 terms is to the sum of the first 16 terms as 36 to 136], which is not at all as 1 to 4. If, then, the stone descends in distinct intervals because the Earth is attracting it by means of material spirits, then these intervals or moments will be nevertheless so small because of the multitude of the particles that their arithmetic proportion will not be sensibly smaller than 1 to 4. The demonstration [by means] of the triangle should therefore be retained.

[Marginal Note:] *The point of equilibrium is sought, that is where the speed of a stone falling in the air is no longer increased.*

In the same way as the space is multiplied [*multiplicatur*], the impediment is multiplied too if you consider it [the fall] in air or water, i.e., in a plenum. The falling object describes an oblong figure, all lines of which are parallel. Since the object falls more quickly in the second hour and traverses more space, the proportion of the figure it describes in the first hour to that which it describes in the second hour is the same as the space traversed in the first hour to that traversed in the second hour. If, therefore, the falling object were not hindered by an impediment, it would encounter as much more air in the second hour as the parallelopiped of the second hour is larger than that of the first hour. But since it is certain that a falling object is impeded by the air – experience showing that the speed of every falling object is not always augmented, but that there is a certain place where, when reached, [the object] will move through the remaining space uniformally – let us see in what way it happens.

The proportion of the triangle appeals to us not because there is really no certain mathematically divisible physical minimum of space in which a minimal physical attractive force moves an object (this force is namely not truly continuous but discrete, and, as one says in Flemish: *sy trect met cleyne hurtkens*, and, therefore, the aforesaid increases consist in a true arithmetical progression); but, I say, it is appealing because this minimum is so small and insensible, that because of the multitude of the terms of the progression, the proportion of the numbers does not differ sensibly from the continuous proportion of the triangle.

This being so, it follows that if an object which falls for one physical minimum moment of time (in which it traverses one physical minimal space) encounters as much air as it itself consists of body, then it will not move more quickly, but will persist in that motion, i.e., if the parallelopiped, which is described in such a moment, contains as much corporeality as the object itself contains, then the attractive force of the Earth will not be able to add to the motion of the object, since the gravity of the body in which it moves, i.e., the air, equals the gravity of the object; but a heavy body in an equally heavy one, and water in water, will not move downward. The motion of a falling object is namely always augmented but in such a manner that if it has to be augmented according to the proportion *ade* to *decb*, it will happen, because of the constantly growing impediment which takes away something of this proportion, that the motion will at last not be augmented anymore; due to the impediment the attractive force is left in the same state, and [the object] retains only so much motion as it had in that last moment. Hereby [the motion] is not diminished because only the attractive force can be destroyed, and once this is taken away, the object continues to move, since what is once moved in vacuum moves [on]. And since there is no reason why the motion should increase, there is no reason why it should encounter more air and describe a longer parallelopiped in the following moments than that which it described in that moment in which it for the first time contained as much air as the object contains corporeality.

(J I, 260-265; part of translation adapted from Clagett 1959, pp. 417-418. For interpretation see 1.3.2 and 1.4.1.)

5.1.3 Descartes' Diary Note (1618)

It happened a few days ago that I came to know an extremely clever man who posed me the following problem:

A stone, he said, descends from A to B in one hour; it is perpetually attracted by the earth with the same force without losing any of the speed impressed upon it by the previous attraction. According to him, that which moves in a vacuum will move always. He asks in what time such a space will be traversed?

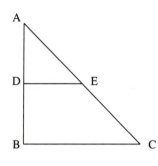

I have solved the problem. In the right isosceles triangle [Fig. 5.4], space ABC represents the motion; the inequality of the space from point A to the base BC [represents] the inequality of the motion. Therefore, AD will be traversed in the time represented by ADE; and DB in the time represented by DEBC; it being noted that the smaller space represents the slower motion. AED is one third of DEBC; therefore [the stone] will traverse AD three times more slowly than DB.

Fig. 5.4 (AT X, 219)

But the problem could be posed in a different way, such that the attractive force of the Earth is equal to that in the first moment: and while the first remains, a new one is produced. In this case the problem would be resolved by a pyramid.

But to lay down the principles of this science [I should say] that an always uniform motion is represented by a line or by a rectangular surface or by a parallelogram or by a parallelopiped; what is augmented by one cause [is represented] by a triangle; by two [causes] by a pyramid, as above; by three [causes] by other figures.

By these [principles] infinitely many questions are resolved. For example, a stone falls in air "gathering strength as it goes": [*viresque acquirit eundo*]² when does it begin to move with equal [i.e. uniform] speed? This is solved thus [Fig. 5.5]. This line represents the gravity of the stone in the first instant; the curvature of the lines AEG and CFH [represents] the inequality of the motion; at

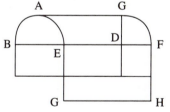

Fig. 5.5 (AT X, 220)

point E, F [the body] would begin to move equally since AEG is a curve only from A to E; from E to G it is a straight line.

(AT X, 219-220; Cogitationes Privatae; part of translation adapted from Mepham, Koyré 1978, pp. 81-82. For interpretation see 1.3.3)

---

² Vergil, *Aeneid*, IV, 175; for Galileo's use of this quotation see p. 235 above.

5.1.4 Descartes' Reasoning on the Compression of Water in Containers (Nov./Dec. 1618)

In order plainly to put forward my thoughts on the proposed questions, many of my fundaments of mechanics should be prefaced, and since there is still time, I will endeavor to explain them in short.

And first, of the various modes of gravitation, all of which we do not have to enumerate here, two should be distinguished, namely the mode in which water that is in a container, presses on the bottom of the same; and the second, the mode in which the whole container with the water in it gravitates. These two are clearly distinct, and it is therefore certain that one can gravitate more or less than the other.

Secondly, in order to understand what the word *gravitate* signifies, we have to imagine that the body, which is said to gravitate, moves downwards and to consider it in the first instant of motion. The force by which it is impelled in the first instant of motion is that which is called gravitation, not that one which moves it downwards during the whole motion, which can be very different from the first. Let us therefore say that gravitation is a force by which the underlying adjacent surface of the heavy body is pressed by the same.

Thirdly, in this principle of imaginable motion attention should also be paid to the imaginable beginning of speed, by which the parts of the gravitating body descend; this [speed] contributes not less to gravitation than the quantity of the body itself. For example, if one atom of water will descend twice as quickly as two other atoms, this single one will gravitate as the other two.

(AT X, 67-68. For context see 1.4.3 and 1.5.5)

5.1.5 Descartes: From a Letter to Mersenne (Nov. 13, 1629)

As to your question concerning the principle on the basis of which I calculate the time in which a weight, which is attached to a chord of 2, 4, 8, and 16 feet, descends, although I will insert it [the principle] into my Physics, I do not want to make you wait for it until then, and I will try to explain it. First I assume that the motion once impressed into a body, remains there always, unless it is removed from it by some other cause, i.e., that which has once started to move in a vacuum moves always and with equal speed. Suppose, then, a weight at A [Fig. 5.6], pushed by its gravity toward C. I say that if its gravity leaves it as soon as it has begun to move, it would nonetheless continue with the same motion until it arrived at C. But it would descend neither more quickly nor more slowly from A to B than from B to C. But since this is not the case, and [the weight] retains its gravity which pushes it downward and which, at each of the moments adds new forces to descend, it happens consequently that it traverses the

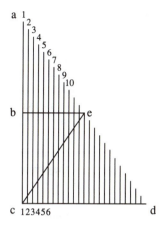

Fig. 5.6 (Paris, Bibl. nat.,
f. fr., nouv. acq. 5160, fol. 48)[3]

space BC much faster than AB, because in traversing it retains all the impetus which moved it through the space AB and, in addition, it adds to this a new [impetus], because of the gravity which propels it anew at each of the moments. As for the proportion in which this speed increases, this is demonstrated by the triangle ABCDE; for the first line denotes the force of speed impressed in the first moment, the second line the force impressed in the second moment, the third the force conferred in the third moment, and so on. Thus is formed the triangle ACD which represents the increase of the speed of motion of the body in its descent from A to C, and ABE, which represents the increase of speed in the first half of the space traversed by this weight; and the trapezium BCDE which represents the increase in speed in the second half of the space traversed by the weight, namely BC. And since, as is clear, the trapezium BCDE is 3 times as large as the triangle ABE, it follows from this that the weight will descend 3 times more quickly from B to C than from A to B: i.e., that if it descends from A to B in 3 moments it will descend from B to C in a single moment; that is to say, in four moments it will traverse twice the distance as in 3 and, consequently, in 12 moments twice as much as in 9, and in 16 moments 4 times as much as in 9, and so on ...

(AT I, 71-73; translation adapted from Mepham, Koyré 1978, pp. 86-87. For interpretation see 1.3.4)

---

[3] Most published versions of this figure contain minor errors and are internally inconsistent. The figure in AT I, 72, which is reprinted in Descartes *Correspondance* (ed. Adam and Milhaud) vol. 1, p. 85, has 3 more lines overall and also one space more in the right half than in the left. Tannery's original publication (1891 p. 531) was differently inaccurate; the diagram given by de Waarde in Mersenne's *Correspondance* (vol. 2, p. 316) is accurate and internally consistent but the triangle is isosceles. Descartes' original drawing is preserved in the National Library in Paris (Paris, Bibl. nat., f. fr., nouv. acq. 5160, fol. 48 recto et verso). We should like to thank Gad Freudenthal for securing us a photocopy of the manuscript.

5.1.6   Descartes: From a Letter to Mersenne (Dec. 18, 1629)

There remain only some things concerning the speed of motion about which you said that Mr. Beeckman had asked you and which will be dealt with best in answering your last [letter] in which you first ask, why I say that the speed impressed by gravity is as one in the first moment and as two in the second moment. I answer, *salva pace*, that I did not understand it this way, but that the speed impressed by gravity in the first moment is as one and again as one in the second moment by the same gravity, etc. But the one of the first moment and the one of the second make two, and with the one of the third they make three, and thus it grows in Arithmetic proportion. I believed that this was sufficiently warranted by the fact that gravity is always joined to the body in which it is; nor can gravity be joined to the body unless it impels it constantly downward. But if we were to suppose, e.g., that a mass of lead falls by the force of gravity and that, after it began to fall in the first moment, God should remove all the gravity from the lead, so that the mass of lead would not be heavier than if it were air or a feather, this mass would nonetheless continue to fall, at least in a vacuum, since it began to move and no reason can be given why it should stop, but its speed will not increase. [Marginal Note: *It should be remembered that we suppose that which is once in motion will always move in the vacuum, and I shall endeavor to prove it in my treatise.*] And if some time later God should restore the gravity in this lead for only one moment of time, and after it had elapsed remove it again, would not the force of gravity impel the lead in this second moment as much as it had in the first moment, and would not therefore the speed of motion be doubled? And the same may be said of the other moments as well.

From whence it certainly follows, that if you let a ball of whatever material fall in a completely empty space of 50 feet in height, it will always employ exactly 3 times as much time for the first 25 feet as for the last 25. But in the air, it is a completely different affair, and to return to Mr. Beeckman, although what he wrote you is false, namely that there is a place such that when a falling body reaches it, it will continue afterwards always with the same speed, it is nonetheless true that after a certain space [of fall] this speed will increase so little that its increase can be judged insensible, and I want to explain to you what he wants to say, since we talked about it once.

He supposes, as I do, that what has begun to move will persist by itself [*sua sponte*] if it is not impeded by some external force, and that it moves always in a vacuum, but in the air it is impeded somewhat by the resistance of the air. He supposes, furthermore, that the force of gravity which exists in the body impels the body anew in the single moments which can be imagined, so that it falls, and that therefore the speed of motion in a vacuum always increases in the proportion which I mentioned above and which I sought eleven years ago when the issue was presented to me and noted in my notebooks at that time. But he [Beeckman] adds of himself the following, namely that the quicker a certain body falls the more does the air resist its motion: this seemed doubtful to me hitherto, but now, having examined the matter carefully, I recognize it to be true. But from this he concludes the following: Since the force producing speed grows always equally, namely by a unit in each moment, but the impeding resistance of the air always grows unequally, namely in the first moment it is less than a unit, but it increases a bit in the second moment and in the following moments; therefore, he says, it will necessarily come that the resistance will be equal to the impulse of gravity, and will remove as much from the speed as the force of gravity adds to it. But from this moment on when this takes place, it is certain, so he says, that the weight will not descend quicker than in the preceding moment, but the speed will neither be augmented in the following moments nor diminished, since the resistance of the air will remain equal – since its inequality arose from the inequality of the speed which is now removed – and the force of gravity always impels equally.

(AT I, 88-91; part of translation adapted from Mepham, Koyré 1978, pp. 88-89. For context see 1.3.4)

5.1.7 Descartes: From a Letter to Mersenne (Oct./Nov. 1631)

You ask me, thirdly, how a stone moves *in vacuo*; but because you have for-gotten to insert the figure, which you suppose to be in the margin of your letter, I could not well understand what you propose, and it does not seem at all that the proportions which you put forward agree with those which I once sent you, where instead of etc. as you have written me, I put $1/3$, $4/9$, $16/27$, $64/81$, etc., and this yields very different consequences. But in order that what I have sent you concerning this [matter] applies, I not only supposed the void, but also that the force which makes that stone move acts always equally, and this is obviously contrary [*repugne apertement*] to the laws of nature: because all natural forces [*puissances*] act more or less accordingly as the object is more or less disposed to receive their action; and it is certain that a stone is not equally disposed to receive a new movement or an augmentation of speed when it already moves very quickly as when it moves very slowly. But I believe that now I could well determine in what proportion the speed of a descending stone augments, not at all *in vacuo* but in this real air. Nevertheless, because my mind is now full of other thoughts I cannot amuse myself by looking for it and there is not much gain in it ...

(AT I, 230-231. For context see 1.3.4 and 1.5.4.2.)

5.1.8 Descartes: From a Letter to Mersenne (Aug. 14, 1634)

Mr. Beeckman came here on Saturday evening and brought me the book of Galileo [the *Dialogo*]. However, since he took it [with him] to Dort this morning, I had it in my hands for about only 30 hours. I did not fail to leaf through the whole book, and I found out that he philosophizes quite well about motion, although there are only very few things he says which I find completely true; what I did notice is that he is more deficient where he follows accepted views than where he departs from them ... I would like, however, to say that I found in his book some of my own thoughts, among others two of which I believe I have written to you before. The first is that the spaces traversed by heavy bodies when they fall, are to one another as the squares of the times which they employ to descend, i.e., that if a ball employs three moments to descend from A to B [Fig. 5.7], it will employ but one to continue from B to C, etc. And I say that with many reservations, since in fact it is never completely true as he thinks he has demonstrated.

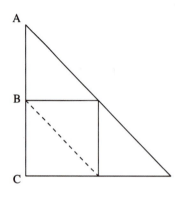

Fig. 5.7 (AT I, 304)

(AT I, 303-305. For interpretation see 1.3.5.)

5.1.9 Descartes: Note on Acceleration in a Vacuum (from the *Excerpta Anatomica*, 1635 or later)

If any body is activated or impelled into motion always with an equal force, i.e., a force imparted to it by a mind [*mens*] (no other such forces can exist), and moves *in vacuo*, it always takes three times as long to travel from the beginning of its motion to the middle as it does from the middle to the end, and so on. But because no such vacuum can be granted, there being space [*spatium*] of some sort, it always resists in some way, and does so in such a way that the resistance always increases in geometric proportion to the speed of the motion, so that eventually it reaches a point where the speed no longer increases appreciably; and one can determine a particular finite speed which it will never equal.

As for those bodies impelled by the force of gravity, since gravity does not always act equally, like a mind [*anima*], but there is some other body already in motion, never can it impel a heavy body as fast as it itself moves. But even *in vacuo* the impulse will always diminish in geometric proportion. Indeed a decrease due to two or more causes in geometric proportion is equivalent to the decrease that arises when all these are considered together as a single cause also in geometric proportion, and the result will still be the same. And similarly, if another cause restrains [the body] with an arithmetic force, a diminution in geometric proportion will arise. But if some other force impels constantly in

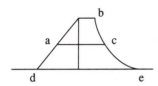

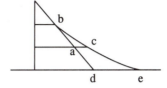

Fig. 5.8 (a) (AT XI, 631)   Fig. 5.8 (b) (AT XI, 631)

geometric proportion simultaneously with that which diminishes geometrically, the geometric force will eventually cease and only the arithmetic will remain and increase the motion which, as was said above, is what a mind [*anima*] does *in vacuo*. Finally, if the impulse increases geometrically and [at the same time] decreases or even increases arithmetically, the speed will increase indefinitely [*in infinitum*] in the composite proportion. This can be explained geometrically by taking the proportional [increases in speed which] the triangle and the curved line [represent], and combining them this way [Fig. 5.8(a)] for addition, and this way [Fig. 5.8(b)] for subtraction, so that the speed during the first [interval] of time is to the speed during the second as the area *abc* is to the area *aced*.

(AT XI, 629-631; translation adapted from Gabbey 1985, pp. 17-18. For interpretation see 1.3.5.)

5.1.10 Descartes: From a Letter to de Beaune (April 30, 1639)

As for gravity [*Pesanteur*], I do not imagine anything else than that all the subtle matter which is between here and the moon and which turns very quickly around the Earth, chases toward it all the bodies which cannot move as quickly. Hence it pushes them with more force when they have not yet begun to descend than when they already descend. Finally, when it happens that they [the bodies] descend as quickly as it [the subtle matter] moves, it will not push them at all anymore, and if they [the bodies] descend more quickly, it [the subtle matter] will resist them. From here you may see that there are many things to be considered before anything concerning the speed can be determined, and it is this which has always distracted me; but many things can be also explained by these principles which were not grasped before.

(AT II, 544. For interpretation see 1.4.3.)

5.1.11 Descartes: From a Letter to Mersenne (March 11, 1640)

I do not doubt at all that many small blows of a hammer will finally have as much effect as a strong blow: I say as much in quantity, although they may be different *in modo*, because there is no quantity that is not divisible into an infinity of parts: and Force, Movement, Impact etc. are species of quantity.

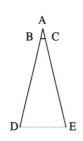

A
B   C

D   E

Fig. 5.9
(AT III, 37)

I cannot determine the speed with which each heavy body descends in the beginning, because this is purely a question of fact, and it depends on the speed of the subtle Matter, which subtracts in the beginning as much of the proportion of the speed with which the bodies descend as the small triangle ABC [Fig. 5.9] subtracts from the triangle ADE, if you take the line BC as [representing] the first moment of speed and DE the last. From this you can easily calculate the force of Impact as compared with Gravity, *positis ponendis*. And because on the basis of these suppositions one can be led far away from the truth, since it is all a matter of fact, I will not engage in it, if you do not mind ...

The subtle matter pushes the descending body in the first moment and gives it one degree of speed; then, in the second moment, it pushes it somewhat less and gives it only almost one degree of speed, and so on in the other [moments]; this occurs in almost double proportion [*fere rationem duplicatam*] at the beginning of the descent of bodies. But this proportion is completely lost when they have fallen several fathoms, and then the speed does not augment any more, or hardly any more.

(AT III, 36-38. For context see 1.4.3.)

5.1.12 Descartes: From a Letter to Mersenne (June 11, 1640)

The reason why I say that falling bodies are pushed less by the subtle matter at the end of their motion than at the beginning is that there is less difference between their speed and that of this subtle matter. For example [Fig. 5.10], if the body A, being without motion, is hit by the body B, which tends to move towards C with a speed such that it could cover one league in one quarter of an hour, it would be pushed more by the body B than it would if it were itself already moving towards C with a speed such that it could travel one league in

A   B   C

half an hour, and it would not be pushed by it at all if it were already moving just as fast [as B], i.e., such that it could cover one league in a quarter [of an hour].

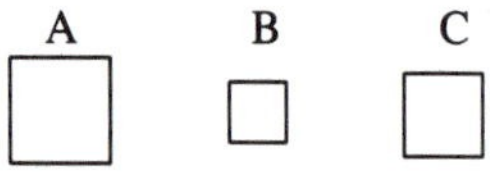

Fig. 5.10 (AT III, 79)

(AT III, 79; translation adapted from Mepham, Koyré 1978, p. 123. For context see 1.4.3.)

5.2 Documents to Chapter 2: Conservation and Contrariety

5.2.1 Leibniz on Measurement by Congruence and Equipollence: From a Letter to de l'Hospital (Jan. 15, 1696)

I remain in agreement with you that a body acts through its mass and its speed; also it is only through these things that I determine the motive force. But it does not at all follow from this that forces are in a proportion composed of the masses and the speeds. Right cones are determined by the height and the base of the triangle that generates them, but they are not in a compound proportion of these two quantities. However, just as two of these cones are equal in size when the generating triangles have the same base and the same height, it is also true that two bodies are equal in force when their masses and their speeds are equal. From which I infer that, given a body AB [Fig. 5.11], having speed H, and a body BCD, double body AB, having speed M equal to speed H, the force of the double body BCD will be double that of the simple body AB when their speeds M and H are equal. For BCD, having two parts BC and CD, each equal to AB, and each part of BCD having its speed equal to that of the whole, that of BC, namely L, will be equal to M and consequently to H, and likewise that of CD, namely N, will be equal to M or to H. Thus, the case of BC with speed L is precisely congruent to the case of AB with speed H and consequently equipollent; likewise the case CD with speed N. Thus, the case BCD with the speed M contains precisely two times the case AB with speed H and consequently it also contains double its force; or a double body is double in force to a simple body of the same speed. This is only all too clear, you say, Sir. However, this is the foundation of my Dynamics, and even of all mathematical estimation [*estime*] and measurement – on condition that one adds here the single principle that the entire effect is equipollent to its cause. For it is the relation between the two that is being dealt with here, since the force is known by the action. And as measurement [*l'estime*] is made by the repetition of the measure, there are two [kinds of] repetitions, a formal repetition which I call *congruence*, when the same subject in which the force is located is repeated; the other [repetition is] virtual, which I call *equipollence*, when this formal repetition or congruence is not located in the subjects themselves which one compares but rather in their full causes or in their entire effects. But neither by the principle of congruence nor by that of equipollence can one demonstrate that the simple body DE with the double speed P is exactly double in force to the simple body AB with the simple speed H, or rather that the double body BCD with the simple speed M is

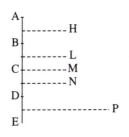

Fig. 5.11 (GM II, Fig. 59)

equal in force to the simple body DE with the double speed P. Congruence there is none at all, and equipollence shows the opposite, for, taking ED with [speed] P, it is true that the speed H is comprehended twice in P but the body AB is not comprehended twice in body DE. Thus there is no repeated congruence at all. And to say that the speed compensates virtually the body, taking the rectangle [i.e. product] of the mass and the speed as the measure of force is to take something not at all demonstrated and in fact just the contrary is demonstrated by the principle of equipollence. Thus, since the case of the two bodies of different speeds cannot be compared by simple congruence or the exact repetition of the same or a congruent [unit], it is necessary to have recourse to the equipollence of cause and effect; that is to say it is necessary to inquire whether there is not a means of producing by a body of double speed an effect which precisely repeats that of a body of simple speed. Now, this can be obtained in a number of ways. For example, if a body of simple speed can raise a pound one foot, a body of double speed can raise precisely four times one pound one foot, either by raising four pounds one foot or by raising one pound four feet; for each, the one and the other, is precisely the fourfold repetition of the elevation of one pound to one foot. So that (to say it in passing) the equality of the raising of one pound four feet and four pounds one foot is also demonstrated by the principle of con-gruence. This proves thus that a body of double speed has four times the force of a similar body with simple speed. And if body A [Fig. 5.12] with a simple speed AQ can tightens a spring Q (which it meets in its path) to a certain degree of tension – without being able to do more – the similar

Fig. 5.12 (GM II, Fig. 60)

body E with a double speed ET would be able to tighten precisely to a similar degree four such springs T, S, R, Q. And what is more: a body of double speed can give the simple speed not only to two but to four bodies which are similar to it in size, as is easy to demonstrate. Thus (according to the principle of the equipollence of cause and effect) a body with a double speed is equipollent to four similar bodies with simple speeds; but (according to the principle of congruence) four equal bodies which have the simple speed have the fourfold force of a single one of them whose speed is simple; finally, a simple body of double speed has the fourfold force of a simple body with a simple speed.

(GM II, 305-307. For interpretation see 2.2.1.)

5.2.2　Descartes: Principles of Philosophy, Bk. II §§36-53 (1644)[4]

36. That God is the primary cause of motion and that he always conserves the same quantity of motion in the universe.

After this consideration of the nature of motion, we must look at its cause. This is in fact twofold: first there is the universal and primary cause – the general cause of all the motions in the world; and second there is the particular cause by which the individual parts of matter acquire motions which they did not have before. Now as far as the general cause is concerned, it seems clear to me that this is none other than God himself, who in the beginning created matter along with motion and rest, and now solely by his ordinary concourse conserves as much motion and rest in this whole as he put there at that time. Admittedly, this motion is nothing in the matter moved but a mode of it. But nevertheless it has a certain and determinate quantity; and this, we easily understand, can be ever the same in the universe as a whole [*in tota rerum universitate*] though it changes in individual parts. Thus namely we should suppose that if one part of matter moves twice as fast as another, and this other is twice as large as the first, there is as much motion in the smaller as in the larger. To the extent that the motion of one part becomes slower, [we should suppose] that the motion of some other part equal to it becomes swifter. For we recognize it as a perfection in God not only that he is immutable in himself but also that he operates in a manner that is always utterly constant and immutable. Thus, except for those changes which plain experience [*evidens experientia*] or divine revelation renders certain and which we perceive or believe to occur without any change in the creator, we should not suppose any other changes in his works lest some inconstancy in him be implied. Thus it follows that it is most consonant with reason to think that from the mere fact that God moved the parts of matter in various ways [*diversimodè*] when he first created them and that he conserves all this matter in completely the same mode and in the same relation [*eodem planè modo eademque ratione*] as he first created it, he also conserves as much motion in it.

---

[4]　The greater part of the translation given here is based on those passages translated by John Cottingham in *The Philosophical Writings of Descartes,* (transl. by J. Cottingham, R. Stoothoff, and D. Murdoch) Cambridge University Press, 1985 (vol. 1, pp. 240-245); it is used with the permission of Cambridge University Press. We have modified it liberally to standardize the terminology as used in our other translations and also to make some renderings more literal (even at the risk of making them on occasion somewhat clumsy). This translation is based on the Latin text; the (contemporary) French translation was consulted – as were various German and English translations – but its renderings were not accepted when they deviated from the Latin. On difficulties with the French version, see Costabel 1967.

37. The first law of nature: that each and every thing, in so far as it can, always continues in the same state; and thus that what is once moved always continues to move.

From this same immutability of God certain rules or laws of nature can be known, which are the secondary and particular causes of the various motions that we notice in individual bodies. The first of these [laws] is that each thing, insofar as it is simple and undivided, always remains in the same state as far as it can and never changes except as the result of external causes. Thus, if a particular part of matter is square, we can be sure without more ado that it will remain square forever, unless something coming from outside changes its shape. If it is at rest, we hold that it will never begin to move unless it is pushed into motion [*ad id*] by some cause. And if it moves, there is equally no reason for thinking that it will ever cease this motion of its own accord and without being checked by something else. Hence we must conclude that whatever moves, so far as it can, always moves. But we live on the Earth, whose constitution is such that all motions occuring near it are soon halted, often by causes hidden from our senses. Hence from our earliest years we have often judged that such motions, which are in fact stopped by causes unknown to us, come to an end of their own accord. And we tend to believe that what we have apparently experienced in many cases holds good in all cases – namely, that by their very nature these [motions] come to an end or tend towards rest. This, of course, is utterly at variance with [*adversatur*] the laws of nature; for rest is the contrary of motion, and nothing can move by its own nature towards its contrary or towards its own destruction.

38. On the motion of projectiles.

Indeed, our everyday experience of projectiles completely confirms this first rule of ours. For there is no other reason why projectiles should persist in motion some time after they leave the hand that threw it, except that things moved continue to move until ithey are slowed down by bodies that are in the way. And it is clear that [projectiles] are normally slowed down, little by little, by the air or other fluid bodies in which they are moving, and that this is why their motion cannot persist for long. The fact that air offers resistance to the motions of other bodies can be proven by our own sense of touch if we beat the air with a fan; it is also confirmed by the flight of birds. And in the case of any other fluid, the resistance offered to the motions of projectiles is even more obvious than in the case of air.

39. The second law of nature: that all motion is in itself rectilinear; and hence that any body moving in a circle always tends to move away from the center of the circle which it describes.

The second law of nature is that each part of matter, considered in itself, always tends to continue moving, not in any oblique lines but only in straight lines.

This is true despite the fact that many parts are often forcibly deflected by their collision with other bodies; and, as I have said above, in any motion a kind of circle is formed of all the matter moving simultaneously. The reason [*causa*] for this rule is the same as that of the preceding, namely the immutability and simplicity of the operation by which God conserves motion in matter. For he always conserves it precisely as it is at the very moment when he conserves it,

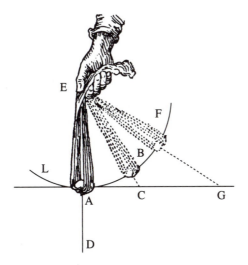

Fig. 5.13 (AT VIII, 64)

without taking any account of the motion which was occurring a little while earlier. It is true that no motion takes place in an instant; but it is manifest that everything that moves is determined [*determinatus esse*] in the individual instants which can be specified as it moves, to continue its motion in a given direction along a straight line, and never along a curved line. Thus, for example [Fig. 5.13], the stone A, rotated in the sling EA in the circle ABF, is indeed, in that instant at which it is at point A, determined [*determinatus est*] to motion in a certain direction, namely along a straight line towards C in such a manner that the straight line AC is a tangent to the circle. It cannot however be assumed to be determined [*determinatum esse*] to any curved motion; for even if it had come before from L to A along a curved line, nothing of this curvedness can be understood to remain within it, when it is in point A. And experience also confirms this, since if [the stone] leaves the sling there, it will not continue to move towards B but towards C. From this it follows that every body moving in a circle constantly tends to recede from the center of the circle which it describes. We feel this in the stone with our hand when we whirl it with a sling. And since we shall often make use of this consideration in what follows, it should be noted carefully, and it will be explicated more fully below.

40. The third law: that a body encountering another, stronger body loses none of it motion; but encountering a weaker body it loses as much [motion] as it transfers to the other body.

The third law is this: when a moving body encounters another, if it has less force to continue in a straight line than the other has to resist it, it is deflected in another direction, and retaining its quantity of motion, it loses [*amittit*] only the determination of the motion. If, however, it has more force, it moves the other body with it and loses [*perdit*] as much of its motion as it gives to the other. Thus we find that when hard projected bodies strike some other hard body, they do not cease to move but rebound in the contrary direction; when, by contrast they encounter a soft body, they are immediately forced to rest because they readily transfer all their motion to it. All the particular causes of the changes which bodies undergo are covered by this third law – or at least those [causes] which are themselves corporeal. I am not here inquiring whether human or Angelic minds have a force to move bodies and of what kind [this force might be], since I am reserving this for a treatise *On Man*.

41. The proof of the first part of this rule.

The first part of this law is demonstrated by the fact that there is a difference between motion considered in itself and its determination in a certain direction; for this determination can be changed while the motion remains unaffected [*integer*]. As I have said above, each single thing that is not composite but simple, as motion is, always persist in being, so long as it is not destroyed by an external cause. Now, in the collision of a hard body it is obvious that there is a cause which prevents the motion of the other body, with which it collides, from remaining determined [*determinatus*] in the same direction; but there is not any [cause] that removes or diminishes the motion itself, since motion is not contrary to motion. Hence it follows that the motion ought not to diminish on that account.

42. The proof of the second part.

The second part [of the law] is also demonstrated from the immutability of the workings of God, who continually conserves the world by the same action by which he originally created it. Since everything is filled with bodies and nevertheless the motion of every single body tends in a straight line, it is clear that God in the beginning, when he created the world, not only moved the different parts of it differently but also at the same time brought it about that the ones impelled the others and transferred their motions to them. Thus, since he now conserves [the world] by the selfsame action and in accordance with the selfsame laws with which he created it, he conserves the motion not as something always fixed in the same parts of matter, but as something transferred [*transeuntem*] from one to the other when they collide with each other. And thus this continual change in created things is itself evidence for the immutability of God.

43. In what the force of any body to act or resist consists.

In this connection we must be careful to note what the force of each body to act on another or to resist the action of another consists in. It consists simply in this, that each single thing tends, so far as it can, to persist in the same state in which it is, as laid down in the first law. Thus what is joined to another thing has some force to resist being separated; and what is separate [has some force] to remain separate; what is at rest [has some force] to persist in its rest and consequently to resist anything that might be able to change this; and what moves [has some force] to persist in its motion, that is, in motion of the same speed and in the same direction. And, this force should be measured not only by the size of the body in which it is and by the surface which separates this body from another, but also by the speed of its motion and by the nature and [degree of] contrariety of the mode [*ac natura et contrarietate modi*] in which different bodies encounter one another.

44. That motion is not contrary to motion but to rest; and that determination in one direction [is contrary] to determination in the opposite direction.

It should be noted that one motion is in no way [*nullo modo*] contrary to another motion that is equally swift; but strictly speaking, only a two-fold contrariety is found here. One opposition is between motion and rest, or even between swiftness of motion and slowness of motion (that is to the extent that this slowness partakes of the nature of rest); the other is between the determination of a body to move in a given direction and the encounter in this direction with a body which is at rest or moving in a different manner; and this contrariety is greater or lesser in accordance with the direction in which the body that collides with the other is moving.

45. How to determine how much the motion of any body is changed by collision with other bodies; this is [calculated] by the following rules.

To enable us to determine, in light of this, in what manner individual bodies increase or diminish their motions or turn in other directions as a result of collision with other bodies, all that is necessary is to calculate the forces of each body to move or to resist motion; we also need to lay it down as a firm principle that that [body] which is the stronger always achieves its effect. And this could easily be calculated if there were only two bodies colliding with each other, and these were perfectly hard and so separated from all remaining bodies that their motions were neither impeded nor assisted by any other surrounding bodies. In this case they would obey the following rules.

46. First [rule]

Fig. 5.14 (AT VIII, 68)

First, if two such bodies [Fig. 5.14], for instance B and C, were completely equal and moved equally swiftly, B from right to left, and C in the direction from left to right, when they collided with each other they would rebound and afterwards continue to move, B towards the right and C towards the left, having lost none of their speed.

47. Second

Secondly, if B were at all greater than C, everything else being posited as before, then only C would rebound and move towards the left with the same speed.

48. Third

Thirdly, if they were equal in size, but B were at all faster than C, both would not only proceed to move towards the left, but also half of the speed by which the former [B] exceeded the latter [C] would be transferred from B to C; that is, if there were six degrees of speed in B and only four in C before, each of them would tend towards the left with five degrees of speed after their collision with one another.

49. Fourth

Fourthly, if body C were completely at rest, and were a little larger than B, whatever the speed with which B moved towards C, it would never move C at all but would be repelled by it in the contrary direction: since a resting body resists a great speed more than a small speed and does this in proportion to the excess of one over the other; and therefore there is always greater force in C to resist than in B to impel.

50. Fifth

Fifthly, if the resting body C were smaller than B, then however slowly B moves towards C, it would move the latter with it, transferring such a part of its motion to it, that both would afterwards move equally fast; thus if B were twice as large as C it would transfer to it a third part of its motion, since that one third part would move body C just as fast as the two remaining [thirds would move] body B which is twice as large. And thus after B collided with C, it would move one third more slowly than before; that is, it would require as much time to move through a space of two feet as it before required to move through as space of three. In the same manner if B were three times greater than C, it would transfer to it one fourth of its motion, and so on.

51. Sixth

Sixthly, if the resting body C were exactly equal to the body B moving towards it, it would in part be impelled by it and would in part repel it in the contrary direction; thus if B came towards C with four degrees of speed, it would communicate to C one degree and rebound with the remaining three in the opposite [*adversam*] direction

52. Seventh

Finally, if B and C moved in the same direction, C more slowly and B following it more swiftly so that they finally touched, and if C were larger than B but the excess of speed in B were greater than the excess of magnitude in C: then B would transfer so much of its motion to C that both would afterwards move equally fast and in the same direction. If, on the contrary, the excess of speed in B is less than the excess of magnitude in C, B would rebound in the contrary direction and would retain all its motion. And these excesses are computed thus: if C were twice as large as B, and B did not move twice as fast as C, it would not push it but rather would rebound in the contrary direction; if however it moved more than twice as fast, it would push it. Thus, if C had only two degrees of speed and B had five, two degrees would be taken from B which, when transferred to C, would effect only one degree more, since C is twice as large as B. Thus it is that the two bodies B and C would afterwards move with three degrees of speed And similar cases are to be judged similarly. These matters do not need proof since they are evident in themselves.

53. The use of these rules is difficult because each body is simultaneously in contact with many others.

But since no bodies in the world can be so separated from all remaining ones and since none around us are normally perfectly hard, the calculation for determining how much the motion of any body is changed by collision with others can be much more difficult. We have to take into account all the bodies which are touching them on every side, and these have quite different effects depending on whether they are hard [*dura*] or fluid. So we must now inquire what this difference consists in. [...]

(AT VIII, 61-70; part of translation adapted from Cottingham, Descartes 1985, vol. 1 pp. 240-245. For interpretation see 2.2.2, 2.3.5, and 2.4.)

5.2.3 Descartes: From a Letter to Clerselier (Feb. 17, 1645)

The reason I say that a body that is without motion could never be moved by another smaller than it, whatever the speed with which this smaller [body] might move, is that it is a law of nature that the body that moves another has more force to move it than the other has to resist. But this "more" depends only on its size; for the one that is without motion has as many degrees of resistance as the other, that moves, has of speed. The reason for this is that, if it is moved by a body that moves twice as fast as another [a third body], it should receive two times as much motion; but it resists these two times as much motion twice as much.

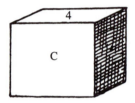 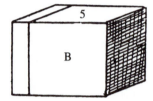

Fig. 5.15 (AT IV, 185)

For example [Fig. 5.15], the body B can only push the body C making it move as fast as it itself moves after having pushed it: namely if B is to C as 5 to 4, B must transfer 4 of the 9 degrees of motion that are in it to C to make it move as fast as itself; this can be done since it has the force to transfer up to 4 and a half (i.e., half of what it has) rather than to reflect its motion in the other direction [*de l'autre costé*]. But if B is to C as 4 to 5, B can only move C if it transfers 5 of its 9 degrees of motion, which is more than half of what it has, and as consequence of this the body C resists more than B has force to act: this is why B should rebound towards the other direction [*refléchir de l'autre costé*] rather than move C. And without this no body would ever be reflected by the collision with another.

For the rest, I am pleased to see that the first and principle difficulty you have found in my *Principles* concerns the rules according to which the motion of colliding bodies changes. For that leads me to think that you found none in what precedes them and that also you will not find much difficulty in the rest, nor in these rules either if you bear in mind that they depend on only a single principle, which is that, *when two bodies collide and have in them incompatible modes, unquestionably there must occur some change of these modes to make them compatible, but this change is always the least possible.* that is to say, *if they can become compatible by changing a certain quantity of these modes, a greater quantity will not change.* And it must be noted that there are in motion [*movement*] two different modes: one is the motion [*motion*] alone, or the speed, and the other is the determination of this motion [*motion*] in a certain direction. Of these two modes, one changes with as much difficulty as the other.

So to understand the fourth, fifth, and sixth rules, in which B's motion and C's rest are incompatible, it must be carefully noted that they can become compatible in two ways: i.e., *if B changes the whole determination of its motion, or, if it changes the rest of body C, transferring to it such a part of its motion that it can push it before it as fast as it itself moves*. And in these three rules I have said nothing other than this: when C is larger than B, it is the first of these two ways that obtains; when it is smaller, it is the second way; and finally when they are equal, the change is made half by one, half by the other. For when C is the larger, B cannot push it before it except by transferring to it more than half its speed, and at the same time more than half its determination to go from right to left, seeing that this determination is joined to its speed. Whereas, rebounding without moving the body C, it [B] changes only the whole of its determination, which is a lesser change than that which would be made up of more than half of the same determination and more than half of the speed. If on the contrary C is less than B, it must be pushed by it, for then B gives it less than half of its speed and less than half of the determination which is joined to it, which makes up less than half of its speed, and less than half of the determination which is joined to it, which makes up less than the whole of the determination, which would have to change if it rebounded.

And this is not at all contrary to experience; for in these rules I understand by a body that is without motion [*mouvement*] a body that is not at all in action to separate its surface from those of other bodies that surround it, and consequently constitutes part of another solid body which is larger. For I have said elsewhere that when the surfaces of two bodies are separated everything positive in the nature of motion can be found just as well in the one that is commonly said not to move at all as in the one said to move; and I later explained why a body suspended in air can be moved by the least force.

(AT IV, 183-187; translation adapted from Gabbey 1980, p. 263. For interpretation see 2.4.2 and 2.4.4.)

5.2.4 Descartes: From Chapter 2 of the *Dioptrics* (1637)[5]

... I shall speak first about reflection, in order to make it easier to understand refraction. Let us suppose that a ball [Fig. 5.16] impelled [by a tennis racket] from A to B meets at point B the surface of the ground CBE, which, preventing its further passage, causes it to be deflected; and let us see in what direction [it will go]. To avoid getting involved in new difficulties, let us assume that the ground is perfectly flat and hard, and that the ball always travels at a constant speed both in its downward passage and in rebounding, leaving aside entirely the question of the power which continues to move it when it is no longer in contact with the racket, and without considering any effect of its weight, size or shape. For there is no point in going into such details here, since none of these factors is involved in the action of light to which the present inquiry must be

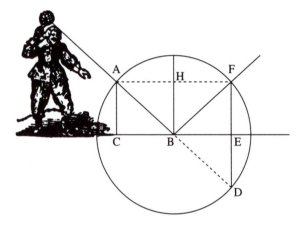

Fig. 5.16 (AT VI, 95)

related. It is only necessary to note that the power, whatever it may be, which causes the ball to continue moving is different from that which determines it to move in one direction rather than another. It is very easy to recognize this from the fact that the movement of the ball depends upon the force with which it has been impelled by the racket, and this same force could have made it move in any other direction as easily as towards B; whereas the ball's tending towards B is determined by the position of the racket, which could have determined the ball in the same way even if a different force had moved it. This shows already that it is

---

[5] The translation given here is by John Cottingham, taken from *The Philosophical Writings of Descartes,* (transl. by J. Cottingham, R. Stoothoff, and D. Murdoch) Cambridge University Press, 1985 (vol. 1, pp. 156-162); it is used with permission of Cambridge University Press. We have made some minor changes to standardize terminology.

not impossible for the ball to be deflected by its encounter with the ground, and hence that its determination to tend towards B could change without any change in the force of its movement, since these are two different things. Consequently we must not imagine as many of our philosophers do, that it is necessary for the ball to stop at point B for a moment before returning towards F. For if its motion were once interrupted by such a halt, no cause could be found which would make it start up again afterwards. Moreover, it must be noted that the determination to move in a certain direction, just as motion itself, and in general any sort of quantity, can be divided into all the parts of which we can imagine that it is composed. And we can easily imagine that the determination of the ball to move from A towards B is composed of two others, one making it descend from the line AF towards line CE and the other making it at the same time go from the left AC towards the right FE, so that these two determinations joined together direct it to B along the straight line AB. And then it is easy to understand that its encounter with the ground can hinder only one of these two determinations, and the other not at all. For it must indeed hinder the one which made the ball descend from AF towards CE, because the ground occupies all the space below CE. But why should it hinder the other, which made the ball advance to the right, seeing that it is not at all opposed to the determination in that direction? So, to discover in precisely what direction the ball must rebound, let us describe a circle, with its center at B, which passes through point A; and let us say that in as much time as the ball will take to move from A to B, it must inevitably return from B to a certain point on the circumference of the circle. This holds in so far as the circumference contains all the points which are as far from B as A is, and the ball is supposed to be moving always equally fast. Next, in order to determine precisely to which of all the points on the circumference the ball must return, let us draw three straight lines AC, HB, and FE, perpendicular to CE, so that the distance between AC and HB is neither greater nor less than that between HB and FE. And let us say that in as much time as the ball took to advance towards the right side from A (one of the points on the line AC) to B (one of those on the line HB), it must advance from the line HB to some point on the line FE. For each of the points on the line FE is the same distance as the others from HB in that direction, and as are those on line AC; and also the ball is as much determined to advance towards that side as it was before. So it is that the ball cannot arrive simultaneously both at some point on the line FE and at some point on the circumference of the circle AFD, unless this is either point D or point F, as these are the only two points where the circumference and the line intersect. Accordingly, since the ground hinders the ball from passing towards D, it is necessary to conclude that it must inevitably go towards F. And so you can easily see how reflection takes place, namely at an angle always equal to the one we call the angle of incidence. In the same way, if a light ray coming from point A falls at point B on the surface of a flat mirror CBE, it is reflected towards F in such a manner that the angle of reflection FBE is neither greater nor less than the angle of incidence ABC.[6]

We come now to refraction. First let us suppose [Fig. 5.17] that a ball impelled from A towards B encounters at point B not the surface of the earth, but a linen sheet CBE which is so thin and finely woven that the ball has enough force to puncture it and pass right through, losing only a part of its speed (say, a half) in doing so. Now given this, in order to know what path it must follow, let us consider again that its motion is entirely different from its determination to move in one direction rather than in another – from which it follows that the quantity of these two factors must be examined separately. And let us also consider that, of the two parts of which we can imagine this determination to be composed, only the one which was making the ball tend downwards can be

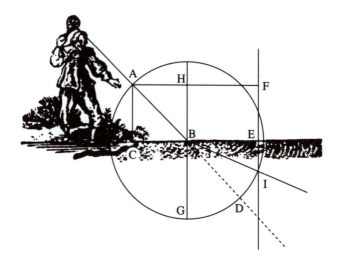

Fig. 5.17 (AT VI, 97)

changed in any way through its colliding with the sheet, while the one which was making the ball tend to the right must always remain the same as it was, because the sheet is not at all opposed to it in this direction. Then, having described the circle AFD with its center at B, and having drawn at right angles to CBE the three straight lines AC, HB, FE so that the distance between FE and HB is twice that between HB and AC, we shall see that the ball must tend towards the point I. For, since the ball loses half its speed in passing through the sheet CBE, it must take twice as much time to descend from B to some

---

6 Descartes uses the terms "angle of incidence" and "reflection" here to mean the angle made with the *surface* not with the normal to the surface; "angle of refraction" refers to the refracted ray's *deviation* from the original direction not its angle to the normal.

point on the circumference of the circle AFD as it took to go from A to B above the sheet. And since it loses none of its former determination to advance to the right, in twice the time it took to pass from the line AC to HB it must cover [*faire*] twice the distance in the same direction, and consequently it must arrive at some point on the straight line FE simultaneously with its reaching some point on the circumference of the circle AFD. This would be impossible if it did not go towards I, as this is the only point below the sheet CBE where the circle AFD and the straight line FE intersect.

Now let us suppose that the ball coming from A towards D does not strike a sheet at point B, but rather a body of water, the surface of which takes away exactly half of its speed, as did the sheet. The other conditions being given as before, I say that this ball must pass from B in a straight line not towards D, but towards I. For, in the first place, it is certain that the surface of the water must deflect it towards that point in the same way as the sheet, seeing that it takes away just as much of its force, and that it is opposed to the ball in the same direction. Then, as for the rest of the body of water which fills all the space between B and I, although it resists the ball more or less than did the air which we supposed there before, we should not say for this reason that it must deflect it

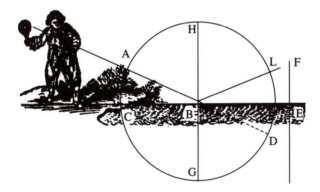

Fig. 5.18 (AT VI, 99)

more or less. For the water may open up to make way for the ball just as easily in one direction as in another, at least if we always assume, as we do, that the ball's course is not changed by its heaviness or lightness, or by its size or shape or any other such extraneous cause. And we may note here that the deflection of the ball by the surface of the water or the sheet is greater, the more oblique the angle at which it encounters it, so that if it encounters it at a right angle (as when it is impelled from H towards B) it must pass beyond in a straight line towards G without being deflected at all. But if it is impelled along a line such as AB, which is so sharply inclined to the surface of the water or sheet CBE that the line FE (drawn as before) does not intersect the circle AD [Fig. 5.18], the ball ought not to penetrate at all, but ought to rebound from it surface B towards the air L, in the same way as if it had struck the earth at that point. People have sometimes experienced this to their regret when, firing artillery pieces towards the bottom of a river for fun, they have wounded those on the shore at the other side.

But let us make yet another assumption here, and suppose that the ball, having been first impelled from A to B, is again impelled at point B [Fig. 5.19] by the racket CBE which increases the force of its motion, say by a third, so that it can then cover [*faire*] as much distance in two seconds as it previously covered in three. This will have the same effect as if the ball were to meet at point B a body of such a nature that it could pass through its surface CBE one-third again more easily than through the air. And it follows manifestly from what has already been demonstrated that if you describe the circle AD as before, and the lines AC, HB, FE so that there is a third less distance between FE and HB than between HB and AC, then point I where the straight line FE and the circular line AD intersect will indicate the position towards which the ball must be deflected when at point B.

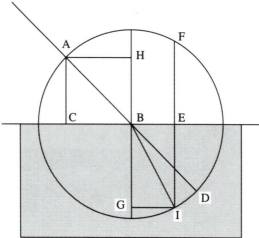

Fig. 5.19 (AT VI, 100)

Now we can also draw the converse of this conclusion and say that since the ball which comes in a straight line from A to B is deflected when at point B and moves towards I, this means that the force or ease with which it penetrates the body CBEI is related to that with which it leaves the body ACBE as the distance between AC and HB is related to that between HB and FI – that is, as the line CB is to BE.

Finally, in so far as the action of light in this respect obeys the same laws as the movement of the ball, it must be said that when its rays pass obliquely from one transparent body into another, which admits [*reçoit*] them more or less easily than the first, they are deflected in such a way that their inclination to the surface between these bodies is always less sharp on the side of the body that admits them more easily, and this [occurs] exactly in proportion as it admits them more easily than does the other. Only it must be noted carefully that this inclination has to be measured [Fig. 5.19] by the quantity of the straight lines CB or AH, EB or IG, and the like) compared to each other, not by that of angles such as

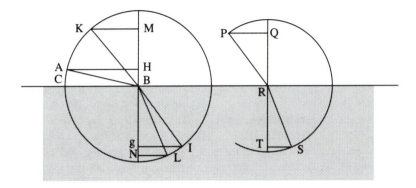

Fig. 5.20 (AT VI, 101)

ABH or GBI, and still less by that of angles like DBI which we call "angles of refraction." For the ratio or proportion between these angles varies with all the different inclinations of the rays, whereas that between the lines AH and IG, or the like, remains the same in all refractions caused by the same bodies. Thus, for example suppose a ray passes through the air from A to B [Fig. 5.20] and, meeting the surface of a glass CBR at point B, is deflected towards I in this glass: and suppose another ray coming from K towards B is deflected towards L, and another coming from P towards R is deflected towards S. In this case there must be the same proportion between the lines KM and LN, or PQ and ST, as between AH and IG, but not the same between the angles KBM and LBN or PRQ and SRT, as between ABH and IBG. ...

(AT VI, 93-101; translation by Cottingham, Descartes 1985, vol. 1, pp. 156-162. For interpretation see 2.5.1.)

5.2.5   Fermat: From a Letter to Mersenne (April or May, 1637)

First of all I doubt, and rightly so it seems to me, whether the inclination to motion must follow the laws of motion itself, for there is as much difference between the two as between potency and act. Furthermore, there seems to be a particular disparity in the fact that the motion of a ball is more or less violent to the extent that it is impelled by different forces, whereas light penetrates transparent bodies in an instant, and there seems to be nothing successive about it. But it is not geometry's business to probe further matters of physics.

In the figure [see Fig. 5.16] by which he explains the reason for reflection, page 15 of the *Dioptrics*, he says that the determination to move in a particular direction, just like motion and in general any other quantity, can be divided into all the parts of which one can imagine it to be composed; and that one can easily imagine that that [determination] of the ball [Fig. 5.21] which moves from A to B is composed of two others, of which the one makes it descend from line AF to line CE, and the other in the same time makes it move from AC on the left towards FE on the right, so that the two joined together conduct it to B along the right line AB.

This assumed, he infers the equality of the angles of incidence and of reflection, which is the foundation of catoptrics.

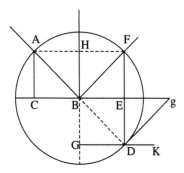

Fig. 5.21 (AT I, 358)   Fig. 5.22 (AT I, 359)

As for me, I cannot admit his reasoning as a proof and a legitimate demonstration. Since, for instance, in the attached figure [Fig. 5.22], in which AF is no longer parallel to CB and where the angle CAF is obtuse: Why can we not imagine that the determination of the ball that moves from A to B is composed of two others, one of which makes it descend from the line AF to the line CE, and the other of which makes it advance along AF? For it is true to say that to the extent that the ball descends in the line AB, it advances along AF; and that this advance is to be measured by the perpendiculars drawn on the line AF from the various points that can be taken between A and B. Whereby it should be obvious that AF makes an acute angle with AB; otherwise, if it were obtuse, the ball would not advance along AF, as is easy to understand. This assumed, we can by the same reasoning of the author [Descartes] infer that the smooth body CE only hinders the first movement, since it is opposed to it only in that sense; and thus the second not being hindered at all, the perpendicular BH being drawn and HF made equal to HA, it follows that the ball should reflect to point F; and thus the angle FBE will be greater than ABC. It is therefore evident that, of all the divisions of the determination to motion, which are infinite, the author has taken only that which can serve to give him his conclusion; and thus he has accommodated his *medium* [i.e. middle term] to his conclusion, and we know as little about it as before. And it certainly seems that an imaginary division that can be varied in an infinite number of ways can never be the cause of a real effect.

(AT I, 357-359. For interpretation see 2.5.2.1.)

5.2.6 Descartes: From a Letter to Mersenne, (Oct. 15, 1637)

You tell me that one of your friends who has seen the *Dioptrics*, finds something there to object to, that he first of all doubts *whether the inclination to motion must follow the laws of motion itself, for there is as much difference between the two as between potency and act.* But I am convinced that he has formed this doubt based on what he imagined that I myself doubted; and because I said on page 8 line 24 : *For it is easy to believe that the inclination to move ought to follow the same laws as motion,* he thought that, when saying that something is easy to believe, I meant it is merely probable. In this he is very far removed from my sentiments. For I hold all that is merely probable almost for false; and when I say that something is easy to believe, I do not mean that it is only probable, but that it is so clear and so evident that there is no need to stop and demonstrate it. As indeed one cannot rightly doubt that the laws followed by motion, which is the act, as he himself says, are observed by the inclination, which is the potency to this act; for although it is not always true that what was in the potency is in the act, it is nonetheless completely impossible that there is something in the act that was not in the potency.

As for what he says afterwards: *that there seems to be a particular disparity in the fact that the motion of a ball is more or less violent to the extent that it is impelled by different forces, whereas light penetrates transparent bodies in an instant, and there seems to be nothing successive about it,* I do not follow his reasoning at all. For he cannot set this disparity in the fact that the motion of a ball can be more or less violent since the action that I take to be light can also be more or less strong; nor in that the one is successive and the other is not, for I think I have made it clear enough by the comparison to the blind man's cane and by that of the wine that descends in a vat, that although the inclination to move is communicated from one place to another in an instant, it does not fail to follow the same path that a successive motion must take, which is all that is in question here.

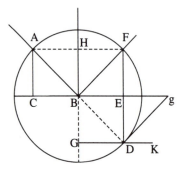

Fig. 5.23 (AT I, 452)

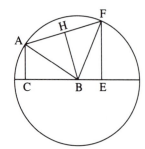

Fig. 5.24 (AT I, 452)

He adds after this a discussion that seems to me to be no demonstration at all. I do not want to repeat his words, since I have no doubt that you have kept the original. But I will say only that from what I wrote, namely that *the determination to move can be divided* (I mean really divided and not at all in the imagination) *into all the parts of which one can imagine it to be composed*, there is no reason at all to conclude that the division of this determination, which is done by the surface CBE [Fig. 5.23], which is a real surface, namely that of the smooth body CBE, is merely imaginary. And he commits a quite manifest paralogism in that, supposing that the line AF [Fig. 5.34] is not parallel to the surface CDE, he wanted us in spite of this to imagine that this line designates the direction to which this surface is not at all opposed; without considering that, just as it is only the perpendiculars, not on that [line] AF as wrongly drawn in his imagination, but on [line] CBE which mark the sense in which the surface CBE is opposed to the motion of the ball, so too, it is only the parallels to that same CBE that mark the sense in which it is not at all opposed to [the motion].

(AT I, 451-452. For interpretation see 2.5.2.1.)

5.2.7 Fermat: From a Letter to Mersenne (Nov. 1637)

... 2. I shall cut short our dispute about reflection, about which I could however continue further, and prove that the author [Descartes] has accommodated his medium to his conclusion, of whose truth he was already certain; for if I should deny that his division of the determinations to motion is the one that must be taken, since we have an infinite number of them, I would force him to prove a proposition that would be very unpleasant to him. But since we do not doubt that reflections make equal angles, it is superfluous to dispute about the proof, since we know the truth; and I suppose it would be better without haggling to turn to refraction, which was the purpose of the *Dioptrics*.

3. I recognize with Mr. Descartes that the motive force or power is different from the determination and consequently that the determination can change without the force's changing and *vice versa*. The example of the first case is found in the figure on the 15th page of the *Dioptrics*,[7] where the ball impelled from point A to point B is deflected to point F; thus the determination to move in the line AB changes without change or diminishment of the force, which sustains its motion. We can take the figure on page 17[8] for the second case. If we imagine that the ball is impelled [Fig. 5.25] from point H to point B, since it falls perpendicularly on to the canvas CBE, it is evident that it will proceed in the line BG and thus its motive force will be weakened and its motion retarded without the determination's changing, since it continues its motion in the same line HBG.

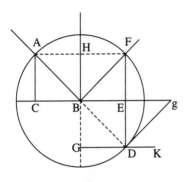

Fig. 5.25 (AT I, 465)

4. I return now to the demonstration of refraction in the same figure of page 17. *Let us consider* (says the author) *that of the two parts of which one can imagine this determination to be composed, it is only the one that makes it tend downward that can be changed in some fashion by the encounter with the canvas; and as for the one that makes it tend towards the right, it ought always to remain the same as it is, because the cloth is not at all opposed to it in this sense.*

5. I remark first that the author has not remembered the difference that he had established between the determination and the motive force or the speed of motion. For it is certainly true that the cloth CBE weakens the motion of the ball but does not hinder it in continuing its determination downward; and although it is slower than before, one cannot say that because the motion of the ball is weakened, the determination that makes it go downward is changed. On the contrary, its determination to move [Fig. 5.26] in the line BI is just as compounded, in the sense of the author, of that which makes it go downward

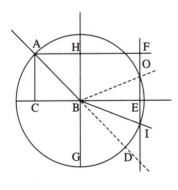

Fig. 5.26 (AT I, 465)

---

7   AT VI, 95; see Fig. 5.16. Figure 5.25 given here was originally published in Clerselier's edition of Descartes' correspondence in 1664.

8   AT VI, 97; see Fig. 5.17.

and that which makes it go from left to right, as is the first determination to move in the line AB.

6. But let us stipulate that the determination along BG or downwards is changed, to speak with the author; we may conclude that the determination along BE, or from left to right, is also changed. For if the determination along BG is changed, it is because in comparison to the first motion the ball which is now deflected and takes the path of BI advances less proportionally along BG than along BE than otherwise would have been the case. However, because we suppose that it advances proportionally less along BG than along BE than it would have otherwise, we may also say that it advances proportionally more along BE than along BG than it would have otherwise. If the first gives us to understand that the determination along BG is changed, the second can just as well let us conceive that the determination along BE is also changed, since change is just as well caused by augmentation as by diminution.

7. But let us again stipulate that the determination downward is changed but not that from left to right, and let us examine the conclusions of the author whose words are: *Since the ball loses nothing at all of the determination that it had to advance toward the right, in twice the time that it took to go from the line AC to HB, it ought to cover twice the distance in the same direction.*

8. See how he falls again into his first error, not distinguishing the determination from the force of motion. And to show you better, let us apply his reasoning to another case. Let us suppose in the same figure [Fig. 5.26] that the ball is impelled from point H to point B, it is certain that it will continue its motion in the line BG and that its determination does not change at all, but also [that] its motion is slower in the line BG than it was before. And nevertheless, if the reasoning of the author were true, we could say: since the ball does not lose any of the determination that it had to advance along HBG (for it is exactly the same), then in the same time as before it would cover the same path. You see that this conclusion is absurd and that to make the argument valid, it would be necessary that the ball lose nothing of its determination and nothing of its force; and thus we have quite manifestly a paralogism.

9. But in order clearly to destroy the proposition we must examine two sorts of compound motions made on two straight lines. Let us consider [Fig. 5.27] the two [lines] DA and AO which make the angle DAO, of whatever size you will; and let us imagine a heavy body [*un grave*] at point A, which descends in the line ACD at the same time that the line advances along AN such that it always makes the same angle with AO, and that the point A of the same line ACD is always in the line AN. If the two motions,

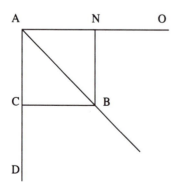

Fig. 5.27 (AT I, 468)

that of the line ACD along AO and that of the same heavy body in the line ACD are uniform, as we may suppose them to be, it is certain that the compound motion will always conduct the heavy body in a straight line such as AB; and if you take a point such as B from which you draw the lines BN and BC parallel to lines DA and AO; then the heavy body will be at point B, in the same time that it would have been at point C if there had only been the motion on ACD and that it would have been at point N if there had been only the other motion alone; and the proportion of the force that conducts it on AD to the force that conducts it along AO is as AC to AN that is to say, as BN to BC. It is this sort of compound motions that was used by Archimedes and the other ancients in the composition of their helices; their principal property is that the two motive forces do not impede each other at all, remaining always the same. But since this motion cannot be applied very well, it is necessary to consider another version and to engage in a particular speculation.

10. Let us suppose in the same figure a heavy body at point A, which is impelled at the same time by two forces, one of which pushes it along AO and the other along AD, with the result that the line of direction of the first motion is AO and that of the second is AD. If there were only the first force alone, the body would always remain on AO, and on AD if there were only the second [force]. But since both forces mutually impede and resist one another, let us suppose (and it should be remembered that we also suppose these motions to be uniform, for otherwise the compound motions would not be carried out in straight lines) that in one minute of an hour, for example, the second force makes the body depart from its direction AO according to the length NB which it must describe parallel to AD; for the body that is transported on AD by the second force, finding itself hindered by the first, will continue on and advance from A toward D by parallels to AD. Let us suppose as well that in the same minute of an hour, the first force makes the body depart from its direction AD according to the length CB, parallel (for the reasons given above) to line AO. It is completely certain that in one minute of an hour the body will be found at point B, which is the intersection of the two lines BN and BC. The compound motion will occur on the line AB and we can say that the body traverses the line AB in one minute.

11. Let us suppose now the angle DAO to be changed [Fig. 5.28] and, for example, to be greater. In the next figure, the same things being posited, I say that in one minute of an hour, as before, the body departs from the direction AO according to the line BN – equal to that to which we have given the same name in the preceding figure. For, since the forces are the same, the second will equally diminish the determination of the first, and will in equal times remove the body from its direction as much as before, because there is always the same resistance.

We may conclude the same thing for line BC.

The compound motion will thus occur here on the line AB, and the line AB will be traversed as before in one minute of an hour. But because in the two triangles ANB of the first and second figures, the sides AN and NB of the first figure are equal to those of the second and because the angles ANB which they comprise are unequal, it follows that the bases AB will be unequal (and consequently the compound motion will be swifter in the second than in the first), and that there will be such a proportion of the speed of the compound motion of the first figure to the speed of the compound motion of the second figure as the length of the line AB in the first figure has to the length of the line AB in the second.

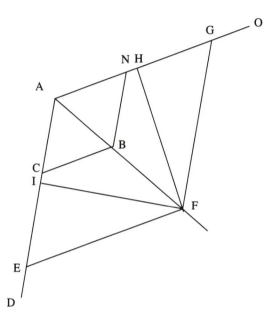

Fig. 5.28 (AT I, 472)

12. I now take an arbitrary point in the line AB [Fig. 5.28], such as F, from which I draw the lines FE, FG parallel to AO and to AD, i.e., FG and BN, as the construction shows us; then FE is to FG as CB to BN. But in the preceding figure the lines BN and BC are equal, respectively, to the lines BN and BC of this second figure (and by the same reasoning we can take an arbitrary point in the line AB of the first figure, in order to draw a conclusion similar to the preceding one). Thus, whatever point that you take in the line AB whether of the first or of the second, the parallels will be to one another as CB and CN, i.e., always in the same proportion. From point F let us draw the perpendiculars FH, FI on the lines AO and AD. In the parallelogram GAEF, the angles AGF and AEF, being opposite [one another], will be equal; thus the triangles GFH and EFI are equiangular; and consequently FI is to FH as EF is to FG. But FI is to FH as the sine of angle DAF is to the sine of angle OAF; and consequently, making, if you will, the same construction in the first figure, you conclude, to

avoid prolixity, that the sine of angle DAB is to the sine of angle OAB in the first figure as the sine of angle DAF to the sine of angle OAF in the second figure.

13. This thus assumed and demonstrated, let us consider the figure on page 20 of the *Dioptrics*,[9] in which the author supposes that the ball, having first been impelled from A to B, being at point B, is impelled in such a manner by the racket CBE, which (doubtless in the sense of the author) pushes along BG. Therefore, from the two motions of which the one pushes along BD and the other along BG a third is made which conducts the ball in the line BI.

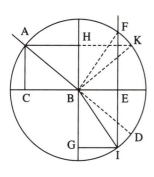

Fig. 5.29 (AT I, 473)

14. Let us now imagine a second figure similar to this one, in which the force of the ball and that of the racket are the same, and that only the angle DBG is greater in this second figure. It is certain by the demonstrations we have just given that there will be the same proportion of sine of the angle GBI to sine of angle IBD in this second figure [Fig. 5.29], which we imagine to be drawn and which we omit to avoid lengthiness. Whereas if the propositions of the author were true, there would be the same proportion of the sine of angle GBD to the sine of angle GBI in the figure of the author that the sine of the angle GBD to the sine of the angle GBI in this second figure which we have imagined. But since this proportion is different from the other, it follows that it cannot hold.

15. Moreover, the principal ground of the author's demonstration is based on the fact that he believes that the motion compounded on BI is always equally swift, even if the angle GBD made by the lines of direction of the two motive forces happens to change; this is false as we have already plainly demonstrated.

16. I do not want to maintain, that in the application of the figure to refraction, which he makes on page 20,[10] one should take my proportion and not his; for I am not sure whether this compound motion should serve as the rule for refraction; on this matter I shall tell you my sentiments another time at more length.

(AT I, 464-474. For interpretation see 2.5.2.1.)

---

[9]   AT VI, 100; see Fig. 5.19.

[10]   AT VI, 100; see Fig. 5.19.

5.2.8 Descartes: From a Letter to Mydorge (March 1, 1638)

First of all, where he [Fermat] says *that I have accommodated my medium to my conclusion and that I would be hard pressed to prove that the division of the determinations that I use is that which must be taken* (after which he quickly moves to other matters), he shows that he has not at all responded to my first letter, in which I clearly proved what he demanded, showing that one must consider in the division of these determinations, not the line drawn the wrong way in his imagination, but the parallel and the perpendicular to the surface where reflection is made.

In the article that begins *I remark first*, he would have it that I supposed such a difference between the determination to move here or there and the speed, that they are not found together and cannot be diminished by the same cause, namely by the cloth CBE: which is contrary to my meaning and contrary to the truth; although this determination cannot be without some speed, nonetheless, the same speed can have various determinations and one and the same determination can be joined to various speeds.

In the following article there is a sophism, or what is the same in matters of demonstration, a paralogism, in these words: *it advances proportionally less along BG than along BE than it would have otherwise, we may also say that it advances proportionally more along BE than along BG than it would have otherwise.* He slips in the word *proportion*, which is not in my book at all, to deceive himself. Since it advances [Fig. 5.30] proportionally less along BG than along BE (i.e., comparing only BG and BE with one another) it also advances proportionally more along BE than along BG, but he concludes that it is true absolutely speaking that it advances more *along BE than it would have otherwise.*

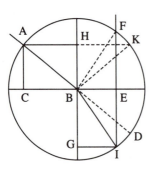

Fig. 5.30 (AT I, 473)

A little later where he says these words: *See how he falls again into his first error*, it is he himself who falls again into his own [first error], believing that the distinction between the determination and the speed or force of motion prevents the one and the other from being able to be changed by the same cause. And he commits a paralogism with these words: *since the ball loses nothing of its determination to speed*, which he has by no means taken from me, for I say nothing of the kind anywhere; and his error is all the greater as he accuses me of committing a paralogism in making it.

Everything that he says after that is only to prepare the reader to accept another paralogism, which consists in the fact that he speaks of the composition of motions in two diverse senses and infers about the one what he has proved about the other. Namely, in the first sense, it is properly only the determination of this motion that is compounded and its speed is not, except insofar as it accompanies this determination as one sees in the second figure, which making AB equal to NA and also to BN, this compound motion, which goes from A to B, is neither more nor less swift than each of the two simple [motions], which go the one from A to N and the other from A to C in the same time; and thus one cannot say that it is its speed that is compounded but only that it is its determination to go from A to B that is compounded of two which are the one to go from A to N and the other from A to C. And, nonetheless, the speed of the motion from A to B can be equal or greater or lesser according as the angle CAN is 120 degrees or more acute or more obtuse; not because it is composed of [the speeds] of the two other motions, but because it must accompany the compound determination and accommodate itself to it. Whereas in his second sense, which is my sense, in the figure on page 20,[11] it is only the speed of the motion that is compounded: namely it is compounded of that which the ball had coming from A to B (for it continues from B toward D) and that which the racket which impels it at point B adds to it. So that, here, it is the speed that follows the laws of composition and not the determination, which is obliged to change in various fashions according to what is required for it to accommodate itself to the speed. And the force of my demonstration consists in the fact that I infer what the determination will be, from the fact that it cannot be other than that which I explain, in order to accord with the speed or better with the force which begins it in B. But his paralogism consists in what he concludes about the composition of the speed after having proved nothing except about the composition of the determination, calling both of them composition of motion.·

And he continues his paralogism up to the end, where he concludes that the compound motion on BI (that is to say the motion whose speed is compounded) is not always equally swift when the angle GBD made by the lines of direction of the two forces (that is to say by the lines that mark how the determinations of the two forces are compounded) is changed; drawing this conclusion from that which he had already proved concerning the motion whose determination – and not whose speed – is compounded that the speed changes when the angle changes. But you can see the faults better than I and if some difficulty in all this should remain which I have not explained enough please oblige me by pointing it out to me.

(AT II, 17-21. For interpretation see 2.5.2.1.)

---

[11] AT VI, 100; see Fig. 5.19.

5.2.9 Descartes: From a Letter to Mersenne (Jan. 21, 1641)

First, he says, that I would have expressed myself more clearly if I had said *determined motion* instead of *determination*. On this matter I do not agree: Although it might be said that the velocity [*velocitas*] of a ball from A to B is

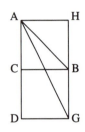

Fig. 5.31
(AT III, 288)

compounded of two others, from A to H and from A to C, I thought one should abstain from such a way of speaking, lest it by chance be so understood that in a motion compounded in this way the quantity of these velocities as well as the proportion of one to the other remains the same; which is not at all true. Because if, for example, we suppose the ball to be carried from A to the right with one degree of speed [*celeritas*] and downwards also with one degree, it would arrive at B with two degrees of speed in the same time in which another [ball] also carried from A to the right with one degree of speed and downwards with two, would arrive at G with three degrees of speed: Whence it would follow that the line AB is to AG as 2 to 3, whereas it is really as 2 to √10 etc.

What he then says about the *ground removing the downward speed* is contrary to the hypothesis: I supposed nothing at all to be taken away from the speed; and it is contrary to all experience: otherwise a ball hitting the ground perpendicularly would never rebound. Therefore my demonstration has no difficulties in any part; but he himself was greatly mistaken in as much as he did not distinguish motion from determination; the motion itself should by no means be diminished, so that reflection occurs at equal angles.

Furthermore, what he assumed, *nothing that cannot be moved [to some extent] by the smallest force can be moved by any force at all*, has no appearance of truth. Who would believe, for example, that a weight of one hundred pounds in a balance would give way a little to a weight of one pound placed in the other pan of the scale. However I willingly concede that the part of the earth which a ball hits gives way a little just as the part of the ball striking the earth is also somewhat bent inwardly, and that since the earth and the ball restitute themselves after the collision, the rebounding of the ball is assisted by this; but I assert that this rebounding is always more impeded by the indenting of the ball and the earth than it is assisted by their restitution; and from this it can be demonstrated that the reflection of the ball, and of other bodies in the same way not perfectly hard never occurs at exactly equal angles. But without demonstration it is easy to find out by experience that softer balls do not bounce back as high nor at such great angles as harder balls. It is thus seen how wrong he was to adduce the softness of the earth to demonstrate the equality of the angles; especially since it follows from this that, if the earth and the ball were hard and did not give way at all, there would be no reflection at all. It is also clear how right I was to assume that both the earth and the ball were perfectly hard, so that the matter can be submitted to mathematical examination [...].

But I am surprised that he asserts that my demonstration is not correct since he clearly offers nothing in support of this reproach except that he says some things contradict some experience, which in fact conform to experience and are most true. But he does not seem to attend to the difference that exists between the refraction of a ball or some other body entering water and the refraction of light, a difference that is twofold and very great. First, the one refraction is made towards the perpendicular and the other in the contrary manner [*modo contrario*]. And [secondly,] rays of light pass more easily by a third of their impetus, more or less, through water than through air; the ball on the other hand ought not to be deprived [*mulctari*] of a third part of its velocity by the same water; for there is no connection between these two. Thus a weak light is not refracted at a different angle by the same water than a strong light; but it is clearly different with the ball, which, impelled by a great force in water cannot be deprived by it of some part of its velocity unless it proceeds more slowly. And thus it is not surprising that experience has shown that a lead ball shot out of a muzzle [*sclopeto*] with maximum force enters the water at an elevation of five degrees since it will not be deprived of more than a thousandth part of its velocity.

Furthermore he insinuates that I suppose that all loss of velocity is to be computed in the motion downward: but I unwaveringly said it was to be computed in the motion as a whole taken *simpliciter*.

(AT III, 288-291. For context see 2.5.2.2.)

5.2.10 Hobbes: From a Letter to Mersenne (Feb. 7, 1641)

...You say, thirdly, that where I say *he would have expressed himself more clearly if he had put determined motion in place of determination*, he does not aggree with me but responds with these words: Although it might be said that the velocity of a ball from A to B is compounded of two others, from A to H and from A to C, he thought he should abstain from such a way of speaking, lest it by chance be so understood that in a motion compounded in this way the quantity of these velocities as well as the proportion of one to the other remains the same; which is not at all true. Because if, for example, we suppose the ball to be carried from A to the right with one degree of speed and downwards also with one degree, it would arrive at B with two degrees of speed in the same time in which another [ball] also carried to the right with one degree of speed and downwards with two, would arrive at G with three degrees of speed: Whence it would follow that the line AB is to AG as 2 to 3, whereas it is really as 2 to $\sqrt{10}$.

I reply: Since Mr. Descartes admits that *it might be said* [dici posse] *that the velocity of the ball from A to B is compounded of two others, from A to H and from A to C*, he should also have admitted that it is indeed true; because he maintains that nothing that is not true may be said [*dici posse*] in philosophy by a philosopher. But *he abstained from this way of speaking because something false could seem to follow from it, namely that the ratio of the line AB to the line AG is not as 2 to √10, but as 2 to 3*; but this is not a proper reason for abstaining. Since, if this falsity were not rightfully inferred from that manner of speaking, then he should not have been afraid of paralogisms which others might thereupon commit on their own accounts; but he himself believed this conclusion to be true, which he himself also drew, but by means of fallacious reasoning. Since, *if we suppose that the ball is carried from A to the right with one degree of speed and downwards also with one degree*, it would nevertheless not arrive at B with two degrees of speed. Similarly, *if A were carried to the right with one degree and downwards with two*, it would nonetheless not arrive at G with three degrees as he supposes. Let us suppose two straight lines AB, AC constructed at a right angle [Fig. 5.32], and let the velocity from A towards B have the same ratio to the velocity from A towards C as the line AB has to the line AC; these two velocities compound the velocity which is from B towards C. I say that the velocity from B towards C is to the velocity from A towards C or from A towards B [i.e. from B towards A] as the straight line BC is to the straight line AC or AB. Let AD be drawn from A perpendicular to BC and let the

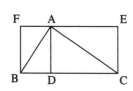

Fig. 5.32[12]

straight line FAE be drawn through A parallel to BC; and likewise also BF and CE, perpendicular to FE. Therefore in as much as the motion from A to B [i.e. from B to A] is composed of the motions from F to A and from F to B [i.e. B to F], the compounded motion AB does not contribute more speed to the motion from B towards C than the components FA, FB can contribute; but the motion FB contributes nothing to the motion from B towards C: this motion is determined downwards and does not at all tend from B towards C. Therefore only the motion FA gives [*dat*] motion from B to C. Similarly it is proved that AC gives motion from D towards C by virtue of AE alone; but the speed which AB draws from FA and by which it acts from B towards C is to the whole speed AB in the proportion FA or BD to AB. Likewise, the speed that AC has by virtue of AE is to the whole speed AC as AE or DC is to AC. Hence the two speeds by which the motion from B towards C occurs joined together are to the speed along AC or along AB [i.e. BA] taken simply as the whole BC is to AC or AB. And therefore referring to the previous figure [Fig. 5.33], the speeds through AB, AG will be as [the

---

[12] Hobbes' original is lost; the figure is reconstructed from the text and from the figure in Descartes' reply (see Fig. 5.37).

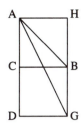

Fig. 5.33
(AT III, 305)

lines] AB, AG themselves, i.e., as √2 to √5, i.e., as √4 to √10, i.e., as 2 to √10, and not as 2 to 3. Hence, no absurdity follows from this way of speaking, as Descartes believed. You see, Father, how even most learned persons are inclined to commit paralogisms because of exaggerated self-confidence.

Fourthly you write that he says, *I should not say that speed is destroyed by the ground because he assumed the contrary, and because it is contrary to experience; otherwise a ball hitting the ground perpendicularly would never rebound.*

I reply, that it certainly was not I in my letter that destroyed his hypothesis, but I said that he himself destroyed it and therefore that he ought not to make use of it (for as far as my opinion in this matter is concerned, I do in fact think that a motion once given cannot ever be removed and thus also not diminished). But that you may judge whether or not he himself destroys the hypothesis, let us take up again the figure [Fig. 5.34]. He supposes A to move towards B by a motion that certainly should never cease but which will not always be in that direction [*in ea determinatione*], as he says. That is, what moves will always move uniformly but not always along the same path [*viam*] or straight line. This I concede. Furthermore, the determination (or path) from A towards B is composed of two other paths (or determinations) of which one is downwards from A towards C, the other lateral from A towards H. This too I concede. From this he thinks he can show that the motion from A to B continues from B to F by the angle FBE equal to the angle ABC, without the destruction of his hypothesis. This I deny. When a ball which moves from A to B arrives at B, it loses the determination (or path) that it had downwards from AH towards CB; thus the determination to the right from AC towards HB remains; but the degree of velocity that it [the ball] had at the beginning is retained; therefore it will go to the circumference of the circle in G. Therefore, he is obliged to demonstrate that, if the entire velocity that it had from A towards B is retained, it is impossible for the

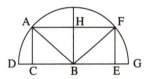

Fig. 5.34 (AT III, 306)

ball to advance farther than E in the same direction [*in eadem determinatione*]; which he could not do, unless the determination from A towards H is taken to be a motion. But he himself seems to understand this determination as a motion since in the demonstration of it he attributes to it a quantity; thus the determination or path of the ball does not have a quantity except in so far as while following it the ball describes a line of such and such a length. However, if these two determinations, the perpendicular and the lateral, are motions, it is manifest that the ball, when it comes to B loses that part of its motion which it had from A towards C. Therefore after impact in B it moves less swiftly than before – which is the destruction of his own hypothesis. He adds that such a diminishment of motion is contrary to experience, since we see that those things that hit the ground perpendicularly rise up again in the perpendicular. But I wonder how it can be known from experience whether the reflection occurs in the perpendicular because there is no loss of motion or because motion is restituted; for the same effect can occur by various means. It is true that experience teaches us that reflection occurs according to equal angles but not by what cause [...].

You write in the tenth place that Mr. Descartes complains *that I insinuate that he computed the whole loss of velocity in the motion downwards, but that he unwaveringly said it was to be computed in the motion as a whole taken* simpliciter.

I reply: I confess that he said straightforwardly that this loss is to be computed in the motion as a whole; but when he said the determination – the perpendicular not the lateral – is diminished in the first penetration by a hard body, he said as a consequence that the perpendicular motion as a whole is diminished; for the determination cannot be diminished unless by determination he means motion. Thus, he does not unwaveringly say that the loss of motion is to be computed in the motion as a whole *simpliciter*. If therefore he has said each of two contradictory propositions, and I ascribe to him the second of them, this is not insinuating anything. Furthermore, if he computes the total loss of velocity in the motion as a whole but computes none in the lateral motion, it is necessary that he computes it all in the perpendicular alone.

(AT III, 303-312. For context see 2.5.2.2.)

5.2.11 Descartes: From a Letter to Mersenne (March 4, 1641)

... Ad 3. I believed that, what I professed *can be said,* can be understood in a sense in which it is true, but it can also be understood in a sense – and in a much more obvious sense – in which it is false; and I therefore have abstained from a way of speaking which is less apt and which would offer the readers an occasion for error; and this reason was completely proper. But it is most improper that he does not admit it to be proper and it is clearly insolent and absurd that he wants to charge me with not rightly understanding the case, whereas it is he himself that does not understand it even now, as will soon become clear. He proposes a completely empy cocoon of a demonstration, so as to deceive the insufficiently attentive. For, in the first place, I would like to know what he supposes when he says: *let the velocity from A towards B have the same ratio to the velocity from A towards C as the line AB has to the line AC; these two velocities compound the velocity which is from B to C.* He cannot suppose that the ball moves from A towards B [Fig. 5.35] and at the same time towards C; this cannot be done. But without doubt he wanted to say *from B towards A* where he said from A towards B; thus namely that we conceive the ball to move from B towards A on line BA while at the same time this line BA moves towards NC, thus that at the same time the ball advances from B to A and line BA to line NC; thus the motion of the ball would describe the line BC. But, whether by chance or purpose, he has confused things so that he seems to say something in what follows, where in truth he says nothing which is not plainly trivial. In order to prove that the velocity from B to C is compounded of the velocities from B to A and from A to C, he divides each of them, saying *in as much as the motion from A to B* (i.e., from B to A)[13] *is compounded out of the motions from F to A and F to B* [i.e. B to F], *the compounded motion AB does not contribute more speed to the motion from B towards C more than FA contributes, nor does AC contribute more than AE etc.* From this he should have inferred that BC is compounded of FA and AE and not of BA and AC, and thus its triviality would also have been apparent: because FA and AE are the same as BC. But he says that the speed BC is compounded of BA and AC because FA and AE are contained in BA and AC, and this is just the same as if he said that an axe is composed of a forest and a mountain because the forest contributes wood for the handle and the mountain [contributes] iron dug up from it. And after these ineptitudes this most urbane person accuses me of having committed a paralogism; but I ask, on what head? I beg of you: precisely where I said that I did not want to make use of an improper way of speaking.

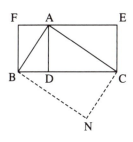

Fig. 5.35 (AT III, 323)

Ad 4. Here he shows that he errs in the very issue on which he said a little before that I should not have been afraid of the paralogisms which others commit on their own accounts. But he himself commits a paralogism in that he considers a determined motion instead of the determination which is in the motion. To understand this it should be noted that a determined motion is to the determination itself of the motion as a flat body to the flatness or surface of that body. For in the same way as it does not follow that, when one surface changes, the others change, too, or that more or less body is attached to them even though they all are in the same body and could not be without it; so too, it does not follow that, when one determination changes the other [also] changes or that more or less motion is attached to it, even though neither of them could be without a motion. If our friend had understood this issue he would not have said that *I was obliged to demonstrate that if the entire velocity that it had from A towards B is retained* [Fig. 5.36], *it is impossible for the ball to advance farther than E in the same direction* [in eadem determinatione]. He would have seen that this is

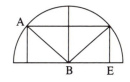

Fig. 5.36 (AT III, 325)

---

13 Descartes' insertion.

demonstrated precisely by demonstrating that the determination to the right is not changed because the motion cannot be increased or decreased in that direction without at the same time adding to or taking away from the determination [in that direction], just as a body cannot be changed in its surface without the surface's changing, too. Nor would he furthermore have said: *But if these determinations are motions etc.* They are no more motions than surfaces are bodies; and he was mistaken and himself committed a paralogism in that he considered a determined motion instead of the determination of a motion; as I promised that I would prove.

(AT III, 322 -326. For interpretation see 2.5.2.2.)

5.2.12 Hobbes: From a Letter to Mersenne (March 30, 1641)

As concerns the dispute over the difference between *determined motion* and *the determination of motion*, I see that it is necessary that I present my opinion more overtly and explicitly than was hitherto the case.

First, it should be known that just as every man is either Peter or Socrates or some other individual, and yet this word *man* is generic term [*vox communis*] (namely one of the five predicables [*vocum*] which Porphery listed in his *Isagoge*) thus, too, every motion is either this or that motion, namely, a motion determined by its *termini a quo* and *ad quem*. As namely Socartes and man are not two men, nor two entities [*res*], but one man under two designations [*appellationes*] (as the same entity is designated by the name [*nomen*] *Socrates* and by the name *man*), so also *motion* and *determined motion* are one and the same motion and one entity under two names.

Secondly, it should be known that what is the efficient cause of some proposed motion is also the efficient cause by which this motion is so determined that the determination of motion taken in an active sense is the action of a movent by which the patient is carried in one directuion rather than in another. But if on other occasions this word *determination of motion* signifies in a passive sense, that is, [signifies] something in the patient, then it means the same as *to be moved thus,* that is, *moved determinately*, and in this sense *determined motion* and *determination of motion* are one and the same.

And Mr. Descartes, at the place where he says that I am mistaken in saying *determined motion instead of the determination which is in the motion*, undertands the determination to be in the body moved and to be passive, and thus *determined motion* and *the determination of motion* are one and the same. But in what manner does he understand this determination to be in the motion? Perhaps as in a subject? This is absurd since motion is an accident, just as it would be absurd to say that whiteness is in color, since whiteness is as much a determination of color as moving to the right or to the left is a determination of motion. But although it is absurd for the *determination* to be *in a motion* as an *acccident* in a *subject*, nevertheless Mr. Descartes has not abstained from [asserting] it. No wonder, since he said that *determined motion* is to the *determination itself of the motion* as a flat body is to the flatness or surface of the same body; but flatness is in a body as in a subject. Thus the comparison should be made in the following way: *determined motion* is to the *determination itself of motion* as a *determined surface* (i.e., a *flat* or *rounded* surface, etc.) to the *determination of the surface* (i.e., to the flatness, roundness, etc.). Since just as much as a flat surface differs from the flatness of the surface so does a *determined motion* differ from the *determination of the motion*. Nor is what you say later at all valid: *In the same way that it does not follow that, when one surface changes the others change, too, thus it does not follow that when one determination changes another also changes.* For as concerns *accidents* in a *subject* (as are two different *surfaces*) one can be lost and another remain. But inasmuch as there is only *one accident* under *two names*, as one motion under the names *determined motion* and *determination of motion*, if what is signified by one name is lost, that which is signified by the other is also lost.

Thirdly, it is to be objected that *one motion* cannot have *two determinations*; for in the figure attached [Fig. 5.37], let A be a body which begins to move towards C, having the straight path AC. If someone tells me that A moves in a straight path to C, he *has determined* that motion for me; I can myself trace the same path as unique and fixed [*unam et certam*]. But if he says A moves along a straight path toward the straight line DC, he has not shown me the *determination* of the motion, since there are infinitely many such paths. Thus the motions from AB to DC and from AD to BC are not determinations of *one*

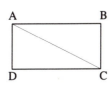

Fig. 5.37 (AT III, 344)

single motion of the body A towards C, but rather the determinations of two motion of two bodies, one of which goes from AB to DC, the other of which from AD to BC.

Fourthly, it should be shown in what way *two determined motion* – of which one is that of the body having the length AB moved perpendicularly to DC, the other being that of the body having the length AD moved laterally to BC – produce the *motion* of the body posited in A determined from A to C. Hence assuming AB to be carried perpendicularly to DC in one minute of time, and in the same way AD carried to BC laterally in the same minute of time, it follows that at the end of this given minute of time the body A is somewhere on CD and also somewhere on BC; it will hence be in C where BC and DC meet. And since AB, AD and A cover in the same minute of time the spaces AD, AB, AC, the velocities by which AB, AD, A are carried will be in the proportion of the lines AD, AB, AC.

Fifthly, it should be noted [Fig. 5.38] that whether A is moved towards C by these two motors AB, AD, as by two winds, or by one motor only, say by a wind that blows from F, it will always be the very same motion carried out from A towards C and will have the same properties.

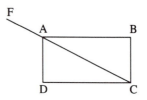

Fig. 5.38 (AT III, 345)

Finally, it should be considered that because the two motions [Fig. 5.39] of the bodies AB and AD give [*conferunt*] from their velocities [*velocitatibus*] the velocity by which the one body A is moved, so that the motion of each of them individually gives part of the speed [*celeritatis*]. It is also clear that the motion of body AB towards DC does not give the body A all its velocity, nor does the motion of the body AD towards BC give all of its velocity; because one prevents the other, from being able to proceed in the next path in which they started, one moving to DC, the other to BC. It should thus be asked, in what proportion the force of each, AB and AD, is diminished. Let the perpendicular DE be drawn from D to AC. I say that the *perpendicular motion* from AB *downwards* gives to the motion of the body A towards C as much speed as suffices to move this body towards C over a space as great as AE; and that the *lateral motion* of the body AD gives as much speed to the motion of the same body A as suffices to carry it over a space as great as EC. And since the speed by which AD is carried laterally is to the speed by which AB is carried perpendicularly, as the straight line AB to the straight line AD as was shown above, the straight lines AD and AB are to one another as AE and EC,[14] and the *lateral* speed of the body AD will be to the *perpendicular* speed of the body AB as EC to AE. And, compounding, as each speed, the lateral and the perpendicular, is to the perpendicular speed alone, so, too, is each straight line AE and EC at the same time to the single straight line AE. Since at the same time each speed, the *lateral* and the *perpendicular*, moves body A through the space AC in one minute, the *perpendicular* speed alone suffices to move the body A in the same minute through precisely as much space as AE; by the same reason the *lateral* speed alone would suffice to move body A over precisely as much space as EC in the same minute [*eodem minuto*

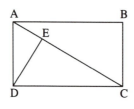

Fig. 5.39 (AT III, 346)

---

[14] Hobbes mistakenly writes "AE and EC"; the correct relation however is not $\dfrac{AD}{AB} = \dfrac{AE}{EC}$ but $\dfrac{AD}{AB} = \dfrac{AE}{DE} = \dfrac{DE}{EC}$; from this follows AE : DE : EC, that is, that DE is the *mean proportional* between AE and EC; thus AE and EC are in the *double proportion* of AD and AB (or AE and DE); therefore $\dfrac{AD^2}{AB^2} = \dfrac{AE}{EC}$.

secundo]. And this is what I meant when I said that the speed of the body A towards C is compounded out of the two speeds AE and EC, undoubtedly diminished in their composition, and not of the entire speeds AD and AB.

And therefore, with this stated, I wanted to demonstrate, why Mr Descartes inferred from my assertion a false consequence, namely, that in a motion compounded this way [Fig. 5.40] the absurd consequence follows that he had inferred. *Let us posit that the ball is carried to the right with one degree of speed and downwards with one degree of speed; it will reach B with two degrees of speed in the same time in which another, which is also carried to the right with one degree of speed and downwards with two, would reach G with three degrees of speed. From this it would follow that the line AB is to the line AG as 2 to 3, whereas it is as 2 to √10.* But from the demonstration just given above it follows clearly enough, I believe, that according to my principles the speed from A to B will not be to the speed from A to G as 2 to 3, but as √2 to √5, which is the proportion of the lines AB to AG themselves, and also the same as the ratio of 2 to √10. Or the speed from A to B is to the speed from A to G not as the speed compounded of AH and HB to the speed compounded of AH and HG, but as the subtenses AB and AG themselves, that is, as the roots of the squares formed on the sides. But I admit that the argument as I wanted to put it in my earlier letter to you dated Paris, February 7, is not correct. I do not at all defend my errors, least of all stubbornly. Unless Mr. Descartes does the same, I will surely be superior in morals. But as to the truth of the matter in dispute between us, what have I not shown sufficiently? What if, knowing the truth of some proposition of Euclid's *Elements* and attempting its demonstration, one does not succeed? Would it therefore be less true given that it has been proved either by others or by myself some other time previously?

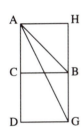

Fig. 5.40
(AT III, 347)

(AT III, 342-348. For interpretation and context see 2.5.2.2.)

5.2.13 Descartes: From a Letter to Mersenne (April 21, 1641)

His last reflections about which you have written me, are as poor as all the others that I have seen of his. For, in the first place, although man and Socrates are not different *supposita*, we nevertheless signify something different by the name of Socrates than by the name Man, namely the individual or particular differences. In the same way a determined motion is not at all different from a motion, but nevertheless the determination is something else than the motion.

Secondly, it is not true that the efficient cause of motion is also the efficient cause of the determination. For example, I throw a ball against a wall; the wall determines the ball to return towards me, but it is not the cause of its motion.

Thirdly, he makes use of a very shallow subtlety asking whether the determination is in the motion as in its subject; as if it were at issue here to know whether motion is a substance or an accident. For it is not at all unseeming or absurd to say that one accident is the subject of another accident, as when we say that quantity is the subject of other accidents. And when I said that the motion was to the determination of motion as the flat body is to its flatness or to its surface, I did not at all mean by this to make a comparison between motion and body as between two substances, but only as between two concrete things in order to show that they were different from those things that can be abstracted.

Finally, his conclusion is very bad that when one determination is changed then the others have to change, too; because, says he, all these determinations are but one accident under different names. If this is so, then according to him it also follows that man and Socrates are but one thing under two different names; and consequently no individual difference of Socrates could be lost, for instance, the knowledge he had of philosophy, unless at the same time he also ceased to be a man.

What he says at the end, namely that a motion has but one determination, is the same as if I said that an extended thing has but one figure only; and this does not prevent this figure from being able to be divided into a number of parts, as the determination also can be.

(AT III, 354-56. For interpretation see 2.5.2.2.)

5.2.14   Descartes: From a Letter to Mersenne (for Bourdin; July 29, 1640)

Whom will he persuade that, when I treated of reflection, I did not know that a mobile which tended in part downward [Fig. 5.41] when moving from A to B, afterwards tends upwards when rebounding from B to F. And what verisimilitude would my reasoning have had, if I had denied this. However, I did not explain this change of determination from downwards to upwards, because it is clear enough of itself [per se est satis nota]; it follows from the fact that when a mobile hits the surface of a hard body perpendicularly it must also rebound perpendicularly; which no one, as far as I know, has ever called into doubt; nor is it my custom in such matters to linger with what is well known and easily grasped. And it would have been least appropriate at that place where I dealt with reflection in passing and in connection with refraction, in which no such change in determination to the contrary is ascertained.

Here he quibbles again and attributes to me incorrect and clearly inept expressions; for it is not the *determination to the right* that *carries the mobile 4 handbreadths* (or *covers 4 handbreadths* as he says equally ineptly further down) but the force as determined to the right, nor can this be inferred from my words, as is seen on page 15 line 2[15] and in all other places where I have dealt with this matter. I said that the determination brings it about not that the mobile moves 4 handbreadths nor that it simply moves, as if it were the cause of motion, but that it moves to the right since it is indeed the reason [*causa*] why the motion occurs towards the right. But the kernel of this quibbling was already introduced above. Saying that the mobile is carried *all the way to B*, he adds *or 5 handbreadths* (which I never wrote) and afterwards he said, *both partial determinations carry the mobile to B*, so that it seems to follows: *therefore 5 handbreadths*. But although, loosely [*improprie*] speaking, it can be said that *the determination carries the mobile to B*, in the sense that it is the cause why it goes in the direction of B, it cannot however be said that it carries the mobile *to B, that is, 5 handbreadths*, since it is not the reason [*causa*] why it goes such a distance. And I wonder that there is anyone in the world who does not blush at attributing to me such things during my lifetime and has no fear lest it be perceived to what extent he deliberately does not to seek the truth.

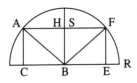

Fig. 5.41 (AT III, 107)

He indeed reveals to us a mighty secret: as if from the fact that I said that figure is to be distinguished from quantity, it is truly necessary for me to be shown that nonetheless the one is not separated from the other and that no extended body can be given which does not have both a quantity and a figure.

He complains that I did not make a mistake and that I did not get into the rut that he himself soon gets stuck in. It should be noted that the collision with the surface CBE divides the determination into two parts but does not divide the force, nor is this surprising, since though a force cannot be without a determination, nonetheless, the same [component] determination can be joined to a greater or lesser force and the same force can remain though the determination changes in some manner. Thus, though a figure does not exist without a quantity, it can be changed without the quantity's changing. And although the surface of a cube is divided into 6 square faces, the cube itself is not thereby divided into 6 parts, but rather the whole body cleaves to each of these faces and corresponds to it.

(AT III, 111-113. For interpretation see 2.5.2.2 and 2.6)

---

15 Page 15 of the *Dioptrics*; AT VI, 94.

5.2.15 Descartes: From a Letter to Mersenne (Aug. 30, 1640)

The principle which I assumed in my *Dioptrics* and which it seems that the quibblings of Father Bourdin have prevented you from noticing, is that the force of motion is not at all changed or diminished by reflection. From this it follows that the determination downwards must necessarily be succeeded by another determination upwards; and thus the ball cannot slide along the surface it arrrives at, unless this surface is so soft that it greatly diminishes the motion; but it is not such surfaces that are being dealt with here, for reflection in such cases does not make an equal angle.

(AT III, 163)

5.2.16 Descartes: From a Letter to Mersenne (Dec. 3, 1640)

But this does not prevent the velitation of Father Bourdin from containing quibblings that were not invented solely by ignorance, but rather by some subtlety that I don't understand. And as for his enclosure, which you say consists in the [objection] that he cannot conceive how the water should not at all slow down the ball from left to right just as well as from up to down, it seems to me I had sufficiently foreseen it in so far as on page 18[16] I considered the refraction by a sheet to show that it is not at all done in the depths of the water but only at its surface. And I explicitly pointed out at the end of page 18, that one should consider only the direction to which the ball is determined when it enters the water because afterwards whatever the resistance of the water it cannot change its determination. Thus, for example if the ball which is impelled from A towards B, when it is at B, is determined by the surface CBE to go towards I whether there is air below the surface or whether there is water, this does not change its determination at all, but only its speed, which is diminished much more in water than in air. But I believe that what perplexes him is the word *determination*, which he wants to consider without any motion, which is chimerical and impossible; in speaking of the determination to the right, I mean all that part of the motion that is determined towards the right. Nonetheless I did not think I ought to mention motion there to avoid embarrassing the reader of this surpising calculation of the velitation, where he says that 3 and 4 are 5 and gives no word of explanation. For it can clearly enough be seen in what I have written that I have tried to avoid superfluous words.

(AT III, 250-251. For an interpretation of Descartes' mistake see 2.6 and 2.5.2)

---

[16] Page 18 of the *Dioptrics*; AT VI, 98-99.

5.2.17 Descartes: From a Letter to Mersenne (April 26, 1643)

The first [principle] is that I assume no real qualities in nature, which would be attached to substance, like little souls to their bodies, and which could be separated by divine power; and thus I attribute no more reality to motion or to all other properties [*varietez*] of substance, which are called qualities, than philosophers commonly attribute to figure, which they do not call *qualitatem realem* but only *modum*. The principle reason for my rejection of the real qualities is that I do not see that the human mind in itself has any notion or particular idea to conceive them by. When mentioning them and maintaining that they exist one mentions a thing that one does not conceive and does not understand oneself. The second reason is that philosophers only assumed these real qualities because they believed they could not otherwise explain all the phenomena of nature; I for my part have found that one can much better explain the phenomena without them.

The other principle is that everthing that is, or exists, always remains in the state it is in unless some external cause changes it; thus I do not believe that there can be any quality of mode that perishes on its own accord. And just as a body that has some shape never loses it unless it is removed from it by collision with some other body, so, too, having a certain motion, it must retain it always unless some cause that comes from elsewhere hinders it. This I prove by metaphysics: for God, who is the author of all things, being completely perfect and immutable, it seems to me contradictory that any simple thing, which exists and consequently has God as its author, has within it the principle of its own destruction. And heat, sound, and other such qualities cause me no difficulty; for they are only motions in the air, where they meet various obstacles which make them stop.

Motion not being a *real quality* but only a *mode*, one cannot conceive that is something other than the change by which one body removes itself from some others, and there are but two properties [*varietez*] of it to be considered; the one is that it can be more or less swift; and the other that it can be determined towards different directions. For although the change can proceed from various causes, it is completely impossible if these causes determine it towards the same direction and make it equally swift, that they give it any diversity of nature [...].

I also believe that it is impossible that a ball perfectly hard, however large it may be, which collides in a right line with a smaller one also perfectly hard can make it move along the same straight line faster than it moves itself. But I add that these two balls must collide in a straight line, that is to say the centers of the one and the other must be in the same straight line along which the motion is made. Thus, for instance [Fig. 5.42], if the large ball B coming from A to D in a straight line encounters the small ball C from the side, which it makes move towards E, there is no doubt whatsoever that even if these balls are perfectly hard, the little one ought to leave more quickly than the large one moves after having encountered it; and constructing the right angles ADE and

CFE, the proportion which holds between the lines CF and CE is the same as holds between the speed of the balls B and C ...

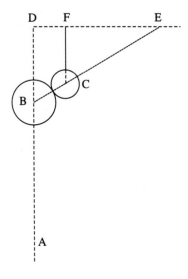

Fig. 5.42 (AT III, 652)

(AT III, 648-652; part of translation adapted from Kenny, Descartes 1970, p. 236. For interpretation see 2.5.3.)

5.3 Documents to Chapter 3: Proofs and Paradoxes

5.3.1 Descartes' Critique of Galileo (from a Letter to Mersenne Oct. 11, 1638)

I shall commence this letter by my observations about Galileo's book [the *Discorsi*]. I find generally that he philosophizes much better than ordinary, in that he avoids as best he can the errors of the Scholastics and undertakes to examine physical matters by mathematical reasonings. In this I accord with him entirely, and I hold that there is no other way to find the truth. But he seems to me very faulty in continually making digressions and never stopping to explain completely any matter, which shows that he has not examined things in order, and that without having considered the first causes of nature he has only sought the reasons of some particular effects, and thus he has built without foundation. Now, insofar as his fashion of philosophizing is closer to the truth, one can the more easily know his faults; as one can better say when those who sometimes follow the right road go astray, than when those go astray who never enter on it [...].

[EN VIII, 116f] Everything he says about the speeds of bodies descending in the void, etc., is built without foundation, for first he should have determined what gravity [*pesanteur*] is, and if he had known the truth, he would have known that it is nothing in the void [...].

[EN VIII, 197f] He supposes that the speeds of falling weights always increase equally, which I formerly believed, as he does, but I now believe I can prove that it is not true.

[EN VIII, 205] He supposes also that the degrees of speed of the same body over different planes are equal when the elevations of the planes are equal, which he does not prove and is not exactly true, and since everything that ensues depends on those two assumptions, one can say that it is entirely built in the air. For the rest, he seems to have written the entire third dialogue only to show that the to's and fro's of the same chord are equal to one another; but he only concludes that weights descend more swiftly along the arc of a circle than along the chords of the same arc, and here, too, he was unable to derive it rigorously from his assumptions.

[EN VIII, 268] He adds another assumption to the preceding, which is no more true; namely, that bodies thrown in the air go uniformly fast following the horizontal, but that in falling their speed[s] increase [as] the double proportion of the space[s]. Now, given this, it is very easy to conclude that the movement of bodies thrown ought to follow a parabolic line; but his assumptions being false, his conclusion can well be very far from the truth.

Fig. 5.43
(AT II, 387)

[EN VIII, 296] It is to be noted that he takes the converse of his proposition without proving or explaining it, i.e., if the shot fired horizontally [Fig. 5.43] from B toward E follows the parabola BD, the shot fired obliquely following the line DE must follow the same parabola DB, which indeed follows from his assumptions. But he seems not to have dared to explain it for fear that their falsity would appear too evident. Yet he makes use only of this converse in all the rest of his fourth discourse, which he seems to have written only to explain the force of cannon shots fired at different elevations. Moreover it is to be noted that in setting forth his assumptions he excludes artillery in order to make them more easily accepted, and yet toward the end it is mainly to artillery that he applies his conclusion. That is to say, in a word, that all is built in the air.

I say nothing about the geometrical demonstrations of which the greater part of his book is full, for I did not have the patience to read them, and I want to believe them all to be true. I only noted, looking at the propositions, that he did not need to be a great mathematician to find them; and glancing at some of them in particular, I noticed that he much fails in them to take the shortest path.

(AT II, 380-387; translation adapted from Drake 1978, pp. 387-391. For interpretation see 3.1 and 3.8.)

5.3.2 Galileo: From Folio 151r

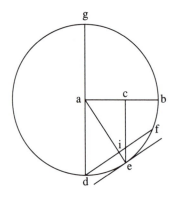

Fig. 5.44 (EN VIII, 378)

Let *gd* be erected [perpendicular] to the horizon, but *df* inclined [to it]: I say that motion takes place from *g* to *d* in the same time as from *f* to *d*.

Thus, the moment along *fd* is the same as that along the tangent in *e*, which is parallel to that same *fd*; therefore, the moment along *fd* is to the total moment as *ca* to *ab*, that is, to *ae*; but as *ca* to *ae*, so is *id* to *da* and also the double [of *id*] *fd* to the double [of *da*] *dg*; therefore the moment along *fd* is to the total moment, namely, along *gd*, as *fd* to *gd*; therefore, the motion along *fd* and *gd* takes place in the same time.

(Folio 151r; EN VIII, 378. For context see 3.3.1)

5.3.3 Galileo: From a Letter to Sarpi (Oct. 16, 1604)

Thinking again about the matters of motion, in which, to demonstrate the phenomena [*accidenti*] observed by me, I lacked a completely indubitable principle to put as an axiom, I am reduced to a proposition which has much of the natural and the evident: and with this assumed, I then demonstrate the rest; i.e., that the spaces passed by natural motion are in double proportion to the times, and consequently the spaces passed in equal times are as the odd numbers from one, and the other things. And the principle is this: that the natural movable goes increasing in velocity with that proportion with which it departs from the beginning of its motion; as, for example, the heavy body [Fig. 5.45] falling from the terminus *a* along the line *abcd*, I assume that the degree of velocity that it has at *c*, to the degree it had at *b*, is as the distance *ca* to the distance *ba*, and thus consequently, at *d* it has a degree of velocity greater than at *c* according as the distance *da* is greater than *ca*.

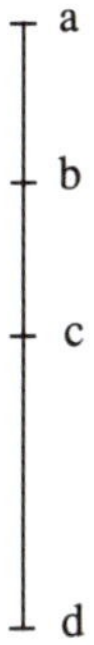

I should like your reverence to consider this a bit, and tell me your opinion. And if we accept this principle, we not only demonstrate (as I said) the other conclusions, but I believe we also have it very much in hand to show that the naturally falling body and the violent projectile pass through the same proportions of velocity. For if the projectile is thrown from the point *d* to the point *a*, it is manifest that at the point *d* it has a degree of impetus able to drive it to the point *a*, and not beyond; and if the same projectile is in *c* it is clear that it is linked with a degree of impetus able to drive it to the same terminus *a*; and likewise the degree of impetus at *b* suffices to drive it to *a*, whence it is manifest that the impetus at the points *d*, *c*, *b* goes decreasing in the proportions of the lines *da*, *ca*, *ba*; whence, if it goes acquiring degrees of velocity in the same (proportions) in natural fall, what I have said and believed up to now is true.

Fig. 5.45
(EN X, 115)

(EN X, 115. Translation adapted from Drake 1969; for interpretation see 3.3.3)

5.3.4 Guidobaldo's Experiment

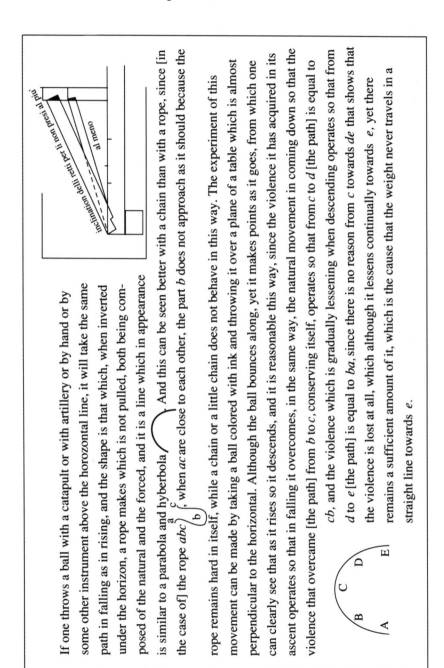

If one throws a ball with a catapult or with artillery or by hand or by some other instrument above the horozontal line, it will take the same path in falling as in rising, and the shape is that which, when inverted under the horizon, a rope makes which is not pulled, both being composed of the natural and the forced, and it is a line which in appearance is similar to a parabola and hyberbola. And this can be seen better with a chain than with a rope, since [in the case of] the rope abc, when ac are close to each other, the part b does not approach as it should because the rope remains hard in itself, while a chain or a little chain does not behave in this way. The experiment of this movement can be made by taking a ball colored with ink and throwing it over a plane of a table which is almost perpendicular to the horizontal. Although the ball bounces along, yet it makes points as it goes, from which one can clearly see that as it rises so it descends, and it is reasonable this way, since the violence it has acquired in its ascent operates so that in falling it overcomes, in the same way, the natural movement in coming down so that the violence that overcame [the path] from b to c, conserving itself, operates so that from c to d [the path] is equal to cb, and the violence which is gradually lessening when descending operates so that from d to e [the path] is equal to ba, since there is no reason from c towards de that shows that the violence is lost at all, which although it lessens continually towards e, yet there remains a sufficient amount of it, which is the cause that the weight never travels in a straight line towards e.

Fig. 5.46 (del Monte MS, p. 236)

Se si tira una palla, o con una balestra, o con artiglieria, o con la mano o con altro instrumento, sopra la linea del' horizonte, il medisimo viaggio fà nel callar, che nel montar, e la figura è quella, che [revoltata][17] sotto la linea horizontale fà una corda, che non stia tirata, essendo l'un è l'altro composto di naturale, e di violento, et è una linea in vista simile alla parabola, et hyperbole ... e questo si vede meglio con una catena, che con una corda perche la corda *abc* ... quando *ac* sono vicini la parte *b* non si accosta come doverebbe percioche la corda resta in se dura che non fà cosi una catena, o catenina. La esperienza di questo moto si pò far pigliando una palla tinta d'inchiostro, e tirandola sopra un piano di una tavola, il qual stia quasi perpendicolare all'horizonte, che se ben la palla va saltando, va però facendo li punti, dalli quali si vede chiaro, che sicome ella ascende, cosi anco descende, et è cosi ragionevole perche la violentia che ella hà acquistata nell'andar in sù, fà, che nel callar vadi medesimamente: superando il moto naturale nel venir in giu', che la violentia che superò da *b* al *c* conservandosi fà che dal *c* al *d* sia eguale a *cb*, e descendendo di mano in mano perdendosi la violenza fà che dal *d* al *e* sia eguale a *ba* essendo che non ci è ragione, che dal *c*[18] verso *de* mostri, che si perda a fatto la violentia, che se ben va continuamente perdendo verso *e*,[19] nondimeno sempre se ne resta, che è causa, che verso *e* il peso non và mai per linea retta.

Translation:

If one throws a ball with a catapult or with artillery or by hand or by some other instrument above the horizontal line, it will take the same path in falling as in rising, and the shape is that which, when inverted under the horizon, a rope makes which is not pulled, both being composed of the natural and the forced, and it is a line which in appearance is similar to a parabola and hyperbola ... And this can be seen better with a chain than with a rope, since [in the case of] the rope *abc* ..., when *ac* are close to each other, the part *b* does not approach as it should because the rope remains hard in itself, while a chain or little chain does not behave in this way. The experiment of this movement can be made by taking a ball colored with ink, and throwing it over a plane of a table which is almost perpendicular to the horizontal. Although the ball bounces along, yet it makes points as it goes, from which one can clearly see that as it rises so it descends, and it is reasonable this way, since the violence it has acquired in its ascent operates so that in falling it overcomes, in the same way, the natural movement in coming down so that the violence that overcame [the path] from *b* to *c*, conserving itself, operates so that from *c* to *d* [the path] is equal to *cb*, and the violence which is gradually lessening when descending operates so that from *d* to *e* [the path] is equal to *ba*, since there is no reason from *c* towards *de* that shows that the violence is lost at all, which, although it lessens continually towards *e*, yet there remains a sufficient amount of it, which is the cause that the weight never travels in a straight line towards *e*.

(del Monte MS, p. 236; tranlation adapted from Naylor 1980a. For interpretation see 3.3.1.)

---

[17] The word is illegible; Libri (1838-1841, vol. 4, p. 398) conjectures "revoltata."

[18] Libri (1838-1841, vol. 4, p. 398) reads "*e*."

[19] Libri (1838-1841, vol. 4, p. 398) reads "*a*."

5.3.5 Galileo: From Folio 147r

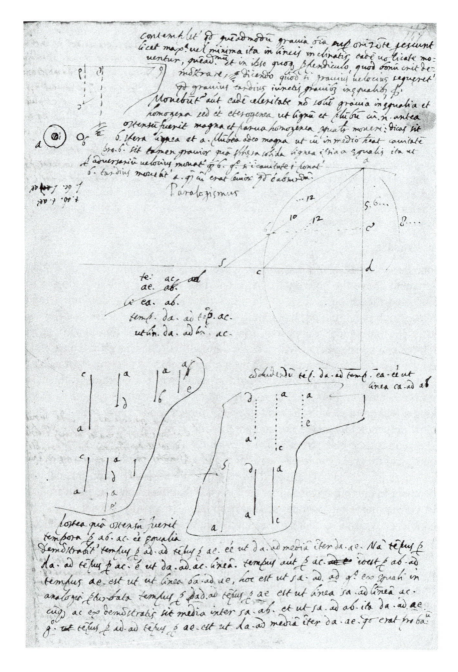

Plate V. On falling bodies (folio. 147r). Published with permission of the Biblioteca Nazionale, Florence.

It is to be considered that just as all heavy bodies rest in the horizontal [plane], the greater [weights] as well as the smaller, so they should move with the same speed [whether heavy or light] along inclined lines just as [they do] in the perpendicular itself. It would be good to demonstrate this, saying that if the heavier were faster, it would follow that the heavier would be slower, unequal heavy bodies having been joined, etc.

Fig. 5.47 (fol. 147r)

Moreover, not only homogeneous and unequal heavy bodies would move at the same speed, but also heterogeneous ones such as wood and lead. Since as it was shown before that large and small homogeneous bodies move equally, you argue: Let *b* be a wooden sphere [Fig. 5.47] and *a* be one of lead so big that, although it has a hollow for *b* in the middle, it is nevertheless heavier than a solid wood sphere equal [in volume] to *a*, so that for the adversary it should move faster than B; therefore if *b* were to be put into the hollow *i*, *a* would move slower than when it was lighter; which is absurd.

Paralogism

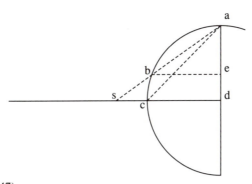

Fig. 5.48 (fol. 147)

After it has been demonstrated that the times [Fig. 5.48] through *ab* and *ac* are equals, it must be shown that the time through *ad* is to the time through *ae* as *da* is to the mean [proportional] between *da* and *ae*. For the time through *da* is to the time through *ac* as line *da* is to *ac*; but the time through *ac* (which is that through *ab*) is to the time *ae* as line *ba* is to *ae*, which is as *sa* is to *ad*. Therefore, by equidistance of ratios in perturbed proportionality [*ex aequali in analogia perturbata*], the time through *ad* is to the time through *ae* as line *sa* is to line *ac*. And since *ac*, as has been demonstrated is the mean [proportional] between *sa* and *ab*, while as *sa* is to *ab*, so *da* is to *ae*, therefore the time through *ad* is to the time through *ae* as *da* is to the mean [proportional] between *da* and *ae*, which was to be proved.

(Folio 147r; EN VIII, 380. See Plate V. Translation adapted from Drake 1978, 94-95. For interpretation see 3.3.2.1.)

5.3.6 Galileo: Folio 182r

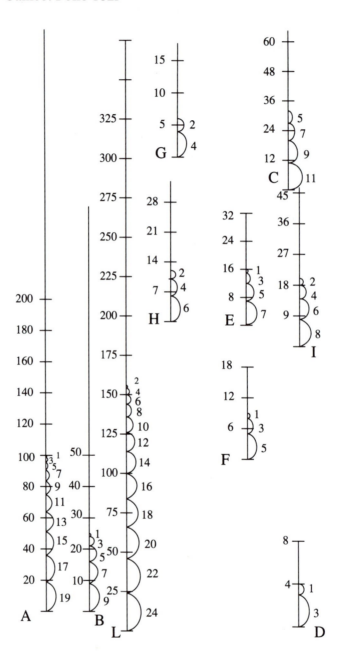

Fig. 5.49 (fol. 182r)

Remarkable for the projectiles for determining how much the natural inclination
downward subtracts from the preternatural
motion of the projection.

If the violent impetus is disposed according
to even numbers, the natural descent takes
away one-half, as it is clear in the examples
[Fig. 5.49] D, F, E, B, C, A; if, however, the
disposition is according to odd numbers, the
natural descent takes so much less as
corresponds to the number of the disposed
parts, as is evident in the examples G, H, I, L.
In G the number of disposed parts – according
to the violent impetus not yet retarded – is 3,
namely 5, 10, and 15, from which in the first,
1 is removed and 4 remain; from the second, 4
having been removed, there remain 6; from the
third, namely from 15, 9 [having been
removed] there remains the same number 6,
which falls short of half of 15 by 3 which is
the number of parts: 5, 10, 15 [i.e., 6 is half
12, which falls short of 15 by 3]. In example
H the number of parts is 4. The subtractions
from the natural motion are 6, 4, 2, which
give 12, whose double [24] falls short of 28
by 4. In example I the subtractions 8, 6, 4, 2
make 20, the double of which [40] falls short
of 45 by 5 which is similarly the number of
parts; and likewise it is clear in L that the
subtractions, namely 156, if duplicated, fall
short of 325 by 13 (which is the number of
parts of the violent motion).

(Folio 182r; EN VIII, 425-426. For context see 3.3.3.1, fn. 94.)

5.3.7 Galileo: Folio 128

I suppose (and perhaps I shall be able to demonstrate this) that the naturally falling heavy body goes continually increasing its velocity according as the distance increases from the terminus from which it parted, as, for example, the heavy body [Fig. 5.50] departing from the point *a* and falling through the line

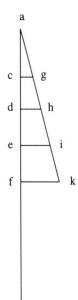

ab. I suppose that the degree of velocity at point *d* is as much greater than the degree of velocity at the point *c* as the distance *da* is greater than *ca*; and so the degree of velocity at *e* is to the degree of velocity at *d* as *ea* to *da*, and thus at every point of the line *ab* it [the body] is to be found with degrees of velocity proportional to the distances of these points from the terminus *a*. This principle appears to me very natural, and one that corresponds to all the experiences that we see in the instruments and machines that work by striking, in which the percussent works so much the greater effect, the greater the height from which it falls; and this principle assumed I shall demonstrate the rest.

Draw line *ak* at any angle with *af*, and through points *c*, *d*, *e*, and *f* draw the parallels *cg*, *dh*, *ei*, *fk*: And since lines *fk*, *ei*, *dh*, and *cg* are to one another as *fa*, *ea*, *da*, *ca*, therefore the velocities at points *f*, *e*, *d*, and *c* are as lines *fk*, *ei*, *dh*, and *cg*. So the degrees of velocity go continually increasing at all points of line *af* according to the increase of parallels drawn from all those same points. Moreover, since the velocity with which the moving body has come from *a* to *d* is compounded from all the degrees of velocity it had at all the points of line *ad*, and the velocity with which it has passed through line *ac* is compounded from all the degrees of velocity that it has had at all points of line *ac*, therefore the velocity with which it has passed line *ad* has that proportion to the velocity with which it has passed the line *ac* which all the parallel lines drawn from all the points of the line *ad* over to *ah* have to all the parallels drawn from all the points of line *ac* over to *ag*; and this proportion is that which the triangle *adh* has to the triangle *acg*, that is the square of *ad* to the square of *ac*. Then the velocity with which the line *ad* is traversed to the velocity with which the line *ac* is traversed has the double proportion that *da* has to *ca*.

Fig. 5.50
(fol. 128)

And since velocity to velocity has contrary proportion of that which time has to time (for it is the same thing to increase the velocity as to decrease the time), therefore the time of the motion along *ad* to the time of the motion on *ac* has half the proportion that the distance *ad* has to the distance *ac*. The distances, then, from the beginning of the motion are as the squares of the times, and, dividing, the spaces passed in equal times are as the odd numbers from unity.

Which corresponds to what I have always said and to experiences observed; and thus all the truths are in accord.

And if these things are true, I demonstrate that the velocity in forced motion goes decreasing in the same proportion with which, in the same straight line, natural motion increases. For let the starting point of the violent motion be the point *b*, and the end point the terminus *a*. And since the projectile does not pass the terminus *a*, therefore the impetus it had at *b* was such as to be able to drive it to the terminus *a*; and the impetus that the same projectile has in *f* is sufficient to drive it to the same terminus *a*; and when the same projectile is in *e, d, c*, it finds itself linked with impetuses capable of pushing it until the same terminus *a*, not more and not less; thus the impetus goes evenly decreasing as the distance of the moving body from the terminus *a* diminishes. But according to the same proportion of the distances from the terminus *a*, the velocity increases when the same heavy body will fall from the point *a*, as assumed above and compared with our other previous observations and demonstrations: thus what we wanted to prove is manifest.

(Folio 128; EN VIII, 373-374. For interpretation see 3.3.3.1.)

5.3.8 Galileo: From Folio 163v

Let the motion from *a* to *b* be made in natural acceleration: I say, if the velocity in all points *ab* were the same as that found in the point *b*, the space *ab* would be traversed twice as fast; because all velocities in the single points of the line *ab* have the same ratio to all the velocities each of which is equal to the velocity *bc* as the triangle *abc* has to the rectangle *abcd*.

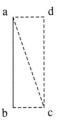

Fig. 5.51

From this it follows, that if there were a plane *ba* inclined to the horizontal line *cd*, and *bc* being double *ba*, then the moving body would come from *a* to *b* and successively from *b* to *c* in equal times: for, after it was in *b*, it will be moved along the remaining *bc* with uniform velocity and with the same with which [it is moved] in this very terminal point *b* after fall through *ab*. Furthermore, it is obvious that the whole time through *abe* is 1$^1/_2$ the time [*sesquialterum*] for *ab*.

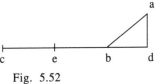

Fig. 5.52

(Folio163v; EN VIII, 383-384; For interpretation see 3.3.3.2.)

5.3.9 Galileo: Folio 85v

I assume that the acceleration of the falling body [Fig. 5.53] along the line *al* is such that the velocity increases in the ratio of the space traversed so that the velocity in *c* is to the velocity in *b* as the space *ca* is to the space *ba*, etc. Matters standing thus, let the line *ax* be drawn at some angle to *al*, and, taking the parts *ab, bc, cd, de,* etc. to be equal, draw *bm, cn, do, ep,* etc. If therefore the velocities of the body falling along *al* in the places *b, c, d, e* are as the distances *ab, ac, ad, ae,* etc., then they will also be as the lines *bm, cn, do, ep.* But because the velocity is successively increased in all points of the line *ae*, and not only in *b, c,* and *d*, which are drawn, therefore all these velocities are to one another as the lines from all the said points of the line *ae* which are generated equidistantly from the same *bm, cn, do.* But those are infinite and constitute the triangle *aep*: therefore the velocities in all points of the line *ab* are to the velocities in all points of the line *ac* as the triangle *abm* to the triangle *acn*, and so for the remaining, i.e., in double proportion of the lines *ab, ac*. But because in the ratio of the increase of [velocity due to] acceleration the times in which the motions themselves occur must decrease, therefore the time in which the moving body traverses *ab* will be to the time in which it traverses *ac* as the line *ab* is to that line which is the mean proportional between *ab* and *ac*.

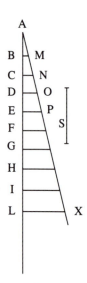

Fig. 5.53
(fol. 85v)

(Folio 85v; EN VIII, 383. Translation adapted from Drake 1978, 98-99 For interpretation see 3.3.3.1.)

5.3.10 Galileo: From Folio 179v

If in the line of natural descent two unequal distances from the starting point of the motion are taken, the moments of velocity with which the moving body traverses these distances are to one another in double proportion of those distances.

Let *ab* be the line of natural descent, in which from the starting point *a* of the motion two distances *ac* and *ad* are taken: I say, that the moments of velocity with which the moving body traverses *ad* are to the moments of velocity with which it traverses *ac* in double proportion of the distances *ad* and *ac*. Draw line *ae* in an arbitrary angle with respect to *ab*. [The note ends here abruptly.]

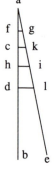

Fig. 5.54
(fol. 179v)

(Folio 179v; EN VIII, 380; translation adapted from Drake 1978, p. 115. For interpretation see 3.3.3.1)

5.3.11 Galileo's *Mirandum* Fragment: From Folio 164v

It has to be seen [*Mirandum*], whether the motion along the perpendicular *ad* is not perhaps faster that along the inclined plane *ab*? It seems so; in fact, equal spaces are traversed more quickly along *ad* than along *ab*; still it seems not so; in fact, drawing the horizontal *bc*, the time along *ab* is to the time along *ac* as *ab* is to *ac*; then, the moments of velocity along *ab* and along *ac* are the same; in fact, that velocity is one and the same which in unequal times traverses unequal spaces which are in the same proportion as the times.

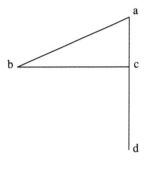

Fig. 5.55 (fol. 164v)

(Folio 164v; EN VIII, 375. For interpretation see 3.4.2)

5.3.12 Galileo: From a Letter to de' Medici (Feb. 11, 1609)

I am now about some questions that remain to me concerning the motion of projectiles, among which are many that bear on artillery shots. And even recently I have found this: that putting the cannon on some elevated place above the plane of the field, and aiming it exactly level, the ball leaves the cannon, driven by much or very little gunpowder, or even just enough to make it leave the cannon; and yet it always goes declining and descending to the ground with the same speed, so that the ball will arrive on the ground at the same time for all level shots, whether the shots are very long or very short, or even if the ball merely emerges from the cannon and falls plumb to the plane of the field. And the same happens with shots at an elevation; these are all completed in the same time provided that they are lifted to the same vertical height. Thus, for example [Fig. 5.56], the shots *aef*, *agh*, *aik*, and *alb*, contained between the same parallels *cd* and *ab*, are all completed in the same time; and the ball consumes the same time in traversing line *aef* as *aik*, or the others – and in consequence, in their halves; i.e., the parts *ef*, *gh*, *ik*, and *lb* are all made in equal times, which correspond to level shots [from *e*, *g*, *i*, and *l*, respectively].

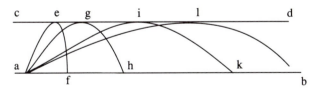

(Fig. 5.56; EN X, 229. Translation adapted from Drake 1973a pp. 303-304. For interpretation see 3.5.1)

5.3.13 Galileo: Folio 152r

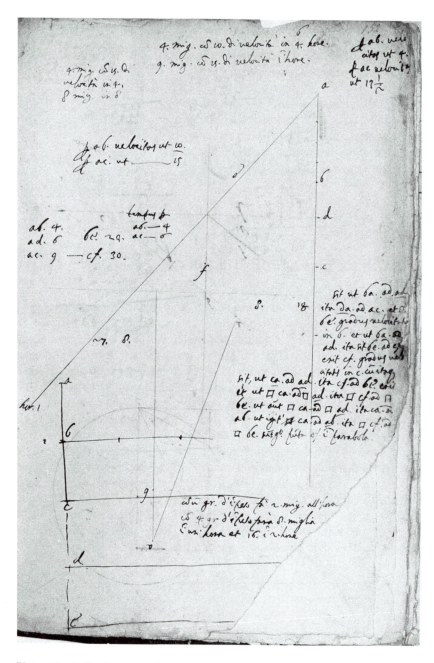

Plate VI. Galileo's reexamination of his proof of the law of fall (folio 152r). Published with permission of the Biblioteca Nazionale, Florence.

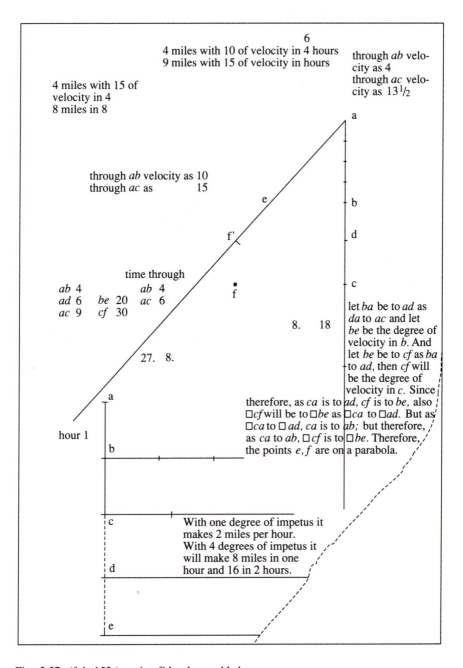

6
4 miles with 10 of velocity in 4 hours
9 miles with 15 of velocity in hours

through *ab* velo-
city as 4
through *ac* velo-
city as 13½

4 miles with 15 of
velocity in 4
8 miles in 8

a

through *ab* velocity as 10
through *ac* as   15

b

d

e

f'

c

time through

ab 4		*ab* 4
ad 6	*be* 20	*ac* 6
ac 9	*cf* 30	

8.   18

f

let *ba* be to *ad* as
da to *ac* and let
be be the degree of
velocity in *b*. And
let *be* be to *cf* as *ba*
to *ad*, then *cf* will
be the degree of
velocity in *c*. Since
therefore, as *ca* is to *ad*, *cf* is to *be*, also
□*cf* will be to □*be* as □*ca* to □*ad*. But as
□*ca* to □ *ad*, *ca* is to *ab*; but therefore,
as *ca* to *ab*, □ *cf* is to □*be*. Therefore,
the points *e, f* are on a parabola.

27. 8.

a

hour 1

b

c

With one degree of impetus it
makes 2 miles per hour.
With 4 degrees of impetus it
will make 8 miles in one
hour and 16 in 2 hours.

d

e

Fig. 5.57. (fol. 152r); point *f'* has been added.

(Folio 152r; EN VIII, 426-427. For interpretation see 3.4.1.1 and 3.4.1.2)

5.3.14 Galileo: Folio 91v

If some moveable is equably moved in double motion, i.e., horizontal and vertical, then the impetus of the movement compounded from both will be equal in the square to both moments of the original motions.

Let some moveable be equably moved in double motion, the ver- tical displacements [*mutationi*] corresponding to space *ab*, and let the horizontal movement carried out in the same time correspond to *bc*. Since spaces *ab* and *bc* are thus traversed in the same time, in equable motions, the moments of those motions will be to one another as *ab* is to *bc*, and the moveable that is moved according to these displacements will describe the diagonal *ac* in the same time in which it makes the displacements along the vertical *ab* and along the horizontal *bc*, and its moment of velocity will be as *ac*. Truly *ac* is equal in the square to *ab* and *bc*; therefore the moment compounded from both moments of *ab* and *bc* is equal in the square to both of them taken together; which was to be shown.

In motion from rest the moment of velocity and the 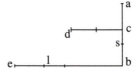 time of this motion are intensified in the same ratio. For let there be a motion through *ab* from rest in *a*, and let an arbitrary point *c* be assumed; and let it be posited that *ac* is the time of fall through *ac*, and the moment of the acquired speed in *c* is also as *ac*, and assume again any point *b*: *I say that the time of fall through* ab *to the time through* ac *will be as the moment of velocity in* b *to the moment in* c. Let *as* be the mean [proportional] between *ba* and *ac*; and since the time of fall through *ac* was set to be *ac*, *as* will be the time through *ab*: it thus has to be shown that the moment of speed in *c* to the moment of speed in *b* is as *ac* to *as*. Assume the horizontal [lines] *cd* to be double *ca*, but *be* to be double *ba*: it follows from what has been shown, that the [body] falling through *ac*, deflected into the horizontal *cd*, will traverse *cd* in uniform motion in an equal time as it also traversed *ac* in naturally accelerated motion; and, similarly, it follows that *be* is traversed in the same time as *ab*: but the time of *ab* itself is *as*: therefore, the horizontal [line] *be* is traversed in the time *as*. But let *eb* be to *bl* as the time *sa* is to the time *ac*; and since the motion through *be* is uniform, the space *bl* will be traversed in the time *ac* according to the moment of speed in *b*: but according to the moment of speed in *c*, in the same time *ac* the space *cd* will be traversed; but the moments of speed are to one another as the spaces, which according to these moments are traversed in the same time: therefore the moment of speed in *c* is to the moment of speed in *b* as *dc* to *bl*. But as *dc* to *be*, so are their halves, i.e., *ca* to *ab*; but as *eb* to *bl*, so *ba* to *as*; therefore, by the same [*ex aequali*], as *dc* to *bl*, so *ca* to *as*: that is, as the moment of speed in *c* to the moment of speed in *b*, so *ca* to *as*, that is, the time through *ca* to the time through *ab*. Which was to be shown.

The impetus in the single points of the parabola *bec* is 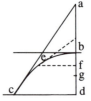 therefore determined by the square [*potentia*] of the moment acquired in the descent along *ab*, which will always be the same and which determines the horizontal impetus, and by the square of the other moment acquired in the descent along the vertical. Thus, for instance in *e* the impetus will be determined by the square root of the sum of the squares of [*linea potente*] *ab* and the mean proportional between *db*, *bf*, which is *bg*.

(Fig. 5.58; fol. 91v; EN VIII, 280, 281-282, and 427. For interpretation see 3.4.1.1 and 3.5.3.3.)

5.3.15 Galileo: Folio 87v

To find the elevation of a parabola descending from which a body describes the given parabola. Let the given parabola be *bf*, whose height is *bi* [Fig. 5.59]and whose amplitude is *if*; if the horizontal *bl* is drawn, assume in the vertical *be* to be equal to *bi*, and connect *elf*, which touches the parabola in *f* and intersects with the horizontal at *l*; let *lb* be to *ba* as *be* to *bl*: I say that *ab* is the elevation falling from which the body, deflected in *b*, will describe the parabola *bf*. For if it is understood that the same *eb* is the time of fall along *eb*, and the same *eb* is the moment of velocity in *b*, then *bl* will be the time and the moment in *b* of the body falling from *a*: hence, falling from *a* to *b*, deflected along the horizon, it will in the time *bl* traverse double *ba*; therefore, in the same motion, in the time *eb*, it will traverse double *bl*: for as the time *eb* is to the time *bl*, double *bl* is to double *ba*. But double *bl* is precisely *fi*; therefore the horizontal *if* will be traversed by the body falling from *a* in the time *be*: but in the same time *eb* the vertical *bi* is traversed from the state of rest in *b*: therefore the body falling from *a* and deflected in *b* traverses in the same time the horizontal *if* and the vertical *bi* from the state of rest in *b*: it will therefore describe the parabola *bf*. It is therefore true, that half the base is the mean proportional between the height of the parabola and the elevation above the parabola, falling from which the projectile describes it.

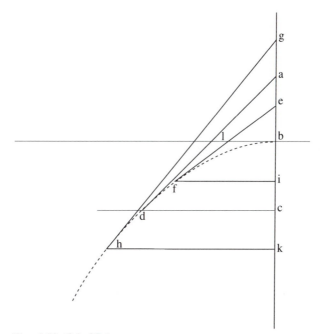

Fig. 5.59 (fol. 87v)

(Folio 87v; EN VIII, 428-429. For interpretation see 3.5.1)

5.3.16 Galileo: Folio 90ᵃr

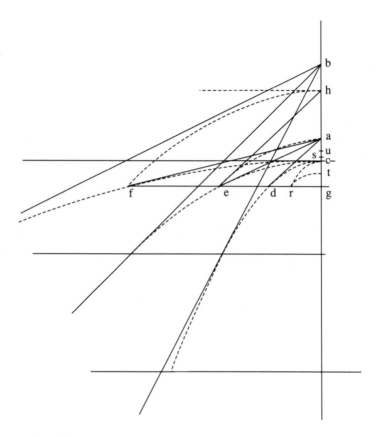

Fig. 5.60 (fol. 90ᵃr)

[A body] falling [Fig. 5.60] from *a* to *c* will – having been deflected – describe the parabola *cd*; but if the moment of velocity in *c* were double, it would describe the parabola *ce*, for which *eg* would be double *gd*: for double the impetus in *c* traverses double the space along the horizontal in the same time. But in order to acquire double the moment in *c*, it is necessary that the fall takes place from four times the height, that is, from *cb*. Similarly, from a height four times *cb* the parabola *cf* will be described, whose amplitude *gf* is twice *ge*. But in *d* the moving body seems to add to the impetus in *c* an impetus acquired through the parabola *cd*, which corresponds to the height *cg*. But in *e* the moving body adds the same moment to the impetus which it had in *c* which was double the impetus of the other body; therefore the impetus of the moving body in *e* seems to be one and a half times the impetus of the moving body in *d*. It is similarly found that the impetus in *f* is to the impetus in *e* as 5 to 3. If therefore the projectile [shot] along the elevation *ea* had an impetus that is one and a half times the impetus in *d*, then projectiles [shot] along the elevation [*eh* and] *da*[20] would be thrown along the parabolas *e*[*a*], *dc* [21]which are parallel to each other, but the distance *eg* would be double that of *dg*.

Let the impetus in *c* of a body falling from *a* be ——— 100.

That of a body falling from *b* will be ——————— 200.

The impetus in *d* [of the body from *a*] will be ——— 200.

The impetus in *e* [of the body from *b*] will be ——— 300.

The impetus in *f* is 500 if [the moving body] comes along the parabola *cf*; but if it comes along the parabola *hf*, the impetus in *f* is 400. From this it is also clear that [the projectile] is hurled farther by the same force along the elevation of half a right [angle] than along a smaller [angle].

The impetus of a body falling from *h* to *a* will be 141 in *a*; deflected along the parabola *ae*, however, the impetus will be doubled in *e*, that is, ——— 282.

It is therefore true that the impetus of the body coming along the parabola *ce* to *e* will be larger than that of the body coming along the parabola *ae*. And if the projectile [shot] from *e*, along the elevation *eh*, has an impetus as 282, it will traverse the parabola *ea*; but along the elevation *ea* the projectile traverses the parabola *ec*, if it has an impetus as 300. Hence it will be thrown farther by the same force along the elevation *eh* of half a right angle than along the elevation *ea*, which is smaller than half a right angle.

The impetus in *c* from *s* will be 50.

In *r* it will be ——————— 150.

But the impetus in *t* from *c* will be about $70^1/_2$.

If [the body] is deflected along the parabola *tr* in *r* it will be 141, namely, smaller than that of a body coming from *s* through *c* to *r*, which was 150. From which it is clear that the projection by the same force along the elevation of half a right angle [resulting in] *rt* is longer than that along the elevation [resulting in] *rc*.

(Folio 90ᵃr; EN VIII, 429-430. For interpretation see 3.5.3.1.)

---

[20] The original ms (and Favaro's transcription) reads simply "*da*."

[21] The original ms (and Favaro's transcription) reads "*ec, dc*"; a number of significant lines, in particular line *eh* and parabola *ea*, are missing from the transcribed diagram on EN VIII, 430. See Plate VII.

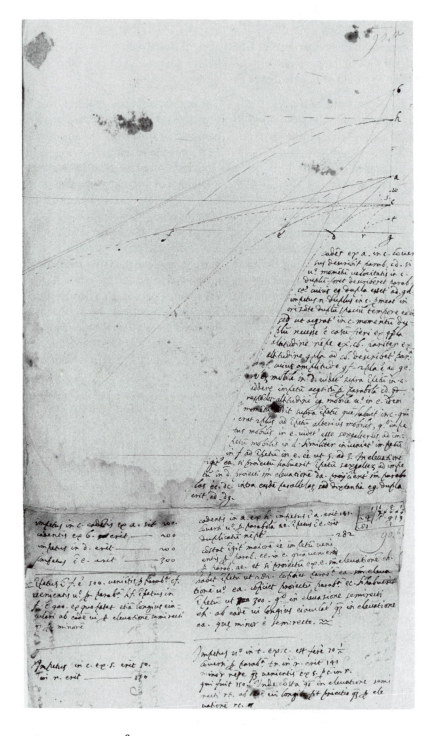

Plate VII. Folio 90^{a}r. Published with permission of the Biblioteca Nazionale, Florence.

5.3.17 Galileo: Folio 80r

In the right triangle *bcd* let the angle *d* be equal to the
angle *cbe* and let *eb* be drawn. The two triangles *dcb*, *ebc*
will thus be similar. Let the whole *dc* be divided in the
middle in *h*; and *hi* be parallel to the line *cb*. Also let *ec*
be divided in the middle in *f* and draw *fg* parallel to *bc*;
and as *dh* is to *hi*, let *hi* be to *hl*, and let *li* be drawn. The
triangle *lih* will be similar to the triangle *dhi*, and for the
same reason also similar to *efg*. But *hi* is equal to *gf* (the
double of both is namely *bc*). Therefore, the remaining
sides *hl, fe* will be equal, so that the third proportional
of these lines *lh, hi*, namely *hd*, will be equal to the
third proportional of the lines *ef, fg*. But the third
proportional of the lines *lh, hi* is *hd*; which is
half the total *dc*. Therefore, the third pro-por
tional of the lines *ef, fg* will be equal to the half
of *cd*, that is, to *ch*. But *ch* is equal to *fl*, be
cause *cf* is equal to *hl* and *fh* is common. Therefore, the third proportional of *ef, fg* will be
fl, which finishes in the point *l*, where the
third proportional of *dh, hi* finishes.
From this it will be shown that the
amplitudes of parabolas are equal
which are made by projectiles along
eleva-tions differing by equal
amounts from half a right angle.

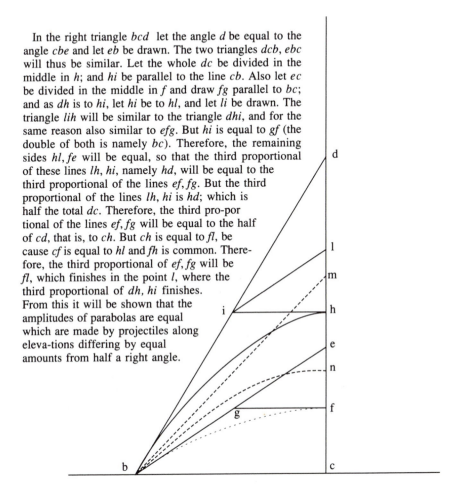

Fig. 5.61 (fol. 80r)

(Folio 80r; EN VIII, 433, table III. For interpretation see 3.5.3.2.)

5.3.18 Galileo: *Discorsi*, Third Day: Naturally Accelerated Motion (1638)[22]

PROPOSITION I. THEOREM I

The time in which a certain space is traversed by a moveable in uniformly accelerated movement from rest is equal to the time in which the same space would be traversed by the same moveable carried in uniform motion whose degree of speed is one-half the maximum and final degree of speed of the previous, uniformly accelerated, motion.

Let line AB [Fig. 5.62] represent the time in which the space CD is traversed by a moveable in uniformly accelerated movement from rest at C. Let EB, drawn in any way upon AB, represent the maximum and final degree of speed increased in the instants of the time AB. All the lines reaching AE from single points of the line AB and drawn parallel to BE will represent the increasing degrees of speed after the instant A. Next, I bisect BF at F and I draw FG and AG parallel to BA and BF; the parallelogram AGFB will [thus] be constructed, equal to the triangle AEB, its side GF bisecting AE at I.

Now if the parallels in triangle AEB are extended as far as IG, we shall have the aggregate of all parallels contained in the quadrilateral equal to the aggregate of those included in triangle AEB, for those in triangle IEF are matched by those contained in triangle GIA, while those which are in the trapezium AIFB are common. Since each instant and all instants of time AB correspond to each point and all points of line AB, form which points the parallels drawn and included within triangle AEB represent increasing degrees of the increased speed, while the parallels contained within the parallelogram represent in the same way just as many degrees of speed not increased but equable, it appears that there are just as many moments of speed consumed in the accelerated motion according to the increasing parallels of triangle AEB, as in the equable motion according to the parallels of the parallelogram GB. For the deficit of moments in the first half of the accelerated motion (the moments

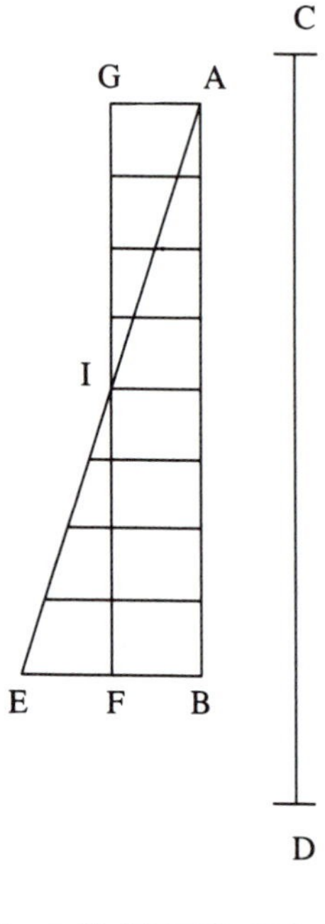

Fig. 5.62 (ENVIII, 208)

22   The translation is taken from Galileo Galilei, *Two New Sciences,* (transl. by Stillman Drake) University of Wisconsin Press, 1974, pp. 165-167; it is used with permission of Stillman Drake, Wisconsin University Press, and Wall & Emerson, Inc.

represented by the parallels in triangle AGI falling short) is made up by the moments represented by the parallels of triangle IEF.

It is therefore evident that equal spaces will be run through in the same time by two moveables, of which one is moved with a motion uniformly accelerated from rest, and the other with equable motion having a moment one-half the moment of the maximum speed of the accelerated motion; which was [the proposition] intended.

PROPOSITION II. THEOREM II

If a moveable descends from rest in uniformly accelerated motion, the spaces run through in any times whatever are to each other as the duplicate ration of their times; that is, are as the squares of those times.

Let the flow of time from some first instant A be represented by the line AB [Fig. 5.63], in which let there be taken any two times, AD and AE. Let HI be

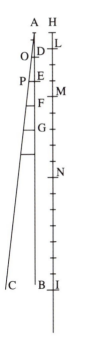

the line in which the uniformly accelerated moveable descends from point H as the first beginning of motion; let space HL be run through in the first time AD, and HM be the space through which it descends in time AE. I say that space MH is to space HL in the duplicate ratio of time EA to time AD. Or let us say that spaces MH and HL have the same ration as do the squares of EA and AD.

Draw the line AC at any angle with AB. From points D and E draw the parallels DO and EP, of which DO will represent the maximum degree of speed acquired at instant E of time AE. Since it was demonstrated above that as to spaces run through, those are equal to one another of which one is traversed by a moveable in uniformly accelerated motion from rest, and the other is traversed in the same time by a moveable carried in equable whose speed is one-half the maximum acquired in the accelerated motion, it follows that spaces MH and LH are the same that would be traversed in times EA and DA in equable motions whose speeds are as the halves of PE and OD. Therefore if it is shown that these spaces MH and LH are in the duplicate ration of the times EA and DA, what is intended will be proved.

Fig. 5.63
(EN VIII, 209)

Now in proposition IV of Book I ["On Uniform "Motion,"] it was demonstrated that the spaces run through by moveables carried in equable motion have to one another the ration compounded from the ratio of speeds and from the ratio of times. Here, indeed, the ratio of speeds is the same as the ratio of times, since the ration of one-half

OD, or of PE to OD, is that of AE to AD. Hence the ration of spaces run through is the duplicate ration of the times; which was to be demonstrated.

It also follows from this that this same ratio of spaces is the duplicate ration of the maximum degrees of speed; that is, of lines PE and OD, since PE is to OD as EA is to DA.

COROLLARY I

From this it is manifest that if [Fig. 5.64] there are any number of equal times taken successively from the first instant or beginning of motion, say AD, DE, EF, and FG, in which spaces HL, LM, MN,, and NI are traversed, then these spaces will be to one another as are the odd numbers from unity, that is, as 1, 3, 5, 7; but this is the rule [ratio] for excesses of squares of lines equally exceeding one another [and] whose [common] excess is equal to the least of the same lines, or, let us say, of the squares successively from unity. Thus when the degrees of speed are increased in equal times according to the simple series of natural numbers, the spaces run through in the same times undergo increases according with the series of odd numbers from unity.

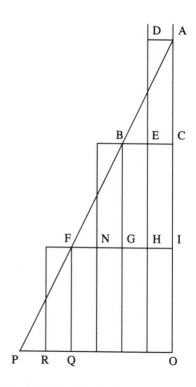

Fig. 5.64 (EN VIII, 211)

(EN VIII, 208-210; translation by Drake, Galilei 1974, pp. 165-167 For interpretation see 3.6.1.)

5.3.19   Galileo: Addendum on Projectile Motion for a Revised Edition of the *Discorsi*

Simpl.: I ask you, before we continue, to put me in a position to understand how to demonstrate the converse, which the Author presupposes as clear and free from doubt; I mean that the projectile which, when it comes from above to below describes a semiparabola, must return to its starting point along the same line if it is hurled conversely from below to above, by following exactly the same traces without having for this another regulator than the direction of the simple straight line that touches the semiparabola already drawn above; if it describes [this line] from above to below, the tranversal, horizontal impetus makes me quietly admit the great curvature at the vertex, but I can neither understand nor see how the impetus generated from below along a straight tangent can reproduce a transversal impetus that would be suited to regulate the same curvature.

Salv.: Signor Simplicio, by mentioning the straight tangent, you admit a condition, namely that the line is tangential and inclined; and just this inclination is sufficient to have the effect that the projectile approaches the axis of the parabola along equal horizontal distances in equal times, as we will perhaps understand further down.

Sagr. But, meanwhile, tell me, Signor Simplicio: do you believe that the line described by a projectile from below upwards along some inclination is really a complete parabolic line? and that is does not matter whether the projection is made from East to West or in the opposite direction?

Simp.: I do believe this, if indeed the elevation is the same and the force of the projector is also the same.

Sagr.: Since you admit this, then if there is a shot in some arbitrary direction, what is it that makes you doubt that the semiparabola from below upwards of the second shot, which will be made contrary to the first, is not the same as the second semi-parabola of the first shot so that the projectile returns along the same path? If that were not the case, then the complete parabola of the second shot would also not be similar to that of the first shot.

Simp.: I have understood and am propitiated; but let us continue.

THEOREM, PROPOSITION VIII

The horizontal amplitudes of parabolas etc. ...

(Cod. B, fol. 14v; EN VIII, 446-447. For interpretation see 3.7.2.2)

5.4 Documents for the Epilogue

5.4.1 Descartes: From a Letter to Huygens (Feb. 18 or 19, 1643)

As for gravity, I also consider that it increases the speed of bodies that it makes fall, almost in the same proportion as the times in which they fall; so that if a drop of water falls for two minutes of an hour, it goes almost twice as fast at the end of the second [minute] as it did at the end of the first; from which it follows that the path it covers is almost in the double proportion [i.e., as the square] of the time; that is to say, if during the first minute it falls from the height of one foot, during the first and the second together it ought to fall from the height of four feet. This is easily explained by the triangle ABC [Fig. 5.65], of which the

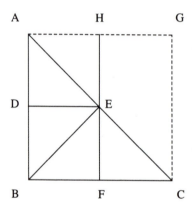

Fig. 5.65 (AT III, 207)

side AD represents the first minute, the side DE the speed that the water has at the end of this first minute, and the space ADE represents the path it meanwhile covers, which is the length of one foot. Then DB represents the second minute, BC the speed of the water in [i.e., at the end of] this second minute, which is double that of the preceding, and the space DEBC the path, which is three time the preceding. And one can also note that if this drop of water continued to move in some other direction with the speed that it acquired by its fall from a height of one foot during the first minute, without gravity's helping it after that, it would cover during one minute the path represented by the rectangle DEFB, which is two feet. But if it should continue to move for two minutes with the speed which it has acquired by descending four feet it would cover the path represented by the rectangle ABCG which is eight feet.

(AT III, 619-620. For interpretation see 4.1.)

5.4.2 Clerselier: From a Letter to Fermat (May 13, 1662)

It is confirmed by experience that, in whatever manner the ball A is pushed at point B by the balls C, D, E, F, G [Fig. 5.66], and whatever are the determinations of which one may suppose their route to be composed, they will always push it towards H.

First of all, for the ball E, it is clear that it would push towards H since ball A is totally opposed to its determination; but what is clear for ball E should be similarly extended to the others which, although they come at a slant towards ball A, only touch it at point B and only push it in so far as they descend towards H, and not at all in so far as they move towards I (or towards K). This is why they cannot impress a different motion on this ball except that of making it go towards H. But since the determinations of the balls D and F are opposed insofar as the one goes to the right and the other to the left, they are not at all opposed insofar as they descend, and thus they ought to produce the same effect on the ball A, which is to push it towards H.

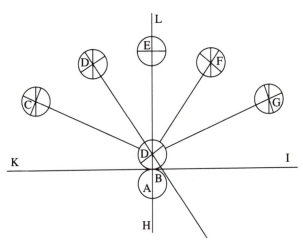

Fig. 5.66 (Fermat 1894, vol. 2, p. 478; figure rotated 180° to agree with text)

But if we suppose that ball A is hard and immobile, all the balls after colliding with it will be constrained to change the determination they have to go towards H into that to go or reflect towards L and to keep the other [determinations] if they have them which it [ball A] cannot cause to change since it is not at all opposed to them in that sense: and this explains the reflection at equal angles.

(*Oeuvres de Fermat*, vol. 2, 477-479. For interpretation see 2.6 and 4.2.)

5.4.3 Pierre Silvain Regis, Law of Reflection (1691)

[Proposition:] That, if the body A [Fig. 5.67] moves obliquely along the line AB and encounters in its path the ground CD, which is assumed to be immoveable, it is reflected along the line BG which is different from the line AB.

To prove this, draw through the points A and B the lines AE and FB perpendicular to CD, and the line AG parallel to the same line CD; having done this, consider in the first place that the body A going towards B approaches at the same time the line CD and FB, that is to say, that its determination from A to B is composed of its determination from A to E and from A to F, or what is the same thing, its determination from top to bottom and from left to right. Consider in the second place that the ground CD is opposed to the determination from A to F and consequently that the body A when it encounters the ground should take a determination completely contrary to that which it had, by which in an equal time it should advance an equal quantity, that is to say that, if in the time of one minute the body A descended from the line A[G] to the line CD, it should in a second minute climb back [*remonter*] from the line CD to the line A[G].[23]

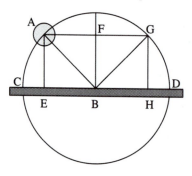

Fig. 5.67
(Regis 1691, vol. 1, p. 352)

Consider in the third place that, as nothing is opposed to the determination from A to G, it should remain the same, and the body A in an equal time should advance according to this determination an equal quantity, that is to say, that if the body A is moved in one minute of time from A to B, and if it has advanced from left to right the entire space comprised between the lines AE and FB, it should advance in a second minute according to the same determination from the line FB to another line as far away from FB as FB is from AE; thus, take BH equal to BE, draw the line GH and say that the body A after two minutes should be found in the line GH.

But in order to determine the point on this line at which body A should be in order to satisfy its determination from left to right, and to determine also at which point in the line AG it should be found in order to satisfy its determination from bottom to top, it is necessary to have recourse to a third antecedent and to consider, that, the body A losing none of its motion in the encounter with the body CD (which we have supposed to be immoveable), the line that it traverses in the second minute will be equal to that which it traversed in the first. Thus describing a circle with the center B and with the interval BA, we should say that the body A, having come from the circumference to the center in the first

---

[23] The original has the misprint "AB" instead of "AG".

minute, should in the second minute go from the center to the circumference. Accordingly, if we join together these three truths we will be obliged to recognize that the body A has arrived at point G where the circle and the lines AG and GH intersect and that it has arrived along the line BG, which makes an angle GBH with the surface of the ground CD, which is called the *angle of reflection*, which is equal to the angle ABE which is called the *angle of incidence* ... Q.e.d.

(Regis 1691, vol. 1, pp. 352-353. For interpretation see 4.2.)

5.4.4 Cavalieri: From *Lo Speccio Ustorio* (1632)

Chapter XXXIX. On the Motion of Heavy Bodies

While, in relation to heavy bodies, there is a great variety of possible considerations which could be made, all fine and all interesting, yet we now will not attempt anything more than to determine the nature of the path along which the heavy body moves, first due to internal gravity, then due to the projector, and finally due to both of them joined together, in order to find out whether conic sections are involved here and, if so, to see which ones.

Hence, I now say that, if we will consider the motion of the heavy body which results from internal gravity alone and without regard to the manner of its action, the body will always be led towards the universal center of heavy things, i.e., towards the center of the earth, and all heavy bodies are universally guided towards this center, since one sees them, if they are not hindered, descending along a line perpendicular to the horizon in all places on the surface of the earth, and it is clear that, when the straight lines perpendicular to the surface of the sphere are prolonged, they all end up in its center; it is, further, most clear that the earth is spherical, whether on the evidence of eclipses or of other phenomena, which demonstrate this to us evidently.

Next, I say that, if one considers a body which is projected towards some given place and if there is no other motive force which pushes it in another direction, then it will go in a straight line towards the place indicated by the projector due to the force also impressed on it along a straight line, from which direction it is not reasonable that the moving body deviate, as long as there is no other moving force, which deflects the body from it, and this is the case, if there is no impediment between the two terminal points [of the motion]; as for example if a cannon ball, which has left the mouth of the cannon, had no other force than that impressed on it by the fire, it would hit point blank the mark posed in the direction of the cannon's barrel, but because there is another motor, which is the internal gravity of the ball, it follows that it will be forced to deviate from this direction to approach the center of the earth.

I say, furthermore, that not only will the projectile go in a straight line towards its target, but that it also would traverse in equal times equal spaces in the same line, as long as the moving body is indifferent to this kind of motion, and also as long as the medium does not exert any resistance to it, so that there would be no cause either for slowing down or for accelerating. The heavy body will thus, due to its internal gravity, go nowhere but towards the center of the earth, while due to its impressed force it can travel in any direction whatever. Hence, if there are two motive forces in the projectile, one being gravity and the other the impressed force, then each of them separately would well make the moving body travel along a straight line, as has been said, but joined together, they will not make it travel along a straight line, except in the following two cases: first, if the heavy body is pushed by the impressed force perpendicular to the horizon; second, if not only the impressed force but also gravity moved the heavy body uniformly, because then the distances moved towards the center of the earth in equal times, as measured from a straight line, would always be equal, as would also be the distances traversed in the same times along this same line along which this heavy body is being pushed; and for this reason the moving body would be always in the same straight line. But if one of the two is not uniform, then the body moved both by gravity and by the impressed force would no longer move in a straight line, but on a curve, the quality and condition of which would depend on the above mentioned uniformity and difformity of the motions joined together. Now gravity is certainly acting on the body, which, separating itself from the projector is directed towards some place or other, being moved for example on a line at an angle to the horizon; but this [gravity] does nothing else than draw the body away from this straight line and has nothing to do with the other motion, except in as much as the body is moved away from the center of the earth, so that, abstracting in the heavy body the inclination towards this center, as well as [the inclination] towards any other place, the body remains indifferent with respect to the motion conferred on it by the projector, and hence, if it were not hindered by the environment, this motion would be uniform. One can therefore reasonably suppose that the heavy bodies pushed by the projector in any direction whatever will travel uniformly due to the impressed force if the impediment of the air is neglected, which, being very fine and very fluid, will possibly allow this uniformity for some noticeable distance.

It now remains for us to reflect upon the body's approach to the center of the earth due to its internal gravity – [a motion] that is called natural motion – and also upon the motion away from it resulting from an impulse conferred on it – which is called violent motion. It is known to all philosophers that the heavy body leaving rest and moving towards the center always increases in velocity [*si vada sempre velocitando*], the more it approaches the center, or better expressed, the more it moves away from its starting point, and that violent motion always slows down away from the center; but in what proportion the natural motion accelerates and the violent one slows down, has been taught to us only recently and only by Mr. Galileo in his *Dialogues* on p. 217,[24] where he says that the increment of the velocity is according to the sequence of odd numbers continued from unity, and the decrease according to the same series, counted backwards [...].

Chapter XL. What Kind of Line do Heavy Bodies Describe in their Motion, when they are Separated from the Projector.

If we consider the heavy body that has to move separately from the projector when it travels conserving its projector [*quando camina di conserva con quello*], so to say, there is no doubt whatsoever that the stone that moves for a little space in the company of the hand or of a slingshot or of some other instrument, is forced to take that path that also its motor takes; as, for example, it will have to go around with the hand or follow any other motion of the hand until it is separated from it. But when it is separated, it is no longer obliged to follow the hand: the impulse then conferred on it at the point of separation is always along a straight line, i.e., along that line, which is in the direction of the motion and which will become the tangent of the curve along which the motion is made. The tangent, I say, at the point of separation – as we are taught in the same way by Mr. Galileo in his *Dialogues* on page 186[25] – is a piece of the same line extended whether the motion of the hand is made along a curved or a straight line. Along this straight line the projectile would therefore go whenever gravity did not continuously deflect it from this line towards the center of the earth. It is true that, if the impulse were along the perpendicular to the horizon, then gravity would also draw the heavy body along the same line, and thus the motion of the projectile would in this case also be along a straight line, as was said in the preceding chapter. But when the impulse is not made along the perpendicular, but according any other straight line, it will follow, that the heavy body will be pulled back towards the center due to its internal gravity and consequently will deviate from this direction in such a way that in equal times it will not descend along equal spaces from this straight line, but rather along unequal spaces, which will grow, as has been said, according to the increase of the odd numbers continued from unity.

Therefore, I say that heavy bodies pushed by the projector in any arbitrary direction other than the perpendicular to the horizon, when they are separated from the projector (and neglecting the impediment of the air), describe a curved line insensibly different from a parabola ...

(Cavalieri 1632, pp. 153-158 and 163-165; part of translation adapted from Mepham, Koyré 238-240. For context see 4.3.)

---

[24] EN VII, 248/221-222

[25] EN VII, 219/193. Here Galileo does indeed explain that the the motion of the projector impresses an impetus upon the projectile to move, when they separate, along a straight line tangent to curve characterizing the motion of the projector at the point of separation, but he does not claim that this motion is continuous.

Bibliography

Agazzi, Evandro, David Gründer, Jaakko Hintikka (eds.) (1981) *Proceedings of the 1978 Pisa Conference on the History and Philosophy of Science*, vol. 1 Dordrecht: Reidel

Aiton, E. J. (1972) *The Vortex Theory of Planetary Motions*, London: Elsevier

Allaire, Edwin B. (1966) "'Tractatus' 6.3751," in: I.M. Copi and R.W. Beard (eds.) 1966, 189-194

Aquinas, Thomas (1963) *Commentary on Aristotle's Physics* (ed. and transl. by R.J. Blackwell, R.J. Spath, W.E. Thirlkel) London: Routledge

Archimedes (1953) *The Works of Archimedes* (ed. by T.L. Heath) New York: Dover

Ariew, Roger (1986) "Descartes as Critic of Galileo's Scientific Methodology," *Synthese 67*, 77-90

Aris, Rutherford and H. Ted Davis (eds.) (1983) *Springs of Scientific Creativity. Essays on Founders of Modern Science*, Univ. of Minnesota Press

Aristotle (1984) *The Complete Works of Aristotle* (ed. by J. Barnes) Princeton Univ. Press

Barbin, Evelyn, Eliane Bonnefon, Michèle Cholière, Gilles Itard, Xavier Lefort, Christiane Lize (1987) *Mathématiques, arts et techniques au XVIIème siècle*, Publications de L' Université Du Maine, no. 4, Le Mans: Équipe I.R.E.M., La Ferté-Bernard: Bellanger

Barbin, Evelyne and Michèle Cholière (1987) "La Trajectoire des projectiles de Tartaglia à Galilée," in: E. Barbin et al. 1987, 40-147

Beeckman, Isaac (1939) *Journal tenu par Isaac Beeckman de 1604 à 1634*, The Hague: Martinus Nijhoff

Benedetti, Giovanni Battista (1585) *Diversarum speculationum mathematicarum, et physicarum liber,* Torino (reissued, Venice 1586, 1599)

Biagioli, Mario (1989) "The Social Status of Italian Mathematicians, 1450-1600," *History of Science 27*, 42-95

Blackwell, Richard J. (1966) "Descartes' Laws of Motion," *Isis 57*, 220-234

Blay, Michel (1990) "Sur quelques aspects de l'évolution du champ conceptuel de la science du mouvement dans la deuxième motié du XVIIème siècle," Lecture given at the International Colloqium on the Concept of Velocity in Astronomy, Mathematics, and Physics from Archimedes to Galileo, Nice, 1990 (typescript)

Buchdahl, Gerd (1969) *Metaphysics and the Philosophy of Science*, Oxford: Blackwell

Buridan, Jean (1942) *Quaestiones super libris quattuor de caelo et mundo* (ed. by E.A. Moody) Cambridge, Mass: Medieval Academy of America

Butts, Robert E. and Joseph C. Pitt (eds.) (1978) *New Perspectives on Galileo* (Papers Deriving from and Related to a Workshop on Galileo held at Virginia

Polytechnic Institute and State University, 1975) *University of Western Ontario Series in Philosophy of Science, 14,* Dordrecht: Reidel

Butts, Robert E. (1978) "Some Tactics in Galileo's Propaganda for the Mathematization of Scientific Experiences," in: Butts and Pitt 1978, 59-86

Cardano, Girolamo (1550) *De Subtilitate libri xxi,* Nürnberg (other editions, Lyons, 1551, 1558, 1580)

Caroti, Stefano (1977) "Nicole Oresme precursore di Galileo e di Descartes?" *Rivista critica di storia della filosofia 1-4,* 11-23, 415-436, 552-563

Caroti, Stefano (1990) "La Vitesse dans les 'Calculationes'," Lecture given at the International Colloqium on the Concept of Velocity in Astronomy, Mathematics, and Physics from Archimedes to Galileo, Nice, 1990 (typescript)

Casper, B.M. (1977) "Galileo and the Fall of Aristotle: A Case of Historical Injustice?" *American Journal of Physics 45,* 325-330

Cassirer, Ernst (1923) *Substance and Function,* New York: Dover

Cavalieri, Bonaventura (1632) *Lo Specchio ustorio,* Bologna

Cavalieri, Bonaventura (1635) *Geometria indivisibilibus continuorum nova quadam ratione promota,* Bologna

Caverni, Raffaello (1895) *Storia del metodo sperimentale in Italia,* 6 vols. Florence (Reprint: New York: Johnson, 1972)

Chalmers, Alan and Richard Nicholas (1983) "Galileo on the Dissipative Effect of a Rotating Earth," *Studies in History and Philosophy of Science 14,* 315-340

Clagett, Marshall (1959) *The Science of Mechanics in the Middle Ages,* Univ. of Wisconsin Press

Clagett, Marshall (1968) "Introduction. Nicole Oresme and the *de configurationibus qualitatum et motuum,*" in: Oresme 1968a

Clagett, Marshall (ed.) (1979a) *Studies in Medieval Physics and Mathematics,* London: Variorum Reprints

Clagett, Marshall (1979b) "Richard Swineshead and Late Medieval Physics. The Intension and Remission of Qualities," in: Clagett 1979a (Reprinted from *Osiris 9,* 1950), 131-161

Clarke, Desmond M. (1977) "The Impact Rules of Descartes' Physics," *Isis 68,* 55-66

Clarke, Desmond M. (1982) *Descartes' Philosophy of Science,* Manchester Univ. Press

Clarke, Desmond M. (1989) *Natural Philosophy in the Age of Louis XIV,* Oxford: Clarendon

Clavelin, Maurice (1968) *La Philosophie naturelle de Galilée: Essai sur les origines et la formation de la mechanique classique,* Paris: Armand Colin

Clavelin, Maurice (1974) *The Natural Philosophy of Galileo* (transl. by A.J. Pomerans) MIT Press

Clavelin, Maurice (1983) "Conceptual and Technical Aspects of the Galilean Geometrization of the Motion of Heavy Bodies," in: Shea 1983a, 23-50

Coffa, José Alberto (1968) "Galileo's Concept of Inertia," *Physis 10,* 261-281

Cohen, I. Bernard (1985) *Revolution in Science,* Cambridge: The Belknap Press of Harvard Univ. Press

Copi, Irving M. and Robert W. Beard (1966) *Essays on Wittgenstein's* Tractatus, London: Routledge

Costabel, Pierre (1960) *Leibniz et la dynamique*, Paris: Hermann

Costabel, Pierre (1967) "Essai critique sur quelques concepts de la mécanique cartésienne," *Archives internationales d'histoire des sciences 20,* 235-252.

Costabel, Pierre (1975) "Mathematics and Galileo's Inclined Plane Experiments," in: Righini-Bonelli and Shea 1975, 175-189

Coulter, Byron L. and Carl G. Adler (1979) "Can a body pass a body falling through the air?" *American Journal of Physics 47,* 841-846

Crombie, Allistair C. (1975) "Sources of Galileo's Early Natural Philosophy," in: Righini-Bonelli and Shea 1975, 157-176

Crombie, Allistair C. (1985) "Sorting Out the Sources," Review of William A. Wallace, *Galileo and His Sources: The Heritage of the Collegio Romano in Galileo's Science, Times Literary Supplement,* London, November 22, 1985

Cronin, Timothy J. (1966) *Objective Being in Descartes and Suarez (Analecta Gregoriana 154)* Rome: Gregorian Univ. Press

Crowe, Michael J. (1967) *A History of Vector Analysis. The Evolution of the Idea of a Vectorial System,* Notre Dame Univ. Press

Degandt, François (1982) "De la vitesse de Galilée aux fluxions de Newton," in: *Penser les mathématiques,* Paris: Le Seuil

del Monte, Guidobaldo *Meditantiunculae Guidi Ubaldi e marchionibus Montis Santae Mariae de rebus mathematicis* (Ms Catalogue No., Lat. 10246) Bibliothèque Nationale, Paris

Descartes, René, Letter to Mersenne Nov. 13, 1629 (Ms Catalogue No., f. fr., nouv. acq. 5160, fol. 48) Bibliothèque Nationale, Paris

Descartes, René (1936-63) *Correspondance* (ed. and transl. by Ch. Adam and G. Milhaud) Paris: Presses Univeritaires de France

Descartes, René (1964-74) *Oeuvres de Descartes* (ed. by Ch. Adam and P. Tannery) Paris: Vrin

Descartes, René (1983) *Principles of Philosophy* (transl. by V.R. and R.P. Miller) Dordrecht: Reidel

Descartes, René (1985) *The Philosophical Writings of Descartes,* 2 vols. (transl. by J. Cottingham, R. Stoothoff, D. Murdoch) Cambridge Univ. Press

Descartes, René (1970) *Philosophical Letters* (ed. and transl. by A. Kenny) Oxford: Blackwell

Dijksterhuis, Eduard J. (1924) *Val en worp. Een bijdrage tot de geschiedenis der mechanica van Aristoteles tot Newton,* Groningen: Noordhoff

Dijksterhuis, Eduard J. (1961) *The Mechanization of the World Picture* (transl. by C. Dikshoorn) Oxford: Clarendon Press

Dijksterhuis, Eduard J. (1970) *Simon Stevin. Science in the Netherlands around 1600,* The Hague: Nijhoff

Dijksterhuis, Eduard J. (1987) *Archimedes* (transl. by C. Dikshoorn) Princeton Univ. Press

Drabkin, I.E. (1960) "A Note on Galileo's *De Motu*," *Isis 51,* 271-277

Drake, Stillman (1958) "Notes and Correspondence," *Isis 49,* 345-346

Drake, Stillman (1964) "Galileo and the Law of Inertia," *American Journal of Physics 32,* 601-608

Drake, Stillman (1968a) "Galileo Gleanings XVI. Semicircular Fall in the *Dialogue*," *Physis 10 ,* 89-100

Drake, Stillman (1968b) "Galileo Gleanings XVII. The Question of Circular Inertia," *Physis 10,* 282-298

Drake, Stillman (1969) "Galileo Gleanings XVIII. Galileo's 1604 Fragment on Falling Bodies," *British Journal for the History of Science 4,* 340-358

Drake, Stillman (1970a) "Galileo Gleanings XIX. Uniform Acceleration, Space, and Time," *British Journal for the History of Science 5,* 21-43

Drake, Stillman (1970b) *Galileo Studies. Personality, Tradition, and Revolution*, Univ. of Michigan Press

Drake, Stillman (1972a) "Galileo Gleanings XX. The Uniform Motion Equivalent to a Uniformly Accelerated Motion from Rest," *Isis 63,* 28-38

Drake, Stillman (1972b) "Galileo Gleanings XXI. On the Probable Order of Galileo's Notes on Motion," *Physis 14,* 55-65

Drake, Stillman (1973a) "Galileo Gleanings XXII. Galileo's Experimental Confirmation of Horizontal Inertia: Unpublished Manuscripts," *Isis 64,* 291-305

Drake, Stillman (1973b) "Galileo's Discovery of the Laws of Free Fall," *Scientific American 228, No. 5,* May, 1973, 84-92

Drake, Stillman (1973c) "Galileo Gleanings XXII. Velocity and Eudoxian Proportion Theory," *Physis 15,* 49-64

Drake, Stillman (1974a) "Impetus Theory and Quanta of Speed before and after Galileo," *Physis 16,* 47-65

Drake, Stillman (1974b) "Galileo's Work on Free Fall in 1604," *Physis 16,* 309-322

Drake, Stillman (1975a) "Free Fall from Albert of Saxony to Honoré Fabri," *Studies in History and Philosophy of Science 5,* 347-366

Drake, Stillman (1975b) "Galileo's New Science of Motion," in: Righini-Bonelli and Shea 1975, 131-157

Drake, Stillman (1975c) "Impetus Theory Reappraised," *Journal of the History of Ideas 36,* 27-46

Drake, Stillman (1976a) "A Further Reappraisal of Impetus Theory: Buridan, Benedetti and Galileo," *Studies in History and Philosophy of Science 7,* 319-336

Drake, Stillman (1976b) "Galileo Gleanings XXIV. The Evolution of *De Motu*," *Isis 67,* 239-250

Drake, Stillman (1978) *Galileo at Work. His Scientific Biography*, Univ. of Chicago Press

Drake, Stillman (1979) "Galileo's Notes on Motion Arranged in Probable Order of Composition and Presented in Reduced Facsimile," *Supplemento agli Annali dell'Istituto e Museo di Storia della Scienza 2, Monografia 3*

Drake, Stillman (1982) "Analysis of Galileo's Experimental Data," *Annals of Science 39*, 389-397

Drake, Stillman (1983) "Comment on the above Note by R.H. Naylor," *Annals of Science 40*, 395-396

Drake, Stillman (1985) "Galileo's Accuracy in Measuring Horizontal Projections," *Annali dell'Istituto e Museo di Storia della Scienza 10*, 1-14

Drake, Stillman (1986) "Galileo's Pre-Paduan Writings: Years, Sources, Motivations," *Studies in History and Philosophy of Science 17*, 429-448

Drake, Stillman (1988) *Galileo. Una Biografia scientifica* (transl. by Luca Ciancio) Bologna: Il Mulino

Drake, Stillman (1989a) "Hipparchus-Geminus-Galileo," *Studies in History and Philosophy of Science 20*, 47-56

Drake, Stillman (1989b) *History of Free Fall: Aristotle to Galileo*, Toronto: Wall and Thomson

Drake, Stillman (1990) *Galileo: Pioneer Scientist*, Univ. of Toronto Press

Drake, Stillman and I.E. Drabkin, (eds. and transl.) (1969) *Mechanics in Sixteenth-Century Italy: Selections from Tartaglia, Benedetti, Guido Ubaldo, & Galileo*, Univ. of Wisconsin Press

Drake, Stillman and James MacLachlan (1975a) "Galileo's Discovery of the Parabolic Trajectory," *Scientific American*, March 1975, 102-110

Drake, Stillman and James MacLachlan (1975b) "Reply to the Shea-Wolf Critique," *Isis 66*, 400-403

Dubarle, D.O.P. (1937) "Remarques sur les règles du choc chez Descartes," in: *Cartesio*, special supplement to *Rivista di filosofia neo-scholastica 19*, 325-334

Duhem, Pierre (1906-1913) *Etudes sur Léonardo de Vinci*, 3 vols. Paris: Hermann

Duhem, Pierre (1987) "An Account of the Scientific Titles and Works of Pierre Duhem," *Science in Context 1*, 333-350

Duhem, Pierre (1906ff) *Le System du monde*, 10 vols., Paris: Hermann

Euclid (1956) *The Thirteen Books of the Elements*, 2nd rev. ed. 3 vols. (ed. and transl. by Thomas L. Heath) New York: Dover

Euler, Leonhard (1846) *Letters to a German Princess on Different Subjects in Physics and Philosophy*, New York

Euler, Leonhard (1665) *Theoria motus corporum solidorum seu rigidorum*, vol. 1 in: *Opera omnia* (second series vol. 3) Berlin and Leipzig: Teubner, 1948

Eustace of St. Paul [Eustacius] (1640) *Summa philosophiae quadrapartita de rebus dialecticis, ethicis, physicis, et metaphysicis*, Cambridge

Favaro, Antonio (1883) *Galileo Galilei e lo Studio di Padova*, 2 vols., Florence

Favaro, Antonio (1983) *Amici e corrispondenti di Galileo* (ed. by P. Galluzzi) 3 vols. Florence: Salimbeni

Feinberg, G. (1965) "Fall of Bodies Near the Earth," *American Journal of Physics 33*, 501-502

Feldhay, Rivka (1987) "Knowledge and Salvation in Jesuit Culture," *Science in Context 1*, 195-213

Fermat, Pierre (1891ff) *Oeuvres de Fermat* (ed. by P. Tannery and C. Henry) Paris

Finocchiaro, Maurice A. (1972) "Vires Acquirit Eundo: The Passage where Galileo Renounces Space-Acceleration and Causal Investigation," *Physis 14*, 125-145

Franklin, Allan (1976) "Principle of Inertia in the Middle Ages," *American Journal of Physics 44*, 529-545

Franklin, Allan (1977) "Stillman Drake's 'Impetus Theory Reappraised'," *Journal of the History of Ideas 38*, 307-315

Fredette, Raymond (1969) *Les De Motu 'plus anciens' de Galileo Galilei: Prolégomènes* (diss.) University of Montreal

Fredette, Raymond (1972) "Galileo's *De Motu antiquiora*," *Physis 14*, 321-348

Freudenthal, Gideon (1985) "'Causa aequat effectum' bei Leibniz und Robert Mayer," *Leibniz – Werk und Wirkung, Proceedings of the 4th International Leibniz Congress (Nov. 14-19, 1983)* Hanover: Leibniz Gesellschaft, 203-210

Freudenthal, Gideon (1986) *Atom and Individual in the Age of Newton*, Dordrecht: Reidel

Gabbey, Alan (1980) "Force and Inertia in the Seventeenth Century: Descartes and Newton," In: Stephen Gaukroger (ed.) (1980) *Descartes, Philosophy, Mathematics, and Physics*, Sussex: Harvester, 230-320

Gabbey, Alan (1985) "The Mechanical Philosophy and its Problems: Mechanical Explanations, Impenetrability, and Perpetual Motion," in: J. Pitt (ed.) (1985) *Change and Progress in Modern Science*. Dordrecht: Reidel (*University of Western Ontario Series in Philosophy of Science, 27*)

Galilei, Galileo (1612) *Discorso al Serenissimo Don Cosimo II ... Intorno alle Cose che stanno in su l'acqua o che in quella si muovono*, Florence

Galilei, Galileo (1632) *Dialogo doue ne i congressi di quattro giornate si discorre sopra i due Massimi Sistemi del Mondo: Tolemaico, e Copernicano*, Florence

Galilei, Galileo (1638) *Discorsi e dimostrazioni matematiche, intorno a due nuove scienze attenenti alla mecanica & i movimenti locali*, Leyden

Galilei, Galileo (1663) *A Discourse Presented to the Most Serene Don Cosimo II Great Duke of Tuscany, concerning the Natation of Bodies upon, and Submerged in, the Water* (transl. by Thomas Salusbury) (new edition with an introduction and notes by Stillman Drake, Univ. of Illinois Press, 1960)

Galilei, Galileo (1960a) *On Motion and On Mechanics* (transl. by Stillman Drake and I.E. Drabkin) Univ. of Wisconsin Press

Galilei, Galileo (1960b) *De Motu*, in: Galilei (1960a)

Galilei, Galileo (1960c) *Le Meccaniche*, in: Galilei (1960a)

Galilei, Galileo (1964-1966) *Le Opere di Galileo; nuova ristampa della edizione nazionale,* 20 vols. (ed. by Antonio Favaro, 1890-1909) Florence: Barbèra

Galileo (1967) *Dialogue Concerning the Two Chief World Systems – Ptolemaic and Copernican* (transl. by Stillman Drake) Univ. of California Press

Galilei, Galileo (1970) *Discours et démonstrations mathématiques concernant deux sciences nouvelles* (transl. by Maurice Clavelin) Paris: Colin

Galilei, Galileo (1974) *Two New Sciences, Including Centers of Gravity and Force of Gravity and Force of Percussion* (transl. by Stillman Drake) Univ. of Wisconsin Press (2nd ed., Toronto: Wall and Thomson, 1989)

Galilei, Galileo, Manuscripts, Div. 2a, Part 5, Vol. II (Mss. Gal. 72), Manoscritti Galileiani Biblioteca Nazionale, Florence

Galluzzi, Paolo (1977) "Galileo contro Copernico," *Annali dell'Istituto e Museo di Storia della Scienza 2,* 87-160

Galluzzi, Paolo (1979) *Momento,* Rome: Ateneo e Bizzarri

Galluzzi, Paolo (ed.) (1983) *Novità celesti e crisi del sapere: Atti del Convegno Internazionale di Studi Galileiani,* in: *Annali dell'Istituto e Museo di Storia della Scienza 2, Monografia 7*

Geach, Peter (1972) *Logic Matters,* Oxford: Blackwell

Giusti, Enrico (1981) "Aspetti matematici della cinematica galileiana," *Bollettino di storia delle scienze mattematiche 1,* 3-42

Giusti, Enrico (1986) "Ricerche galileiane: il trattato *De Motu aequabili* come modello della teoria delle proporzioni," *Bollettino di storia delle scienze mattematiche 6,* 89-108

Golino, Carlo L. (ed.) (1966) *Galileo Reappraised,* Univ. of California Press

Grant, Edward (1971) *Physical Science in the Middle Ages,* New York: John Wiley and Sons

Grant, Edward and John E. Murdoch (eds.) (1987) *Mathematics and its Applications to Science and Natural Philosophy in the Middle Ages,* Cambridge Univ. Press

Grattan-Guinness, I. (ed.) (1987) *History in Mathematics Education* (Proceedings of a workshop held at the Univ. of Toronto, Canada July-August 1983) *Cahiers d'histoire et de philosophie des sciences, nouvelle série, no 21,* Societé française d'histoire des sciences et des techniques, Paris: Belin

Halbwachs, F. and A. Torunczyk (1985) "On Galileo's Writings on Mechanics: An Attempt at a Semantic Analysis of Viviani's Scholium," *Synthese 62,* 459-484

Hall, A. Rupert (1952) *Ballistics in the Seventeenth Century: A Study in the Relations of Science and War with Reference Principally to England,* Cambridge Univ. Press

Hall, A. Rupert (1958) "Galileo's Fallacy," *Isis 49,* 342-344

Hall, A. Rupert (1959) "Another Galilean Error," *Isis 50,* 261-262

Hall, A. Rupert (1965) "Galileo and the Science of Motion," *British Journal for the History of Science 2,* 185-201

Hall, A. Rupert (1990) "Was Galileo a Metaphysicist?" in: Levere and Shea 1990, 105-122

Hanson, Norwood Russell (1972) *Patterns of Discovery,* Cambridge Univ. Press

Heninger, S. K., Jr. (1977) *The Cosmographical Glass. Renaissance Diagrams of the Universe,* San Marino: Huntington Library

Hill, David K. (1979) "A Note on a Galilean Worksheet," *Isis 70,* 269-271

Hill, David K. (1986) "Galileo's Work on 116v: A New Analysis," *Isis 77,* 283-291

Hill, David K. (1988) "Dissecting Trajectories: Galileo's Early Experiments on Projectile Motion and the Law of Fall," *Isis 79,* 646-668

Hobbes, Thomas (1839-45) *The English Works of Thomas Hobbes* (ed. by W. Molesworth) London

Humphreys, W.C. (1967) "Galileo, Falling Bodies, and Inclined Planes," *British Journal for the History of Science 3,* 225-244

Itard, Gilles (1988) "La Réfraction de la lumière, éléments d'histoire de l'antiquité à la fin du XVIIème siècle," in E. Barbin (ed.) 1988, 173-254

Joyce, W.B and Alice Joyce (1976) "Descartes, Newton, and Snell's Law," *Journal of the Optical Society of America 66,* 1-8

Kalmar, Martin (1981) *Some Collision Theories of the Seventeenth Century: Mathematicism vs. Mathematical Physics* (diss.) Johns Hopkins University

Kant, Immanuel (1900ff) *Gesammelte Schriften,* Berlin: Königlich-Preußischen Akademie der Wissenschaften

Kant, Immanuel (1929) *Critique of Pure Reason* (transl. by N. Kemp Smith) New York: St. Martin's Press

Kant, Immanuel (1985) *The Metaphysical Foundations of Natural Science,* in: *Kant's Philosophy of Material Nature* (transl. by James Ellington) Indianapolis: Hackett

Keynes, John Neville (1906) *Studies and Exercises in Formal Logic,* New York: Macmillan

Kneale, William and Martha Kneale (1969) *The Development of Logic,* Oxford: Clarendon

Knudsen, O.M.P. (1968) "The Link between Determination and Conservation of Motion in Descartes' Dynamics," *Centaurus 13,* 183-186.

Koertge, Noretta (1977) "The Development of Galileo's Ideas Concerning the Problem of Accidents," *Journal of the History of Ideas 38,* 389-408

Koyré, Alexandre (1939) *Etudes galiléennes* (2nd ed. 1966) Paris: Hermann,

Koyré, Alexandre (1955) "A Documentary History of the Problem of Fall from Kepler to Newton; *de motu gravium naturaliter cadentium in hypothesi terrae motae,*" *Transactions of the American Philosophical Society 45,* 329-395

Koyré, Alexandre (1957) "La Dynamique de Nicolo Tartaglia," in: *La Science au seizième siècle* (Colloque international de Royaumont 1-4 juillet 1957, Union internationale d'histoire et de philosophie des sciences) Paris: Hermann

Koyré, Alexandre (1978) *Galileo Studies* (transl. by John Mepham) Sussex: Harvester

Kuhn, Thomas S. (1977) *The Essential Tension,* Univ. of Chicago Press

Lefèvre, Wolfgang (1978) *Naturtheorie und Produktionsweise,* Neuwied: Luchterhand

Leibniz, G. W. (1849-1863) *Leibnizens Mathematische Schriften,* Berlin and Halle

Leibniz, G. W. (1875-1890) *Die Philosophischen Schriften,* Berlin

Levere, Trevor H. and William R. Shea (eds.) (1990) *Nature, Experiment, and the Sciences: Essays on Galileo and the History of Science in Honour of Stillman Drake (Boston Studies in the Philosophy of Science, 120)* Dordrecht: Kluwer

Lewis, Christopher (1976) "The Fortunes of Richard Swineshead in the Time of Galileo," *Annals of Science 33,* 561-584

Lewis, Christopher (1980) *The Merton Tradition and Kinematics in Late Sixteenth and Early Seventeenth Century Italy,* Padua: Editrice Antenore

Libri, Guillaume (1838-1841) *Histoire des Sciences mathématique en Italie, depuis la renaissance des lettres jusqu'à la fin du XVIIème siècle,* 4 vols., Paris (Reprint: Hildesheim: Olms, 1967)

Lindberg, David. C. (1968) "The Cause of Refraction in Medieval Optics," *British Journal for the History of Science 4,* 23-38

Lindberg, David C. (ed.) (1978) *Science in the Middle Ages,* Univ. of Chicago Press

Lukasiewicz, Jan (1967) *Aristotle's Syllogistic from the Standpoint of Modern Formal Logic,* Oxford: Clarendon

Lukasiewicz, Jan (1979) "Aristotle on the Law of Contradiction," in: Barnes, J. (ed.) *Articles on Aristotle.* vol. 3, London: Duckworth, 40-62

Mach, Ernst (1942) *The Science of Mechanics* (transl. by Thomas J. McCormack) La Salle: Open Court

Mach, Ernst (1953) *The Principles of Physical Optics* (transl. by J.S. Anderson and A.F.A.Young) New York: Dover

Machamer, Peter (1978) "Galileo and the Causes," in: Butts and Pitt 1978, 161-180

MacLachlan, James (1976) "Galileo's Experiments with Pendulums: Real or Imaginary," *Annals of Science 33,* 173-185

MacLachlan, James (1982) "Note on R.H. Naylor's Error in Analysing Experimental Data," *Annals of Science 39,* 381-384

MacLachlan, James (1990) "Drake against the Philosophers," in: Levere and Shea 1990, 123-144

Mahoney, Michael (1973) *The Mathematical Career of Pierre de Fermat,* Princeton Univ. Press

Maier, Anneliese (1949) *Die Vorläufer Galileis im 14. Jahrhundert,* Rome: Edizioni di storia e letteratura

Maier, Anneliese (1951) *Zwei Grundprobleme der scholastischen Naturphilosophie*, 2nd ed., Rome: Edizioni di storia e letteratura

Maier, Anneliese (1952) *An der Grenze von Scholastik und Naturwissenschaft*, Rome: Edizioni di storia e letteratura

Maier, Anneliese (1955) *Metaphysische Hintergründe der spätscholastischen Naturphilosophie*, Rome: Edizioni di storia e letteratura

Maier, Anneliese (1958) *Zwischen Philosophie und Mechanik*, Rome: Edizioni di storia e letteratura

Maier, Heinrich (1896-1900) *Die Syllogistik des Aristoteles*, Tübingen

Marci, Marcus (1967) *De Proportione motus, seu Regula Sphygmica ad celeritatem et tarditatem pulsuum ex illius motu ponderibus geometricis librato absque errore metiendam, Acta historiae rerum naturalium necnon technicarum. 3.* (Czechoslovak Studies in the History of Science, Special issue 3; commemorating the 300th anniversary of Marci's death) Prague, 1967

Marshall, D.J.J. (1979) *Prinzipien der Descartes-Exegese*, Freiburg: Alber

Matuszewski, Roman (1986) "On Galileo's Theory of Motion: An Attempt at a Coherent Reconstruction," *International Studies in the Philosophy of Science 1,* 124 -141

McMullin, Ernan (ed.) (1967) *Galileo Man of Science,* New York : Basic Books

Mersenne, Marin (1945ff) *Correspondance du P. Marin Mersenne* (ed. by C. de Waard) Paris: Presses Universitaires

Meyerson, Emile (1962) *Identity and Reality,* New York: Dover

Milhaud, Gaston (1921) *Descartes savant*, Paris: Alcan (Reprint: New York: Garland, 1987)

Mittelstrass, Jürgen (1970) *Neuzeit und Aufklärung: Studien zur Entstehung der Neuzeitlichen Wissenschaft und Philosophie,* Berlin: de Gruyter

Mittelstrass, Jürgen (1972) "The Galilean Revolution: The Historical Fate of a Methodological Insight," *Studies in the History and Philosophy of Science 2,* 297-328

Molland, George (1982) "The Atomisation of Motion: A Facet of the Scientific Revolution," *Studies in History and Philosophy of Science 13,* 31-54

Moody, Ernest A. (1951) "Galileo and Avempace: The Dynamics of the Leaning Tower Experiment," *Journal of the History of Ideas 12,* 163-193, 375-422

Mouy, Paul (1934) *Le Développement de la physique cartésienne*, 1646-1712, Paris: Vrin

Murdoch, John E. and Edith Sylla (1978) "The Science of Motion," in: Lindberg (1978) pp. 206-264

Napolitani, Pier Daniele (1988) "La Geometrizzazione della realtà fisica: il peso specifico in Ghetaldi e in Galileo," *Bollettino di storia delle scienze matematiche 8,* 139-237

Nardi, Antonio (1986) "Moto delle acque e gravi in caduta: Descartes, Toricelli, Mersenne, Bernoulli," *Giornale citico della filosofia italiana 6,* 331-365

Nardi, Antonio (1988) "La Quadratura della velocità," *Nuncius 2,* 27-64

Naylor, Ronald H. (1974a) "Galileo and the Problem of Free Fall," *British Journal for the History of Science 7*, 105-134

Naylor, Ronald H. (1974b) "Galileo's Simple Pendulum," *Physis 16*, 23-46

Naylor, Ronald H. (1974c) "The Evolution of an Experiment: Guibaldo del Monte and Galileo's *Discorsi* Demonstration of the Parabolic Trajectory," *Physis 16*, 323-346

Naylor, Ronald H. (1975) "An Aspect of Galileo's Study of the Parabolic Trajectory," *Isis 66*, 394-396

Naylor, Ronald H. (1976a) Review of W. Wisan, "The New Sciences of Motion: A Study of Galileo's *De Motu Locali*," *Annali dell Istituto e Museo di Storia della Scienza 2*, 88-97

Naylor, Ronald H. (1976b) "Galileo: Real Experiment and Didactic Demonstration," *Isis 67*, 398-419

Naylor, Ronald H. (1976c) "Galileo: The Search for the Parabolic Trajectory," *Annals of Science 33*, 153-172

Naylor, Ronald H. (1977a) "Galileo's Theory of Motion: Processes of Conceptual Change in the Period 1604-1606," *Annals of Science 34*, 365-392

Naylor, Ronald H. (1977b) "Galileo's Need for Precision: the 'Point' of the Fourth Day Pendulum Experiment," *Isis 68*, 97-103

Naylor, Ronald H. (1979) "Mathematics and Experiment in Galileo's New Sciences," *Annali dell'Istituto e Museo di Storia della Scienza 4*, 55-63

Naylor, Ronald H. (1980a) "Galileo's Theory of Projectile Motion," *Isis 71*, 550-570

Naylor, Ronald H. (1980b) "The Role of Experiment in Galileo's Early Work on the Law of the Fall," *Annals of Science 37*, 363-378

Naylor, Ronald H. (1982) "Galileo's Law of Fall: Absolute Truth or Approximation," *Annals of Science 39*, 384-389

Naylor, Ronald H. (1983a) "Galileo's Early Experiments on Projectile Trajectories," *Annals of Science 40*, 391-396

Naylor, Ronald H. (1983b) Letter to the Editor, *Annals of Science 40*, 397

Neile, John (1669) "Hypothesis of Motion," in: *The Correspondence of Henry Oldenburg*, vol. 5 (ed. by A.R. and M.B. Hall) Univ. of Wisconsin Press, 1968, 519-524.

Ogawa, Yutaka (1989) "Galileo's Work on Free Fall at Padua: Some Remarks on Drake's Interpretation," *Historia Scientiarum 37*, 31-49

Olschki, Leonardo (1927) *Galileo und seine Zeit*, Halle: Max Niemeyer (Vol. 3 of Olschki, *Geschichte der neusprachlichen wissenschaftlichen Literatur*)

Oresme, Nicole (1966) *De Proportionibus proportionum and Ad Pauca respicientes* (ed. and transl. by Edward Grant) Univ. of Wisconsin Press

Oresme, Nicole (1968a) *Nicole Oresme and the Medieval Geometry of Motions* (ed. and transl. by M. Clagett) Univ. of Wisconsin Press

Oresme, Nicole (1968b) *Le Livre du ciel et du monde* (ed. by A.D. Menut and A.J. Denomy) Univ. of Wisconsin Press

Osler, Margaret J. (1973) "Galileo, Motion, and Essences," *Isis 64*, 504-509

Overmann, Ronald James (1974) *Theories of Gravity in the Seventeenth Century* (diss.) Indiana University

Palter, Robert M. (ed.) (1969) *Toward Modern Science,* 2 vols., New York: Dutton

Paul of Venice [Paulus Venetus] (1503) *Summa philosophie naturalis,* Venice (Reprint: Hildesheim: Olms, 1974)

Piaget, Jean and Garcia Rolando (1989) *Psychogensis and the History of Science* (transl. by. H. Feider) Columbia Univ. Press

Pitt, Joseph C. (1978) "Galileo: Causation and the Use of Geometry," in: Butts and Pitt 1978, 181-196

Planck, Max (1913) *Das Princip der Erhaltung der Energie,* Leipzig: Teubner

Régis, Pierre Silvain (1691) *Cours entier de philosophie, ou système général selon les principes de M. Descartes,* Amsterdam (Reprint: New York: Johnson, 1970)

Renn, Jürgen (1984) *Galileo's Theorem of Equivalence, Part I: Galileo's Calculations,* Preprint Freie Universität Berlin: FUB/HEP 84/5

Renn, Jürgen (1988) "Galileo's Manuscripts on Mechanics: The Project of an Edition with Full Critical Apparatus of Mss. Gal. Codex 72," *Nuncius 3,* No. 1, 193-241

Renn, Jürgen (1989) "Einige Paradoxien des Geschwindigkeitsbegriffes bei Galilei," in: W. Muschik and W.R. Shea (eds.) *Philosophie, Physik, Wissenschaftsgeschichte: Ein gemeinsames Kolloquium der TU Berlin und des Wissenschaftskollegs zu Berlin, 24-25, April 1989,* 30-67 Berlin: TUB-Dokumentation

Renn, Jürgen (1990) "Galileo's Theorem of Equivalence: The Missing Keystone of his Theory of Motion," in: Levere and Shea 1990, 77-104

Renn, Jürgen (1990b) "Die Rolle von Zeitfaktoren und Generationsdynamik für die wissenschaftlichen Erfolge Galileis und Einsteins," in: P.H. Hofschneider, and K.U. Mayer (eds.) *Generationsdynamik und Innovation in der Grundlagenforschung,* Max-Planck-Gesellschaft, *Berichte und Mitteilungen* No. 3/90, 1990, pp. 225-242

Righini-Bonelli, M.L., and W.R. Shea (eds.) (1975) *Galileo's New Science of Motion: Reason, Experiment, and Mysticism in the Scientific Revolution,* New York: Science History Publications

Rohault, Jacques (1672) *Traité de Physique,* Amsterdam

Romo Feito, José (1985) *La Fisica di Galileo. La problemática en torno a la ley de caida de los cuerpos,* Seminario de Historia de las Ciencias, Bellaterra: Universidad Autónoma de Barcelona

Rose, Paul-Lawrence (1968) "Galileo's Theory of Ballistics," *British Journal for the History of Science 4,* 156-159

Russell, Bertrand (1900) *A Critical Exposition of the Philosophy of Leibniz,* London: Allen and Unwin

Russell, Bertrand (1903) *The Principles of Mathematics,* London: Allen and Unwin

Sabra, A. I. (1967) *Theories of Light from Descartes to Newton*, London: Oldbourne

Schmitt, Charles (1969) "Experience and Experiment: A Comparison of Zabarella's View with Galileo's in *De Motu*," *Studies in the Renaissance 16*, 80-138

Schuster, John (1977) *Descartes and the Scientific Revolution, 1618-1634: An Interpretation* (diss.) Princeton University

Scott, Joseph F. (1976) *The Scientific Work of René Descartes (1596-1650)*, London: Taylor and Francis

Scotus, Johannes Duns (1891) *Opera omnia*, Paris (Reprint: Westmead: Gregg, 1969)

Segré, Michel (1980) "The Role of Experiment in Galileo's Physics," *Archive for History of Exact Sciences 23*, 227-252

Segré, Michel (1983) "Torricelli's Correspondence on Ballistics," *Annals of Science 40*, 489-499

Settle, Thomas B. (1961) "An Experiment in the History of Science," *Science 133*, 19-23

Settle, Thomas B. (1966) *Galilean Science: Essays in the Mechanics and Dynamics of the "Discorsi"* (diss.) Cornell University

Settle, Thomas B. (1967) "Galileo's Use of Experiment as a Tool of Investigation," in: McMullin 1967, 315-338

Settle, Thomas B. (1983) "Galileo and Early Experimentation," in: Aris et al. 1983, 3-20

Settle, Thomas B. (1987) "The Tartaglia Ricci Problem: towards a Study of the Technical Professional in the 16th Century," in: *Cultura, scienze e tecniche nella Venezia del cinquecento* (Atti del convegno internazionale di studio Istituto Veneto di Scienze, Lettere ed Arti) Venice, 217-226

Settle, Thomas B. (1988) "On Accepting Problematical Evidence," Brooklyn, Polytechnic Institute (typescript)

Shapere, Dudley (1974) *Galileo. A Philosophical Study*, Univ. of Chicago Press

Shapiro, Alan E. (1973) "Kinematic Optics. A Study of the Wave Theory of Light in the Seventeenth Century," *Archive for History of Exact Sciences 11*, 134-266

Shea, William R. (1972a) *Galileo's Intellectual Revolution: Middle Period, 1610-1632*, New York: Science History Publications

Shea, William R. (1975) "Trends in the Interpretation of Seventeenth Century Science," Introduction to Righini-Bonelli and Shea 1975, 1-18

Shea, William R. (1977) "Galileo and the Justification of Experiments," in: Robert E. Butts and Jaakko Hintikka (eds.) *Historical and Philosophical Dimensions of Logic, Methodology and Philosophy of Science*, Dordrecht: Reidel, 1977 (Part Four of the Proceedings of the Fifth International Congress of Logic, Methodology and Philosophy of Science, London, Ontario, Canada, 1975) *University of Western Ontario Series in Philosophy of Science 12*, 81-92

Shea, William, R. (1978) "Descartes as a Critic of Galileo," in: Butts and Pitt 1978, 139-159

Shea, William R. (ed.) (1983a) *Nature Mathematized* (papers deriving from the Third International Conference on the History and Philosophy of Science, Montreal, Canada, 1980) Dordrecht: Reidel (*Western Ontario Series in the Philosophy of Science 20*)

Shea, William R. (1983b) "Do Historians and Philosophers of Science Share the Same Heritage?" Introduction to Shea 1983a, 3-20

Shea, William R. (1983c) "The Galilean Geometrization of Motion: Some Historical Considerations," in: Shea 1983a, 51-59

Shea, William R. and Neil S. Wolf (1975) "Stillman Drake and the Archimedean Grandfather of Experimental Science," *Isis 66*, 397- 400

Sigwart, Christoph (1904) *Logik*, Tübingen: Mohr

Smith, A. Mark (1987) "Descartes Theory of Light and Refraction: A Discourse on Method," *Transactions of the American Philosophical Society 77*, No. 3, 1-92.

Sorabji, Richard (1988) *Matter, Space and Motion Theories in Antiquity and Their Sequel*, Cornell Univ. Press

Soto, Domingo de (1554) *Summulae*, Salamanticae (Reprint: Hildesheim: Olms, 1980)

Souffrin, Pierre (1986) "Du Mouvement uniforme au mouvement uniformément accéléré," *Bollettino di storia delle scienze matematiche 6*, 135-144

Souffrin, Pierre (1987) "La Quantification du mouvement chez les scolastiques Oresme et l'intensité des formes," University of Nice (typescript)

Souffrin, Pierre (1988) "Etudes sur les acceptations Galiléennes de *velocitas/ velocità*," Lecture given at the Conference "350 Years Galileo's *Discorsi*," Florence, December 1988

Souffrin, Pierre and Jean-Luc Gautero (1989) "Note sur la démonstration "mécanique," du théorème de l'isochronisme des cordes du cercle dans les *Discorsi* de Galilée," University of Nice, January 1989 (typescript)

Souffrin, Pierre and J.P. Weiss (1988) "Le Traité des Configurations des Qualités et des Mouvements: remarques sur quelques problèmes d'interprétation et de traduction," in: P. Souffrin and A. Ph. Segonds (eds.) (1988) *Nicolas Oresme. Tradition et innovation chez un intellectuel du XIVème siècle*, Paris: Les Belles Lettres, 125-134

Specht, Rainer (1966) *Commercium mentis et corporis. Über Kausalvorstellungen im Cartesianismus*, Stuttgart-Bad Cannstatt: Frommann-Holzboog

Spinoza, Baruch (1925) *Opera omnia*, Heidelberg: Winters

Strawson, Peter (1952) *Introduction to Logical Theory*, London: Methuen

Suarez, Franciscus (1856ff) *Opera omnia*, Paris

Sylla, Edith (1973) "Medieval Concepts of the Latitude of Forms: The Oxford Calculators," *Archives d'histoire doctrinales et littéraire du moyen âge 40*, 223-283

Sylla, Edith (1986) "Galileo and the Oxford *Calculatores:* Analytical Languages and the Mean-Speed Theorem for Accelerated Motion," in: Wallace 1986, 53-108

Szabó, István (1977) *Geschichte der mechanischen Prinzipien,* Basel: Birkhäuser

Tannery, Paul (1891) "Neuf lettres inédites de Descartes à Mersenne," *Archiv für Geschichte der Philosophie 4,* 529-556

Taliaferro, R. Catesby (1964) *The Concept of Matter in Descartes and Leibniz,* Notre Dame Mathematical Lectures 9, Notre Dame

Tartaglia, Niccolò (1537) *La Nova scientia ...,* Venice

Tartaglia, Niccolò (1546) *Quesiti, et inventioni diverse ...,* Venice

Tartaglia, Niccolò (1556a) *La Prima parte del general trattato di numeri, et misure ...,* Venice

Tartaglia, Niccolò (1556b) *La Seconda parte del general trattato di numeri, et misure ...,* Venice

Taton, René (1975) "The Mathematical Revolution of the Seventeenth Century," in: Righini-Bonelli and Shea 1975, 283-290

Tenneur, Jacob Alexandre (1649) *De Motu naturaliter accelerato tractatus...,* Paris

Thuillier, Pierre (1983) "Galilée et l'éxperimentation," *La Recherche 14,* no. 143, 442-454

Toletus, Franciscus (1589) *Commentaria, una cum Quaestionibus in Universam Aristotelis Logicam,* in *Opera omnia,* Cologne, 1616-17 (Reprint: Hildesheim: Olms, 1985)

Torricelli, Evangelista (1644) *De Motu Gravium,* in: *Opera di Evangelista Torricelli,* vol. 2 (ed. by G. Loria and G. Vassura) Faenza: Montanari, 1919

Torrini, Maurizio (1979) *Dopo Galileo. Una polemica scientifica (1684-1711),* Florence: Olschki

Vergil [Publius Vergilius] (1972-1973) *The Aeneid of Vergil* (ed. by R.D.Williams) London: Macmillan

Wallace, William A. (1971) "Mechanics from Bradwardine to Galileo," *Journal of the History of Ideas 32,* 15-28

Wallace, William A. (1977) *Galileo's Early Notebooks: The Physical Questions,* Univ. of Notre Dame Press

Wallace, William A. (1978) "Galileo Galilei and the Doctores Parisienses," in: Butts and Pitt 1978, 87-138

Wallace, William A. (1981) *Prelude to Galileo. Essays on Medieval and Sixteenth-Century Sources of Galileo's Thought,* Dordrecht: Reidel (*Boston Studies in the Philosphy of Science* 62)

Wallace, William A. (1984) *Galileo and His Sources. The Heritage of the Collegio Romano in Galileo's Science,* Princeton Univ. Press

Wallace, William A. (ed.) (1986a) *Reinterpreting Galileo,* Catholic Univ. of America Press (*Studies in Philosophy and the History of Philosophy 15*)

Wallace, William A. (1986b) "Reinterpreting Galileo on the Basis of His Latin Manuscripts," in: Wallace 1986a, 3-28

Wallace, William A. (1990) "The Dating and Significance of Galileo's Pisan Manuscripts," in: Levere and Shea 1990, 3-50

Watanabe, Hiroshi (1974) "On the Divergence of the Concept of Motion in the Collaboration of Beeckman and Descartes," in: *Proceedings no. 2, XIVth International Congress of the History of Science, Tokyo and Kyoto, 19-27, August, 1974,* Science Council of Japan, 338-340

Weizsäcker, Carl Friedrich von (1952) *The World View of Physics,* London: Routledge

Wertheimer, Max (1945) *Productive Thinking,* New York: Harper

Westfall, Richard S. (1966) "The Problem of Force in Galileo's Physics," in: C.L. Golino (ed.) *Galileo Reappraised,* Univ. of California Press, 65-95

Westfall, Richard S. (1971) *Force in Newton's Physics: The Science of Dynamics in the Seventeenth Century,* London: Elsevier

Westfall, Richard S. (1972) "Circular Motion in Seventeenth-Century Mechanics," *Isis 63,* 154-189

Wigner, Eugene P. (1967) *Symmetries and Reflections,* Bloomington: Univ. of Indiana Press

Wisan, Winifred Lovell (1974) "The New Science of Motion: A Study of Galileo's *De Motu Locali,*" *Archive for History of Exact Sciences 13,* 103-306

Wisan, Winifred Lovell (1977) "Mathematics and Experiment in Galileo's Science of Motion," *Annali dell'Istituto e Museo di Storia della Scienza 2,* 149-160

Wisan, Winifred Lovell (1978) "Galileo's Scientific Method: A Reexamination," in: Butts and Pitt 1978, 1-58

Wisan, Winifred Lovell (1981) "Galileo and the Emergence of a New Scientific Style," in: Evandro Agazzi, David Gründer, Jaakko Hintikka (eds.) *Proceedings of the 1978 Pisa Conference on the History and Philosophy of Science,* vol. 1 Dordrecht: Reidel

Wisan, Winifred Lovell (1983) "Galileo's *De Systemate Mundi* and the New Mechanics," in: Paolo Galluzzi (1983) 41-47

Wisan, Winifred Lovell (1984) "Galileo and the Process of Scientific Creation," *Isis 75,* 269-286

Wittgenstein, Ludwig (1929) "Some Remarks on Logical Form," in: Copi and Beard 1966

Wittgenstein, Ludwig (1972) *Tractatus Logico-philosophicus* (ed. and transl. by D.F. Pears and B.F. McGuiness) New York: Humanities Press

Wohlwill, Emil (1883-1884) "Über die Entdeckung des Beharungsgesetzes," *Zeitschrift für Völkerpsychologie und Sprachwissenschaft 14* (1883) 365-410 and *15* (1884) 70-135, 337-387

Wohlwill, Emil (1909) *Galilei und sein Kampf für das kopernikanische Weltbild,* Hamburg: Leopold Ross

Wolff, Michael (1978) *Geschichte der Impetustheorie,* Frankfurt: Suhrkamp

Wolff, Michael (1987) "Impetus Mechanics as a Physical Argument for Coperni-
    canism. Copernicus, Benedetti, Galileo," *Science in Context 2*, 215-256

Wolff, Michael (1988) "Hipparchus and the Stoic Theory of Motion," in J.
    Barnes and M. Mignucci (eds.) *Matter and Metaphysics* (Fourth Symposium
    Hellenicum) Bibliopolis, 471-545

Wunderlich, Herbert (1977) *Kursächsische Feldmeßkunst, artilleristische Richt-
    verfahren und Ballistik im 16. und 17. Jahrhundert,* Berlin (GDR): VEB
    Deutscher Verlag der Wissenschaften

Wundt, Wilhelm (1906-08) *Logik; eine Untersuchung der Prinzipien der
    Erkenntnis und der Methoden der wissenschaftlichen Forschung*, 3 vols., 3rd
    ed., Stuttgart: Enke

Wundt, Wilhelm (1910) *Die Prinzipien der mechanischen Naturlehre* (originally
    1866 as, *Die Physikalischen Axiome und ihre Beziehung zum Kausal-
    prinzip*). Stuttgart: Ferdinand Enke

Yokoyama, Masahiko (1972) "Origins of the Experiment of Impact with
    Pendulums," *Japanese Studies in the History of Science 11*, 67-72

Young, John (1967) "A Note on Falling Bodies," *The New Scholasticism 41*,
    465-481

Index

Index locorum

Galileo

Collected Works

Manuscripts

Beeckman

Journal